# Management and Welfare of Farm Animals

## UFAW FARM HANDBOOK
### 4th edition

Edited by
R Ewbank
F Kim-Madslien
C B Hart

Established 1926

Universities Federation for Animal Welfare
The Old School, Brewhouse Hill, Wheathampstead, Herts AL4 8AN, UK

© 1999

Universities Federation for Animal Welfare
The Old School
Brewhouse Hill
Wheathampstead, Herts AL4 8AN, UK
Tel: 01582 831818   Fax: 01582 831414
E-mail: ufaw@ufaw.org.uk

**Publishing history**

1st edition 1971    *The UFAW Handbook on the Care and Management of Farm Animals*
                    Edited by UFAW. Published by Churchill Livingstone, Edinburgh
2nd edition 1978    *The Care and Management of Farm Animals*
                    Edited by W N Scott. Published by Baillière Tindall, London
3rd edition 1988    *Management and Welfare of Farm Animals. The UFAW Handbook*
                    Published by Baillière Tindall, London
                    Reprinted by UFAW in 1996 and 1997.

ISBN 1 900630 00 1

Printed by: Halstan & Co Ltd, Amersham, Bucks, UK

# Contributors

**E Ruth I BISWAS** PhD BSc NDD                                              Guineafowl
Retired Senior Lecturer- Animal Production
19055 Berlin, Germany

**R EWBANK** MVSc MRCVS FIBiol                                          Animal welfare
Formerly Director of UFAW

**A J HANLON** BSc MSc PhD                                                        Red deer
Lecturer in Animal Behaviour and Welfare
Faculty of Veterinary Medicine, University College, Dublin, (address at time of writing:
Macaulay Land Use Research Institute, Aberdeen, Scotland)

**R R HENRY** BVMS MRCVS                                                            Ducks
(formerly of Cherry Valley Farms Ltd)
Sudbrooke, Lincoln

**B HODGETTS** DPI MIBiol CBiol                                          Quail production
Buckeye International Ltd, South Petherton, Somerset

**J O L KING** PhD MVSc BSc(Agric) FRCVS FIBiol                                  Rabbits
Emeritus Professor of Animal Husbandry
Faculty of Veterinary Science, University of Liverpool

**I J LEAN** BSc PhD CBiol MIBiol                                                    Pigs
Senior Lecturer in Animal Production
Wye College, University of London

**J D LEAVER** BSc PhD CBiol FIBiol FRAgS                                     Dairy cattle
Professor of Agriculture
Wye College, University of London

**S A LISTER** BSc BVetMed CertPMP MRCVS                                        Turkeys
Private Veterinary Practitioner
Crowshall Veterinary Services, Attleborough, Norfolk

**A MOWLEM** FIAT CBiol MIBiol                                                      Goats
Consultant
Water Farm Goat Centre, Stogurfey, Bridgewater, Somerset

**A J F RUSSEL** BSc (Hons Agric) MAgSc(Hons) PhD NDA — Goats
c/o Macaulay Land Use Research Institute, Aberdeen, Scotland

**D W B SAINSBURY** MA PhD BSc MRCVS — Broiler chickens
Emeritus Fellow, Wolfson College, University of Cambridge

**A E WALL** BVM&S CertVOpthal MSc MRCVS — Fish farming
Veterinarian
Fish Vet Group, Inverness, Scotland

**A J F WEBSTER** MA VetMB PhD MRCVS — Beef cattle and veal calves
Professor of Animal Husbandry
School of Veterinary Science, University of Bristol

**R G WELLS** BSc(Agric)Hons MSc(Poultry Sci) NDA — Laying hens
Senior Lecturer in Animal Production
Harper Adams University College, Newport, Shropshire

**H LL WILLIAMS** BSc(Agric) MSc PhD FRAgS HonAssocRCVS — Sheep
Emeritus Reader in Animal Husbandry
Royal Veterinary College, University of London

# Editors

**R EWBANK** MVSc MRCVS FIBiol
Formerly Director of UFAW

**F KIM-MADSLIEN** BSc MSc
Formerly UFAW Scientific Officer now Research Student, School of Veterinary Science,
University of Bristol

**C B HART** BSc MRCVS FIBiol
Member of UFAW Council

# Preface

The Universities Federation for Animal Welfare was formed in 1938 in direct succession to the University of London Animal Welfare Society, which had been founded in 1926. From the outset it promoted the better treatment of all animals, working on the principle that one of the best ways of doing this was through education, ie by lectures, symposia and workshops and, more generally, by technical publications.

The 1st edition of this Handbook was published in 1971, and followed upon UFAW's involvement in submitting evidence to the UK Government's committee of enquiry – the so-called Brambell Committee – which reported in 1965 on the welfare of animals kept under intensive livestock husbandry systems.

The intentions in this, the 4th edition, are the same as those originally put forward by Charles Hume – UFAW's Founder - in his Foreward to the 1st edition.

> *'This book . . . seeks to put the case for the humane treatment of farm animals in as rational way as possible. UFAW does not necessarily support all the procedures that are described, and hopes that as knowledge accumulates some alternative and more humane methods will be developed. It is acknowledged, however, that farming is a business and that, in the face of intensive competition, improved and more humane techniques can be brought about only if consumers will pay more for the food that they buy.'*

Hume's last comment on the cost of food is most pertinent to the economic situation that currently applies in farming in the UK and many other countries. A surplus of many foods of animal origin have forced down the price that marketing organisations are willing to pay for animal products. In this situation the value of the individual live animals is low, necessitating the introduction of economies which may not be in the best interests of the animals. For example, more animals may be cared for by fewer stockmen, veterinary care may be less readily sought and routine preventive measures neglected, and economies made in housing and feeding, so that the animals come to be at risk - and all this at a time when the consumer is being more demanding about the standards under which food producing animals are kept.

The writers of the individual chapters of this edition have been asked to emphasise the welfare aspects of the husbandry systems they describe. Mention is made, in many cases, of the economic background to the keeping of the various species and the authors have often attempted to look ahead to some of the problems - welfare and otherwise - which may be met in the various sections of the food animal producing industry in the next few years.

We - the editors - are most grateful to the contributors to the 4th edition, who have given of their time and effort (in some cases over a somewhat extended period); to the members of UFAW staff concerned with the production of the book and especially to Samantha Griffin, who word-processed the manuscript into camera-ready format.

It is hoped that this new edition will continue to have a role in keeping to the fore the welfare needs of the farmed animals.

May 1999

R Ewbank
F Kim-Madslien
C B Hart

# Contents

# 1 Animal welfare

## R Ewbank

Man uses animals in enormous numbers. He uses them for companionship, for work, for scientific purposes, for sport, for food and for the production of wool, hides and hair.

Over the last 200 years farmers have increasingly applied technological methods to agriculture with a resulting rise in productivity. During the last few decades world economic forces have changed the structure of farming so that many livestock units in the 'developed' world are now large in size and intensive in nature. At the same time, ie over the last 30–40 years, many of the people living in the urban and suburban parts of the 'developed' world have become separated from the realities of rural life and are now increasingly critical of the ways in which animals are used.

These criticisms, which range from the practical to the philosophical, have led to the general public, the scientists and the farming community taking a much greater interest in the whole subject of animal welfare.

## WELFARE DEFINITIONS AND EXPLANATIONS

### What is 'animal welfare'?

Animal welfare is a complex issue and many attempts have been made to define the term. There is, as yet, no one universally agreed definition. The 1993 New Shorter Oxford English Dictionary considers welfare as 'happiness, well-being, good health or fortune, (of a person, community etc); successful progress, prosperity'. In common usage welfare is perceived as a satisfactory or positive state. In the scientific and technical literature however, there is a range of definitions. Some have argued, eg Dawkins 1980, 1983; Duncan and Petherick 1991 that an animal's welfare is only impaired if it is experiencing an unpleasant mental state ie it is suffering. The definitions that only acknowledge the mental aspects of welfare rule out certain criteria. For example, a tumour that an animal cannot as yet feel would be regarded as just a health problem and not of welfare concern. Other definitions acknowledge more the underlying biological mechanisms: 'Welfare on a general level is a state of complete mental and physical health where the animals is in harmony with its environment (Hughes 1976) and: 'The welfare of an individual is its state as regards its attempts to cope with its environment, (Broom 1986). Both of these wordings refer, in a general way, to the biological balance that should exist between an animal and its surroundings. Hughes's definition additionally introduces the concept of health. It also considers welfare as one positive state while Broom's embraces a scale of welfare: position on the scale depending on how well the animal is coping. Animals which are not coping are considered as being in a state of poor, bad or negative welfare.

There is practical merit in simply replacing the word 'welfare' by the term 'health and well-being'. Health is more than the mere absence of disease and well-being is more than just the absence of discomfort and emotional distress. This positive approach is to be welcomed, because, hopefully, it encourages high standards and it is also in tune with the natural pride of the good stockman in having contented, thriving and productive animals.

It is the stockperson, the veterinary surgeon and the government inspector who will be looking at the animals and deciding whether or not they are showing signs of health and well-being.

Animals that are not showing these well recognised signs are generally in a state of poor welfare. We must always remember, however, that it is a human being who is making a value judgement as to what he or she believes to be the welfare status of the animal. Value judgements, to some extent, depend on the beliefs of the judgement-maker and they may change over time as the individual's beliefs change. Hill and Sainsbury's (1990) definition of welfare incorporates this idea: it proposes that animal welfare is a state '–which is considered by human observers to be consistent with the consensus of current human knowledge of the best interests of the animal(s).'

It seems that there is to be no easy or widely acceptable way of defining or measuring animal welfare – see article by Mason and Mendl (1993) – yet most people seem to agree that an animal in a poor state of welfare is probably suffering from some combination of pain, discomfort, emotional distress or lasting harm.

### Possible causes of suffering in animals

*Ill-treatment* is a term that covers those actions (or inactions) by humans that cause animals to suffer from pain, discomfort, emotional distress or lasting harm. It is possible to divide ill-treatment into three main forms: abuse, neglect and deprivation (see Table 1.1). Some people use the term cruelty as an alternative to the word abuse.

*Abuse and neglect* result in suppression of the signs of health and well-being of animals. There is individual biological inefficiency on the part of the animal (lowered growth rate, lowered milk yield, food conversion inefficiency etc) and financial loss to the farming enterprise. The stockkeeper's traditional

claim that his animals cannot be suffering because they are producing so well, is probably valid as long as only abuse and neglect are being considered.

*Deprivation* relates to situations where animals cannot fulfill their physiological and/or behavioural needs (see below). An absence of the facilities necessary to meet such needs is characteristic of several of the modern intensive systems, eg battery cages, sow tethers and stalls, veal crates, barren unbedded fattening pens. Animals in such accommodation sometimes show behavioural abnormalities, compared with those in more extensive systems. Although such alterations in behaviour may or may not be signs of suffering, they are fairly certainly indicators that the animals are having to make an extreme behavioural adaptation to their environment. Deprivation is sometimes linked to depression in biological production – however, some so-called deprivation systems are highly productive, especially if measured solely in financial terms. High production alone is not necessarily a full defence against an accusation of deprivation!

It has been argued that some modern breeds of livestock have been genetically selected to have such high levels of production, eg daily weight gain, numbers of eggs, milk yield, and are kept under such challenging nutritional and environmental conditions that they are in danger of over-stretching their physiological and anatomical limits and will start to suffer from the so-called production or over-production diseases.

**Table 1.1**       **Consequences of ill-treatment of animals.**

| Type | Symptoms/signs | Effect on production |
|---|---|---|
| **Abuse**<br>(deliberate) | Fear, injury, pain, distress etc (ie suffering) | Individual biological inefficiency and financial loss |
| **Neglect**<br>(occasional) through idleness, ignorance or overwork | Malnutrition, disease, distress etc (ie suffering) | Individual biological inefficiency and financial loss |
| **Deprivation**<br>(built into some husbandry systems) of facilities to fulfill behavioural and/or physiological needs | Changes in behaviour, occasional abnormal behaviours etc (suffering?) | Uncertain |

*(Modified from Ewbank 1985)*

In assessing the effects of husbandry systems on animal production, it is vital to distinguish between the effects on the biological production of individual animals and the financial return from the whole enterprise. Although, for example, increases in stocking rates will usually lead to a lowering of individual daily live weight gains, or of productivity in terms of, say, the number of eggs produced per bird, the financial return from the enterprise – measured both as a total and on an individual animal basis – may increase. This is possibly due to a sharing of the capital cost of the same building across a larger number of animals. Another and perhaps more important economy results from the greater number of animals being looked after by the same or even a smaller number of human attendants.

**Stress, overstress and distress**

Animals respond to challenges in their environment by employing a variety of interlocking physiological, biochemical and behavioural adaptation mechanisms. Potentially harmful stimuli that activate these mechanisms are called *stressors* and the response itself has been termed the *stress* or *stress response*. These changes are largely adaptive but the animal usually makes them at a biological cost (see Figure 1.1). In acute or prolonged stress the side-effects of some of the responses can be actually harmful to the

animal. Salpolsky (1994) gives an excellent if somewhat human orientated general account of stress, while Broom and Johnson (1993) and Wiepkema and Koolhaas (1993) discuss in detail the relationship between stress and animal welfare.

Stress is often accompanied by an increase in the adrenaline and the corticosteroid hormones secreted into the blood stream by the adrenal glands. Rises in the levels of these hormones in the blood are used as one of the main measurable indicators of stress. These indicators are even more reliable if they are accompanied by evidence of a rise in metabolic rate and/or of a suppression of the animal's immune response (Barnett 1987). The body also releases other hormones eg glucagon, prolactin, vasopressin and the endogenous morphine-like substances called endorphins and encephalins which, amongst other functions, help blunt pain perception in some stressful situations. Some researchers have suggested that a particular mix of these hormones may be characteristic of a particular stressor.

Fraser *et al* (1975) strongly believe in using the term stress only when an animal has to make an abnormal or extreme adjustment in its physiology or behaviour to cope with adverse aspects of its environment. This argument has considerable merit, but a case can be made (Ewbank 1985) to subdivide this extreme-adjustment stress response into *over-*

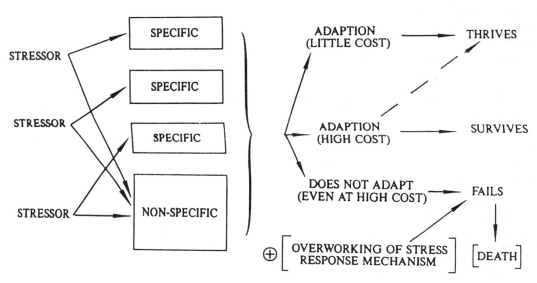

**Figure 1.1**     **An overview of the stress response and its role in adaption.**
*(from Ewbank 1992)*

*stress* and *distress*. Over-stress describes the adaptive, medium-level response that an animal makes at a relatively low biological cost, but which may be accompanied by some damage to the animal's biological systems. *Distress* is the high-level response that has a high biological cost, is damaging to the animal, is probably sensed by the animal as unpleasant (ie causing it suffering). An individual will outwardly express distress by recognizable changes in its behaviour.

The term *distress* also fits in with the legal terminology of the *Agriculture (Miscellaneous Provisions) Act 1968* (see below for details of this legislation) where the main offence is to cause 'unnecessary pain or distress'. Pain is an internal physical and/or psychological process causing suffering. A human observer will notice that an animal is in pain because he/she can see the outward signs of distress. It is also likely that animals can also suffer from emotional distress which is not associated with pain. Most stockkeepers would probably agree that their animals feel and probably suffer from fear. If farm animals experience one emotion, could they not have others? Separating a ewe from her lamb causes her to show physiological and behavioural changes indicative of distress. How does the ewe's reaction compare with the response of a human mother forcedly separated from her child? There are as yet no real answers to these questions. However, practical experience, common sense and humanity suggest that, in the absence of positive proof to the contrary, animals should be given the benefit of the doubt. If an animal is in a situation that would clearly induce unpleasant emotions in humans and shows behavioural signs of distress, then it is probably undergoing an unpleasant emotional process. This animal is probably suffering. For many years the UK law has recognised emotional suffering in animals. The *Protection of Animals Act 1911* stipulates that it is an offence – amongst other things – to 'infuriate or terrify any animal' (see below for further details of legislation).

**Physiological and behavioural (ethological) needs**
The *European Convention for the Protection of Animals Kept for Farming Purposes* (Council of Europe 1976) states several times that farm animals must be housed (managed, fed etc) according to, among other things, 'their physiological and ethological needs'.

The concept of physiological needs is straightforward. If, for example, an animal does not have adequate quantities of a suitable food, it will probably not grow properly or, in the case of a mature animal, it will be difficult to breed from; it may show signs of malnutrition or vitamin or mineral deficiencies. If kept long enough on this inadequate diet it may even die. There is a physiological need for sufficient amounts of the correct kind of food. Preventing animals from exercising can cause pathological changes to occur in their joints, bones and muscles – it is, therefore, possible to argue that there is a physiological need for exercise.

Behavioural (ethological) needs are less readily understood. The concept, however, is central to many ideas in farm animal welfare. In 1964, the UK government, in response to the public concern for animal welfare that had been brought to a head by the publication of the outstandingly important book *Animal Machines* (Harrison 1964), appointed a Technical Committee (Brambell 1965) to enquire into the welfare of animals kept under conditions of intensive husbandry. The deliberations of the committee were strongly influenced by contemporary behavioural ideas.

As early as the 1940s one of the main areas of interest for European ethologists was the study of animals' internal motivational states. The early ethologists took the view that most, if not all, behavioural patterns were largely controlled by genetically determined internal drive mechanisms. They believed that over a period of time, the level of the internal drive state would steadily rise until it exceeded a threshold and the animal would then perform the appropriate behaviour. External stimuli from the environment might alter (usually lower) the threshold; they might be required for the full expression of the resulting behaviour but they were not necessary either for the build-up of motivation or for triggering the behaviour. If it was made difficult for the animal to carry out the behaviour (eg by not giving it the required physical facilities, such as, in the case of the domestic fowl, nesting material for the pre-egg laying behaviour) then the animal might be expected to show so-called intention movements and/or inappropriate behaviours or even overtly abnormal behaviour. Many people considered these types of responses to be signs of frustration and a possible form of emotional distress. These ideas were largely accepted by the Brambell Committee, which further held that some intensively kept livestock housed in seemingly barren surroundings did indeed show behavioural signs of frustration.

More recently, however, some ethologists have been pointing out that the internal drive theory is too simple an explanation of why animals perform certain behaviours at certain times and places. They are suggesting that needs would be better termed biological, and that animals satisfy them through a mixture of physiological and behavioural mechanisms. For a critical yet sympathetic review of some of the modern ideas on behavioural needs, see Dawkins (1983).

The idea of behavioural needs, however, has been taken up by the legislators. The Council of Europe's 1976 *European Convention for the Protection of Animals Kept for Farming Purposes*, took on board many of the points, including the concept of behavioural needs, contained in the Brambell Report. The UK has signed and ratified the Convention and has thus indicated not only that the government of the day agreed with the spirit of the document but that it also believed that its domestic legislation was sufficient to implement all the conditions (including the meeting of behavioural needs) laid down by the Convention. It seemed, from the point of view of the UK lawyers, that the important word was 'need'. A need is not a luxury or an option but a necessity. An animal deprived of a necessity will, in the view of most people, suffer – showing signs of pain, discomfort and/or distress. Such suffering could be the basis of a prosecution under the *Protection of Animals Act 1911* or under the *Agriculture (Miscellaneous Provisions) Act 1968*. The lawyers, therefore, argued that the UK law indirectly ensures that animals' physiological and behavioural needs are likely to be met. In practice, however, many intensively kept animals are not able to satisfy all their needs and yet their owners/keepers are not taken to court.

**Table 1.2    The FAWC Five Freedoms.**

The welfare of an animal, whether on farm, in transit, at markets or at place of slaughter should be considered in terms of 'five freedoms'. These freedoms define ideal states rather than standards for acceptable welfare. They form a logical and comprehensive framework for analysis of welfare within any system together with the steps and compromises necessary to safeguard and improve welfare within the proper constraints of an efficient livestock industry.

1.    Freedom from hunger and thirst
           - by ready access to fresh water and a diet to maintain full health and vigour
2.    Freedom from discomfort
           - by providing an appropriate environment including shelter and a comfortable resting area
3.    Freedom from pain, injury or disease
           - by prevention or rapid diagnosis and treatment
4.    Freedom to express normal behaviour
           - by providing sufficient space, proper facilities and company of the animal's own kind
5.    Freedom from fear and distress
           - by ensuring conditions and treatment which avoid mental suffering.

These freedoms will be better provided for if those who have care of livestock practise:-
   - caring and responsible planning and management
   - skilled, knowledgeable and conscientious stockmanship
   - appropriate environmental design
   - considerate handling and transport
   - humane slaughter

*(from FAWC Press Notice 92/7)*

Another approach to ensuring, as far as is reasonable, that animals are able to fulfill their biological needs has been to draw up lists of basic husbandry requirements. The earliest, if somewhat restricted, example was put forward in the Brambell Report as long ago as 1965. 'An animal should at least have sufficient freedom of movement to be able, without difficulty, to turn round, groom itself, get up, lie down and stretch its limbs.' This statement on the size of the physical environment for animals came to be known as the 'five freedoms' (*Brambell Five Freedoms*).

The concept has been extended and developed over the years and is now put forward as the Farm Animal Welfare Council's Five Freedoms (*FAWC Five Freedoms*) (Table 1.2). The FAWC is the UK government's advisory body on farm animal welfare - see Winter (1998). This approach implies that animals probably have 'interests' (eg an interest not to suffer; an interest in being able to find sufficient food) and animal users have a duty; as far as possible, to safeguard these interests.

## Animal rights

There are different ethical views as to how human animals should treat other animals. These vary from the concept that we have an absolute dominion over the other creatures, through a middle position that we have a duty to give the animals we use a good life and a quick humane death, to the suggestion that animals have a total right to be left alone and that this right must be fully respected.

Whether we think it is right to use animals in various ways largely depends on our concept of animals compared with humans. The philosophical arguments concerning animal rights have their roots in human ethics, morals, religion and history and in questions about animal intelligence, awareness, self-awareness and capacity to suffer.

The spread of animal rights ideas in the UK over the last 10-20 years is largely due to the influence of three semi-popular but serious books written by moral philosophers (Singer 1976, Rollin 1981 and Regen 1982). The arguments put forward are often convincing within their own philosophical framework but at first reading may seem to bear only a limited relevance to the practical situation.

Animal rights and animal welfare are sometimes considered to be synonymous. The viewpoint and aims of the extreme 'animal rights movement' differs fundamentally from those of most animal welfare organisations. The animal welfare approach is largely based on the idea that humans are responsible for a stewardship over animals – a responsibility that encompasses humane care. The aims of most animal welfare organisations are to improve the health and well-being of animals within the bounds of captivity, domestication and the care of the environment. The extreme animal rights movement aims to stop all use of animals by man, although it must be admitted that some animal rights supporters take a somewhat less extreme stance.

Some animal rights campaigners tend to have very fixed views and be intolerant during discussions. This is a pity, because some of their ideas can contribute to an understanding of the complex ethical relationships that exist between man, animals and the environment.

Many animal users are happier with the concept that it is permissible to use animals as long as they are given a good life while they are alive and a quick, humane death when they have to be killed. It must be remembered, however, that this statement incorporates value judgements and is influenced by beliefs as to how man should interact with the natural world.

## LEGISLATION IN THE UK

### General animal welfare legislation

This chapter has already touched upon the ways in which the present UK law is supposed to be adequate to implement the conditions that the 1976 *European Convention for the Protection of Animals Kept for Farming Purposes* lays down for the fulfilment of animals' physiological and behavioural needs.

### *The Protection of Animals Act 1911*

This is the principal statute relating to the protection of both domesticated and captive animals. The main offence is to inflict cruelty. Actions amounting to cruelty include:

1. Cruelly to beat, kick, ill-treat, over-ride, over-drive, over-load, torture, infuriate or terrify any animal
2. To cause unnecessary suffering by doing, or omitting to do, any act

3. To convey or carry any animal in such a manner as to cause it unnecessary suffering
4. To perform any operation without due care and humanity (in relation to which the provisions of the *Protection of Animals (Anaesthetics) Acts* are particularly relevant – see below)
5. The fighting or baiting of any animal or the use of any premises for such a purpose
6. The administering of any poisonous or injurious drug or substance to any animal.

This Act is commonly grouped with several other related Acts and Amendments and is known collectively as the *Protection of Animals Acts 1911-1964*. Although these Acts do not apply to Scotland, there is separate legislation of a very similar nature applying specifically to that country.

The *Agriculture (Miscellaneous Provisions) Act* 1968. This statute was largely enacted in response to some of the recommendations made in the 1965 Brambell Committee Report. It is the main piece of legislation that applies specifically to the welfare of farm animals, and has four main points:

1. It makes it an offence to cause unnecessary pain or unnecessary distress to livestock being kept for farming purposes on agricultural land
2. It gives authority for veterinary officers of the State Veterinary Service to inspect, on welfare grounds, farms where livestock are being kept
3. It empowers the appropriate Minister to introduce Regulations to improve the welfare of livestock. There are currently three such regulations:

   (1) *The Welfare of Livestock (Prohibited Operations) Regulations 1982* as amended in 1987. (see below)
   (2) *The Welfare of Livestock Regulations 1994.* These regulations make specific provision for the welfare of laying hens in battery cages, calves and pigs, and general provisions for other livestock
   (3) *The Welfare of Livestock (Amendment) Regulations 1998* these regulations change considerably the provisions relating to the welfare of calves.
4. It authorizes the Agricultural Ministers to prepare *Codes of Recommendations for the Welfare of*

*Livestock.* Codes have been produced for cattle, sheep, pigs, rabbits, goats, farmed deer, domestic fowls, turkeys, ducks and pigs (MAFF 1983-1997). These Codes are of an advisory nature (unlike the Regulations listed above), and failure to observe them is not in itself an offence. However, they do have a certain legal standing: a person cannot claim ignorance of them, and he/she disregards their advice at his/her own risk. A prosecution under the 1968 Act for causing unnecessary pain or distress to livestock on agricultural land, may use failure to observe the provisions of the Codes as evidence.

The Preface to most of the *Codes of Recommendations* list a number of necessary provisions that, if adopted, should go a long way towards ensuring that the animals are being kept in accordance with the concept of the *FAWC Five Freedoms*. The provisions include:

1. Comfort and shelter
2. Readily accessible fresh water and a diet to maintain the animals in full health and vigour
3. Freedom of movement
4. The company of other animals, particularly of like kind
5. The opportunity to exercise most normal patterns of behaviour
6. Light during the hours of daylight, and lighting readily available to enable the animals to be inspected at any time
7. Flooring which neither harms the animals, nor causes undue strain
8. The prevention, or rapid diagnosis and treatment, of vice, injury, parasitic infestation and disease
9. The avoidance of unnecessary mutilation
10. Emergency arrangements to cover outbreaks of fire, the breakdown of essential mechanical services and the disruption of supplies.

These Prefaces can be seen as guidelines that set the basic standards for the keeping of animals and which establish the attitudes that people should take towards their stock. The MAFF Codes of Recommendations themselves are detailed suggestions as to how these standards are to be practically implemented.

**Transport of animals**

The rules and regulations governing the transport of farm animals are complex and are mainly laid down in *The Welfare of Animals (Transport) Order 1997*. For details see the official booklet '*Guidance on the Welfare of Animals (Transport) Order 1997*' (Ministry of Agriculture Fisheries and Food 1998a).

**Control of minor surgery**

The whole question as to which minor surgical operations (for example, castration, docking, debeaking, dehorning) are allowed by legislation to be performed on farm animals, who can do them, and at what age it is necessary to use anaesthetics is complex (see MAFF 1995). The situation is covered by three interacting groups of legislation: *The Welfare of Livestock (Prohibited Operations) Regulations 1982*, as amended in 1987; the *Protection of Animals (Anaesthetics) Acts 1954 and 1964*, with the various Orders made under them; and various Orders under the *Veterinary Surgeons Act 1966*.

*The Welfare of Livestock (Prohibited Operations) Regulations 1982* (as amended in 1987) further prohibits a number of operations, unless they are being carried out as first aid measures, or by a veterinary surgeon in the treatment of injury or disease. These are:

1. Penis amputation and other penial operations

2. Freeze dagging of sheep

3. Short-tail docking of sheep, unless sufficient tail is retained to cover the vulva in the case of female sheep and the anus in male sheep

4. Tongue amputation in calves

5. Hot branding of calves

6. Tail docking of cattle

7. Devoicing of cockerels

8. Castration of a male bird by a method involving surgery

9. Any operation on a bird with the object or effect of impeding its flight, other than feather clipping

10. Fitting any appliance which has the object or effect of limiting vision to a bird by a method involving the penetration or other mutilation of the nasal septum

11. Tail docking of a pig unless the operation is performed by the quick and complete severance of the part of the tail to be removed and either:

    (i) the pig is less than eight days old, or

    (ii) the operation is performed by a veterinary surgeon who is of the opinion that the operation is necessary for reasons of health or to prevent injury from the vice of tail biting

12. Removal of any part of the antlers of a deer before the velvet of the antlers is frayed and the greater part of it has sheared

13. Tooth grinding of sheep is totally prohibited (even by a veterinary surgeon) by the 1987 amendment.

The *Protection of Animals (Anaesthetics) Acts 1954 and 1964* aim to prevent the infliction of unnecessary suffering during an operation. It is legal to perform certain minor operations without the use of an anaesthetic. They include, among others, the giving of injections, emergency first aid and the castration of a male animal before it has reached a specified age (see individual chapters for details).

The *Veterinary Surgeons Act 1966* stipulates that, after a certain age, male animals may be castrated only by a veterinary surgeon. In the case of farm animals this age coincides with that after which the operator must, by legislation, use an anaesthetic/analgesic.

The age component in this legislation is largely based upon the assumption that newborn or very young animals do not feel pain or at least feel a reduced level of pain compared with adults. Recent research, however, is revealing that this assumption may be wrong. At the practical level, however, the giving of an anaesthetic/analgesic to these young animals may cause them more pain and distress than the actual operation itself.

For further details of the legislation covering farm animal welfare, see Brookman and Legge (1997); Cooper (1987); Ministry of Agriculture Fisheries and Food (1995) and Porter (1991a,b).

**ASSESSING ANIMAL WELFARE**

We are unable to communicate directly (talk) with animals and thus have no immediate means of assessing their needs, feelings and welfare status. We largely have to depend upon one or more of the following groups of indirect indicators.

Behavioural – presence and/or absence of normal and abnormal behaviours, results of preference tests

Physical – presence of cuts, bruises, disease lesions - changes in the immediate surround (ie state of faeces, rub marks on fences, etc)

Physiological – changes in heart rate, blood parameters (cortisols, adrenaline, ß endorphins), immune responses

Production – changes in growth rate, food conversion efficiency, eggs and milk yields, conception rates.

These four groups of indicators and their examples are not an exhaustive list. The actual details will vary between species, they may differ at different times of the year and they may be characteristic of a particular husbandry system.

There are advantages and disadvantages with each of the indicator groups. Behavioural indicators are non-invasive and are often obvious at a distance; a good stockkeeper will often know at a glance if there is something wrong with his/her animals. Changes in behaviour are often the first signs of disease and the main signs of distress. An important general point is that the study of the behaviour of animals in their environments tends to produce a broad ecological/behavioural picture by which it is often possible to judge whether a particular husbandry system is biologically sound. The study of animal behaviour should be encouraged amongst animal keepers, agriculturalists and veterinarians. The best practical introduction to farm animal behaviour is probably the book by Kilgour and Dalton (1984). A fuller, more recent and more general text is Fraser and Broom (1990). Physical indicators can also be observed in the field situation and their number and distribution can be quantified. Physiological changes usually require blood or saliva samples to be taken and then examined/analysed in a laboratory. The results can be quantified. It is often possible to assess production criteria from an examination of farm records. The conclusions have to be treated with care: a sudden fall in production is possibly indicative of a welfare problem; the maintenance of production, however, does not necessarily meant that the welfare of the animals is good.

There is often a total pattern of signs that the stockperson recognizes (sometimes, it seems, nearly subconsciously) as being normal, denoting the healthy and productive state. As the stockperson works with the animals he/she tends to monitor this pattern and any deviation from the norm will alert the handler to trouble. Once alerted, the good stockperson will undertake a thorough inspection of the animals. This will probably reveal that they are either normal (healthy) or abnormal (unhealthy). If the stockkeeper cannot at this time come to any real conclusion, he/she will then take action (if appropriate) and wait to see what happens and/or seek professional advice.

Once the welfare status of an animal (or group of animals) has been determined, the husbandry system as a whole can then be viewed against a perceived positive/negative welfare scale (see Figure 1.2).

If the welfare is poor, ie the enterprise lies in the negative part of the scale, it is likely that obvious signs of pain, discomfort or distress and that action should be taken to remedy the situation. An inspecting officer might decide that there is sufficient evidence of prolonged and/or unrelieved suffering to institute a prosecution. There will obviously be a variable, indeterminate 'grey area' between the clearly positive and negative zones, but in most cases it will be obvious that the enterprise lies in one zone or the other.

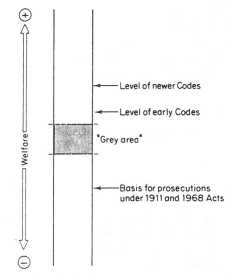

**Figure 1.2 Perceived positive/negative welfare scale.**

It can be argued that the original, 1971-77, *Codes of Recommendation for the Welfare of Livestock* aimed at a welfare standard just above the 'grey zone' and that the more recent and revised Codes (1990 and later) are set at a higher level. As the welfare of an animal moves up the positive side of the scale there could be an increase in individual biological production (growth rates, etc). This may result in an improvement in the profitability of the system. Another scale of reference on which it can be helpful to base attitudes and welfare standards is one proposed by McInerney (1994) (See Figure 1.3). Improvements in the profitability of a system will probably only occur if the animal is on the A to B part of this perceived animal welfare/livestock productivity relationship graph. Most modern animal units are probably on the B to D part of the curve, where increased production and a cheaper product to the consumer is being bought at the expense of lowered welfare. Improved animal welfare in these cases can probably only result from lowering productivity and with the end product (eggs, meat, milk) costing more to the consumer.

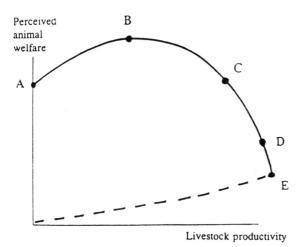

**Figure 1.3    The relationship between perceived animal welfare and livestock productivity. (McInerney 1994).**

## STOCKMANSHIP

Within any system the quality of the stockperson is the most important determinant of welfare. The criteria for good stockmanship are: a proper understanding of animals, a sense of caring and sufficient experience based on proper training.

Stockmanship involves stock sense (a knowledge of, rapport with and ability to observe and understand the ways of animals) and skill in stock tasks, namely the practical aspects of handling, care and manipulation of animals. A good stockperson should be observant, patient, informed about animals and their needs, competent in stock tasks and able to recognize health and disease states. They should be knowledgeable about the workings of environmental control equipment and the measures to take when it fails. Most farmers and stockkeepers take a positive pleasure in having contented thriving animals under their care. This natural inclination should be respected and encouraged. One of the criticisms of modern intensive systems is that far too many animals are being looked after by far too few stockmen. It may be possible to perform much of the routine work with mechanized equipment but there may still be little time for individual attention to the animals themselves.

It is therefore essential that stockpeople are well-trained, provided with the equipment and facilities to perform their work and have sufficient time to quietly and regularly observe their charges.

In the UK, training courses and proficiency tests for stockpersons are run by Lantra National Training Organisation Ltd (previously – Agricultural Training Board Landbase), local authorities and the National Federation of Young Farmers' Clubs. A person with a liking for rural life and a sympathy for farm animals requires experience and proper, formal training before they can become a good stockperson.

Management backing is vital; there must be clear lines of communication. In large units the subdivision of responsibility can be a problem. In old-fashioned, small enterprises the stockperson, the manager and the owner were often the same individual. In modern large establishments, these three functions are often in separate hands and may be operating from different sites. The stockkeepers can only perform their work well if they are provided with the correct physical facilities and if they have the continued support of the manager/owner.

For further full discussion of stockmanship see Seabrook (1987) and English *et al* (1992), and for an account of human/farm animal relationships, see Albright (1986) and Hemsworth and Coleman (1998).

## CONTROL OF DISEASE

Disease can lead to substantial suffering in animals and is an important cause of economic loss. Its control demands an intelligent cooperation between the farmer's own veterinary surgeons, specialist advisers and the stockkeeper. It is crucial to think of disease control in the planning stage of any new animal enterprise. Well-planned preventive measures are not only more humane, in that the diseases do not develop and therefore the animals do not suffer, they are also more efficient in terms of productivity and costs. For further information, see Moss (1992) and Sainsbury (1998).

## HUMANE KILLING

It may be necessary at times to kill diseased, injured, deformed or surplus animals on the farm. The veterinary surgeon and licensed slaughterman may do this in the course of their routine duties but, in an emergency, ie to relieve acute suffering, a farmer or stockkeeper may personally have to kill the animal(s). This emergency slaughter must be carried out as humanely as possible and it is essential that the farmer or stockman should give some thought in anticipation of the problem and prepare accordingly. This will involve consulting with a veterinary surgeon and making sure that the appropriate equipment is always available, and that the farm staff have been trained in its use.

A welfare problem can arise, however, over the disposal of diseased or injured stock that are not suffering acutely. Sending them alive (as a so-called casualty animal) with an appropriate veterinary certificate to a slaughterhouse that is willing to accept them, will probably give the farmer a greater financial return than if they had been killed on the farm and the carcase disposed of through a knacker's yard. There is, therefore, a temptation for some farmers to transport animals that should, on humane grounds, have been slaughtered on the farm. The

Ministry of Agriculture (MAFF 1998b) has issued a most useful booklet *Guidance on the Transport of Casualty Food Animals*. The welfare problem of casualty animals will probably only be overcome by a change in attitude of farmers, veterinary surgeons and slaughterhouse owners. Also vital is the firmer implementation of the laws that make it an offence to cause unnecessary pain, suffering or distress to animals. The adoption of farm insurance schemes by which the financial loss resulting from the killing of an animal on humane grounds could be claimed for, would encourage the humane treatment of hopelessly diseased or badly injured stock.

For practical details of humane killing, see under the appropriate species chapter.

## EXTENSIVE AND INTENSIVE SYSTEMS

Much of the concern about the welfare of farm animals centres around so-called intensive husbandry systems (the modern dairy industry, battery cages, fish farming). For a recent compilation of critical articles and photographs, see Animal Welfare Institute (1987). Curtis (1986) presents a strong supportive case for intensive systems. Concurrently with the criticism of intensive systems has been an interest in the development of new extensive/alternative/organic methods of animal keeping (eg Boehncke and Molkenthin 1991 and Lampkin 1994).

Many intensive systems are restrictive and in some cases may not provide animals with their basic biological needs. It is therefore tempting to suggest that extensive systems which give animals more freedom must be better.

It is possible to classify husbandry systems as intensive, intermediate and extensive. Figure 1.4, for example, illustrates the range of housing systems available for laying birds. Generally, as systems become less intensive, they are more costly to run, and the end-product is likely to be more expensive to the consumer (although not as expensive as some people would like to imagine). Such non-intensive systems demand a higher level (both in quality and quantity) of stockmanship, the animals are more exposed to the weather and there is sometimes a higher level of disease. The animals, however, do have more freedom, although at times this can appear as 'freedom to get into trouble'!

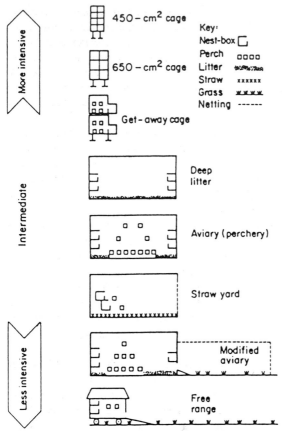

**Figure 1.4** **Diagrammatic representation of housing system for laying hens (not to scale).**
*(Modified from Ewbank 1981)*

Extensive systems are not necessarily 'better' and, conversely, intensive systems are not necessarily 'worse'. Each type has its good and bad points (see individual chapters for details). Rather we should realize that there are both good and bad units in each category, as measured by health, well-being and productivity. The aim should be to encourage good welfare, whatever the system.

## KEEPING ANIMALS IN GOOD HEALTH AND WELL-BEING

To keep animals in good health and well-being there should be:

1. A full understanding of their biological (physiological and behavioural) needs

2. The provision of facilities that allow the animals to fulfill these needs

3. The provision of round-the-clock back-up services so that, in case of breakdown of equipment, accidents or outbreaks of disease, the animals do not suffer unduly

4. Sufficient time allowed for the stockpeople to give attention to their charges

5. An attitude on the part of the stockperson, the manager and the owner which ensures that the animals are given a good life when alive and a humane death

6. A willingness on the part of the consumer to pay for humanely produced animal products.

## THE WAY AHEAD

Some progress has been made since Ruth Harrison, with her book *Animal Machines*, set in train the current interest in farm animal welfare – but it is still believed eg Harrison 1988, Johnson 1991 that much still remains to be done. There are, for example, several areas of agriculture and intensive farming that face heavy criticism:

1. The export of dairy calves to traditional (single animal) crate-rearing systems for veal production

2. Confinement of laying birds in battery cages

3. Beak trimming in poultry as a means of preventing featherpecking and cannibalism

4. Production/overproduction diseases which can lead to such conditions as leg damage in broilers and lameness in dairy cows

5. Transport and slaughter conditions for animals that have little individual value eg spent hens

6. The production and disposal of young unwanted animals eg male calves of some dairy breeds, male chicks of laying strains of poultry.

7. The genetic make up of some breeds such that they have difficulty in performing some of their normal activities, eg the double muscling which results in difficult births in some breeds/strains of beef cattle.

It has been suggested that regular official inspection and possibly the annual licensing of intensive farm animal units would be one way of improving welfare. This may be so – the manpower and administrative costs could be high but inspection and licensing might help to allay or remove some of the worst and sometimes unfounded fears of the general public. The voluntary quality assurance schemes being run by various animal welfare organisations, supermarket chains and agricultural producer groups often have a welfare inspection component. These schemes should increase the welfare status of the animals being produced and should improve the image of the agricultural industry but, of course, they are voluntary and do not cover all animal production.

Criticism is also being levelled at some aspects of extensive farming. For example, most free range laying hens are routinely beak trimmed to control feather pecking and they are sometimes accommodated in large sheds where many of the birds have little real chance of going outside. Outdoor sows often have rings pierced through their nose to prevent them rooting and, in winter, some are kept on outdoor sites somewhat reminiscent of World War I battlefields. Hill sheep are left exposed in distant areas where farmers are not easily able to check them and give assistance when emergencies arise. Nevertheless there is a feeling amongst many farmers, scientists and veterinary surgeons that well run extensive systems do have the potential to provide better welfare conditions for their animals than do many of the intensive enterprises.

Real progress, however, mainly comes through changes in the attitude of those involved in the use of animals and we must remember that the main user of animals is the consumer. Modern intensive methods of animal husbandry have given the developed world large quantities of wholesome, relatively cheap, animal protein-rich foods. The relative cheapness has in many ways been bought at the expense of good husbandry. A change to less intensive methods will often involve a financial cost. Who, then, should bear this cost? As Webster (1994) states in his book on animal welfare '– since it is the consumer who has benefited most from the intensification of animal production it is the consumer who will have to concede the most in order to ensure that farm animals get a fair deal.'

The European Union has recently adopted a protocol (European Union 1997) obliging member states to pay full regard to animal welfare in forming and in implementing relevant policy and legislation. This legally binding protocol, which, in effect amends the Treaty of Rome - the agreement which was set up by original European Economic Community, the forerunner of the present European Community/Union - should ensure that farm animals are treated as sentient beings and not just as mere agricultural products. All this may, in due course, help to raise the level of farm animal welfare across Europe but it will do little to counteract the possible negative animal welfare influence of the liberalisation of world trade. The rules of the World Trade Organisation which now incorporates the Articles of the General Agreement on Tariffs and Trade (GATT) make it difficult to ban the import of animal products on welfare grounds. This means that products produced outside Europe under poor welfare conditions and therefore probably produced more cheaply will, under free market-forces, have to be accepted for sale within Europe. They will compete with the more expensive European high welfare products and probably come to occupy a high proportion of the market. For further details of this complex and worrying development see Eurogroup for Animal Welfare (1998).

It is vital both for the health and well-being of the animals involved and for the economic future of the farming industry that an increasing and critical interest should be taken in the whole complex subject of farm animal welfare. The literature on the subject is immense and is perhaps best approached through the following important and comprehensive books: Appleby and Hughes (1997); Broom and Johnson (1993); Dawkins (1980); Fox (1984); Fraser and Broom (1990); Rollin (1995); Sainsbury (1986) and Webster (1994).

As society changes its attitudes to animal usage, the legislation will probably change and the keeping of farm animals may become more regulated. More importantly, the attitudes of farmers, stockkeepers, veterinary surgeons and agriculturalists, who are themselves members of a changing society, will change and they will then acquire the motives, knowledge and skills necessary to further advance the welfare of the animals in their charge.

# REFERENCES AND FURTHER READING

**Albright J L** 1986 Human/farm animal relationship. In Fox M W and Mickley L D (eds) *Advances in Animal Welfare Science 1986/87* pp 51-56. Humane Society of the United States: Washington DC, USA

**Animal Welfare Institute** 1987 *Factory Farming: the Experiment that Failed*. Animal Welfare Institute: Washington DC, USA

**Appleby M C and Hughes B O** 1997 *Animal Welfare*. CAB International: Wallingford, UK

**Barnett J L** 1987 The physiological concept of stress is useful for assessing welfare. *Australian Veterinary Journal 64:* 195-196

**Boehncke E and Molkenthin V** (eds) 1991 *Alternatives in Animal Husbandry*. (Proceedings of International Congress, University of Kassel at Witzenhausen) Witzenhousen: Agrarkultar Verlag

**Brambell F W R** (Chairman) 1965 *Report of the Technical Committee to Enquire into the Welfare of Animals Kept Under Intensive Livestock Husbandry Systems*. Cmnd 2836. HMSO: London, UK

**Brookman S and Legge D** 1997 *Law Relating to Animals*. Cavendish Publishing: London, UK

**Broom D M** 1986 Indicators of poor welfare. *British Veterinary Journal 142:* 524-526

**Broom D M and Johnson K G** 1993 *Stress and Animal Welfare*. Chapman and Hall: London, UK

**Cooper M E** 1987 *An Introduction to Animal Law*. Academic Press: London, UK

**Council for Agricultural Science and Technology** 1981 *Scientific Aspects of the Welfare of Food Animals*. Report No 91. CAST: Ames, Iowa, USA

**Council of Europe** 1976 *European Convention for the Protection of Animals Kept for Farming Purposes, European Treaty Series No 87*. Council of Europe: Strasbourg

**Curtis S E** 1986 The case for intensive farming of food animals. In: Fox M W and Mickley L D (eds) *Advances in Animal Welfare Science 1986/87* pp 245-255. Humane Society of the United States: Washington DC, USA

**Dawkins M S** 1980 *Animal Suffering*. Chapman and Hall: London, UK

**Dawkins M S** 1983 Battery hens name their price: consumer demand theory and the measure of ethological 'needs'. *Animal Behaviour 31:* 1195-1205

**Duncan I J H and Petherick J C** 1991 The implications of cognitive processes for animal welfare. *Journal of Animal Science 69:* 5017-5022

**English P, Burgess G, Segundo R and Dunne J** 1992 *Stockmanship. Improving the care of the pig and other livestock*. Farming Press: Ipswich, UK

**Eurogroup for Animal Welfare** 1998 *Conflict or Concern? Animal Welfare and the World Trade Organisation*. RSPCA: Horsham, UK

**European Union** 1997 *Protocol on Protection and Welfare of Animals*. European Union Treaty of Amsterdam

**Ewbank R** 1981 Alternatives: definitions and doubts. In: *Alternatives to Intensive Husbandry Systems* pp 5-9. UFAW: Potters Bar, UK

**Ewbank R** 1985 Behavioral responses to stress in farm animals. In: Moberg G P (ed) *Animal Stress* pp 71-79. American Physiological Society: Bethesda, USA

**Ewbank R** 1992 Stress: a general overview. In: Phillips C and Piggins D (eds) *Farm Animals and the Environment* pp 255-262. CAB International: Wallingford, UK

**Fox M W** 1984 *Farm Animals: Husbandry, Behavior and Veterinary Practice*. University Park Press: Baltimore, USA

**Fraser A F and Broom D M** 1990 *Farm Animal Behaviour and Welfare*, 3rd edition. Baillière Tindall: London, UK

**Fraser D, Ritchie J S D and Fraser A F** 1975 The term 'stress' in a veterinary context. *British Veterinary Journal 131:* 653-662

**Harrison R** 1964 *Animal Machines*. Stuart: London, UK

**Harrison R** 1988 *Farm animal welfare, what, if any, progress?* (6th Hume Memorial lecture). UFAW: Potters Bar, UK

**Hemsworth P H and Coleman G C** 1998 *Human-Livestock Interactions: The Stockperson and the Productivity and Welfare of Intensively Farmed Animals*. CAB International: Wallingford, UK

**Hill J R and Sainsbury D W B** 1990 *Farm Animal Welfare, Who Cares? How?* Cambridge Centre for Animal Health and Welfare: Cambridge, UK

**Hughes B O** 1976 Behaviour as an index of welfare. In: *Proceedings 5th European Poultry Conference, Malta.* II: 1005-1012

**Johnson A** 1991 *Factory Farming*. Blackwell: Oxford, UK

**Kilgour R and Dalton C** 1984 *Livestock Behaviour. A Practical Guide*. Granada: London, UK

**Lampkin N** 1994 *Organic Farming* Amended Edition. Farming Press: Ipswich, UK

**Mason G and Mendl M** 1993 Why is there no simple way of measuring animal welfare? *Animal Welfare 2:* 301-309

**McInerney J P** 1994 Animal welfare: an economic perspective. In: Bennett R M (ed) *Valuing Animal Welfare* (Proceedings of a workshop) pp 9-25. University of Reading: Reading, UK

**Ministry of Agriculture Fisheries and Food** 1983-1997 *Codes of Recommendations for the Welfare of Livestock: Cattle, Sheep, Rabbits, Goats, Farmed Deer, Domestic Fowls, Turkeys, Ducks, Pigs.* MAFF Publications: London, UK

**Ministry of Agriculture Fisheries and Food** 1995 *Summary of the Law Relating to Farm Animal Welfare, (reprinted 1998).* MAFF Publications: London, UK

**Ministry of Agriculture Fisheries and Food** 1998a *Guidance on the Welfare of Animals (Transport) Order 1997.* PB 3766. MAFF Publications: London, UK

**Ministry of Agriculture Fisheries and Food** 1998b *Guidance on the Transport of Casualty Farm Animals.* (First published in 1993 but updated in 1998). MAFF Publications: London, UK

**Moss R** (ed) 1992 *Livestock Health and Welfare.* Longmans Scientific and Technical: Harlow, UK

**Porter A R W** 1991a Animal Welfare. In: *Legislation Affecting the Veterinary Profession in the United Kingdom* pp 17-45. Royal College of Veterinary Surgeons: London, UK

**Porter A R W** 1991b Practice of veterinary medicine and surgery. In: *Legislation Affecting the Veterinary Profession in the United Kingdom* pp 119-124. Royal College of Veterinary Surgeons: London, UK

**Regen T** 1982 *All That Dwell Therein.* University of California Press: Berkeley, USA

**Rollin B E** 1981 *Animal Rights and Human Morality.* (Revised edition, 1992) Prometheus Books: Buffalo, NY, USA

**Rollin B E** 1995 *Farm Animal Welfare. Social, Bioethical and Research Issues.* Iowa State University Press: Ames, USA

**Sainsbury D** 1986 *Farm Animal Welfare.* Collins: London, UK

**Sainsbury D** 1998 *Animal Health*, 2nd edition. Blackwell Science: Oxford, UK

**Sapolsky R M** 1994 *Why Zebras Don't Get Ulcers.* Freeman: New York, USA

**Seabrook M** (ed) 1987 *The Role of the Stockman in Livestock Productivity and Management.* Report EUR 10982EN. Commission of the European Communities: Luxembourg

**Singer P** 1976 *Animal Liberation.* (2nd edition, 1990). Chapman and Hall: London, UK

**Webster J** 1994 *Animal Welfare: A Cool Eye Towards Eden.* Blackwell Science: Oxford, UK

**Wiepkema P R and Koolhaas J M** 1993 Stress and animal welfare. *Animal Welfare 2:* 195-218

**Winter A C** 1998 The role of the Farm Animal Welfare Council. In: Michell A R and Ewbank R (eds) *Ethics, Welfare, Law and Market Forces: the Veterinary Interface*, pp 25-39. UFAW: Wheathampstead, UK

# 2 Dairy cattle

## J D Leaver

## THE UK INDUSTRY
### Trends in the industry

Since the early 1960s, the UK dairy industry has changed rapidly. The trend towards fewer but larger herds accelerated in the 1960s and 1970s; however, national herd size remained constant (Table 2.1). The introduction of milk quotas in 1984 led to a decline in the total number of cows nationally. By 1995 producers kept over 40 per cent of dairy cows in herds of over 100.

**Table 2.1**     **UK herd structure 1965-1994.**

| Year | UK milk producers | Dairy cows (millions) | Average herd size |
|------|------|------|------|
| 1965 | 124688 | 3.186 | 26 |
| 1975 | 76827 | 3.242 | 42 |
| 1985 | 48827 | 3.133 | 64 |
| 1995 | 36583 | 2.600 | 71 |

*Source: Dairy Facts and Figures, 1995, published by the Federation of Milk Marketing Boards.*

The traditional system was to house and milk in cowsheds. It was a labour intensive system, not ideal for cow welfare because they could not exercise properly. Increasing herd sizes, and the swing to silage as the main forage, has led to a predominance of loose housing systems. Cubicle systems are now the most common type of housing.

Manufacturers use the milk that is surplus to liquid consumption. In 1965, 70 per cent of milk that UK farmers produced was sold as liquid milk. In 1995, the public consumed only 50 per cent in this form (Table 2.2).

**Table 2.2**     **Production and use of milk in the UK, 1965-1994 (millions of litres).**

| Year | Total milk supplied from farms | Liquid sales of milk | Manufact- ured milk |
|------|------|------|------|
| 1965 | 10710 | 7459 | 3251 |
| 1975 | 13167 | 7761 | 5407 |
| 1985 | 15242 | 6926 | 8301 |
| 1995 | 14008 | 6931 | 6964 |

*Source: Dairy Facts and Figures, 1995, published by the Federation of Milk Marketing Boards.*

The consumption of liquid milk declined from 2.8 litres/week/capita in 1965 to 2.2 litres/week in 1995. Also, in the 1980s, there was a swing to skimmed and semi-skimmed milk, stimulated by the 'health lobby', at the expense of full-fat milk. The consumption of butter fell similarly.

Expanding production, declining consumption of milk and its products, and the static population size, has led to surpluses in the UK and EU generally.

### Structure of the industry

Most milk producers are in the wetter, western regions, where farms are smaller than in the major arable areas. Dairying relies largely on grass and its conserved products that are plentiful in these areas. In the UK farmers house cows for 4-8 months of the year depending on geographical location.

Between 1933 and 1994, the Milk Marketing Boards (MMB) in England and Wales, Northern Ireland and Scotland controlled the marketing of milk. These were producer organizations that had the exclusive right to buy all milk from producers. They had the power to equalize the price paid (pool price), irrespective of the Board region in which units were situated, or of the intended use of the milk.

In late 1994 the Government abolished this monopoly and replaced it with a free market system. This allows producers to sell to dairy companies or cooperatives either individually, or through selling groups. Wholesale producers supply most milk, which the buyer collects from their farms. Producer-retailers and producer-processors produce the remainder. They sell their milk through retail and wholesale outlets, and to farmhouse cheesemakers.

The price that wholesale producers receive for their milk is dependent on the value of milk sales for liquid consumption and for manufacture. There are monthly variations in price, these generally aim to depress production levels in spring and increase them in late summer. This is because milk production reaches a peak nationally in May after farmers turn out cows from their winter housing onto spring grass. The price is also subject to additions or subtractions that reflect the milk's compositional quality (in particular fat and protein content). Buyers pay a premium for good hygienic quality, which has a low total bacterial count (TBC). The detailed standards differ for each buyer, but premium payments are generally for milk with a TBC count below 20,000 bacteria/ml. Buyers also monitor somatic cell contents (SCC), which reflect the level of mastitis in a herd; they generally pay premiums for milk under 150,000 cells/ml of milk. Antibiotic contamination and adulteration with water results in deductions and penalties.

EU policy for dairying aims to manage the Community markets to obtain a target price for milk. The Council of Agricultural Ministers decide this target, but they are unable to guarantee the price. Each EU Government attempts to secure product prices through a complex system of intervention buying of butter and skimmed milk powder, and of subsidies and levies. These methods were successful in maintaining prices to farmers, but stimulated surplus production that accumulated in stores of butter and skimmed milk powder. This situation led, in 1984, to the imposition of milk quotas in all EU countries, with super levies on overproduction. Initially, quotas were well above the levels of consumption, but ever since they have fallen. Milk quotas operate both at national level and at farm level.

## BREEDING AND GENETICS
### Breeds
*Bos taurus*, from mainly temperate regions, which includes the European breeds, is one of the two major species of domesticated cattle. The other is *Bos indicus*, which occurs in tropical regions.

Dairy cattle have a conformation well adapted for milk production. Cattle breeders have always been careful not to select on performance alone but also for conformation. One objective of breeding management is to produce cows with strong suspensory ligaments to hold the udder, and sound hind legs and feet. However, there is a conflict of objectives because cows with better (wider) udders usually have worse back legs. For high yields of milk, the cow needs to mobilize body fat in early lactation and convert nutrients to milk rather than to body fat. Consequently, high yielding cows are often quite thin.

Since the 1950s the popularity of the Dairy Shorthorn and Ayrshire breeds has declined, while that of the Friesian and Holstein breeds has increased. The Dairy Shorthorn was the dominant pre-war breed. It is a dual purpose animal with a good beef carcase and dairy attributes. In Scotland, unlike the rest of Britain, the brown and white Ayrshire was the dominant breed until the 1970s. Its poor beefing potential, and the failure of those farming it to use artificial insemination (AI) technology, contributed to its decline.

Channel Islands breeds, namely the Jersey and Guernsey, produce milk of a high fat and protein content. Many Channel Island herds still remaining produce milk for direct processing on the farm.

The Friesian originated in the Netherlands. In Britain, producers selected the breed for its dual-purpose characteristics. This led to a small to moderate sized black and white animal, often with large amounts of subcutaneous fat. North American farmers selected for milk yield and size. The result was a much taller and more dairy-like animal, the Holstein. Today the continuing trend is to cross the traditional British Friesian to produce the Holstein Friesian containing an increasing proportion of genes from North America. Black and white breeds now account for over 90 per cent of dairy cows in the UK. The higher yield of these breeds (Table 2.3) has been a contributory factor to the overall increase in milk yield per cow.

The milk yield potential of cows within breeds has also risen through selection. Progeny testing of bulls and the use of AI have contributed significantly to higher production levels. The greater understanding of nutrition and management of cows has also been effective in improving yields per cow.

Table 2.3     **Production figures for major breeds in recorded herds in England and Wales 1995-96.**

| Breed | 305-day yield (kg) | Milk fat (g/kg) | Milk protein (g/kg) |
|---|---|---|---|
| *Ayrshire* | 5822 | 40.5 | 33.0 |
| *Holstein Friesian* | 6638 | 40.4 | 32.4 |
| *Dairy Shorthorn* | 5587 | 38.5 | 32.5 |
| *Guernsey* | 4703 | 48.0 | 35.3 |
| *Jersey* | 4491 | 55.5 | 38.5 |

*Source: National Milk Records, Annual Report 1995/96*

### Genetic improvement

For the past 200 years there has been a gradual improvement in the performance of dairy cows. Over the last 40 years AI and progeny testing of bulls has accelerated progress within breeds. So far the development of cross-breeding schemes to exploit hybrid vigour, as in other species of farm animals, is small.

Breeders improve genetics by selection. Most milk production traits are quantitative, as many genes are responsible for their control. Those genes controlling milk yield, for example, are concerned with food intake, digestion and metabolism, blood supply to the udder, the amount of secretory tissue and udder size. It is not surprising therefore that there is considerable variation between cows in their ability to produce milk.

The heritability of a trait determines the rate at which selection can improve it. Milk, and milk fat and protein yields have moderate heritabilities of about 0.25. Milk composition (fat, protein and lactose content), however, has a higher heritability of about 0.50. Traits concerned with viability and health, such

as fertility, disease resistance and longevity, have low heritabilities of less than 0.10. This makes them difficult to improve by selection. In these cases only management can bring significant improvements.

Only about 6 per cent of genetic progress comes from female selection and this emphasizes the overwhelming importance of sire selection. Currently, breeders select young potential AI sires from the offspring of cows of high genetic merit and progeny tested bulls. They then progeny test potential candidates. After using 750 straws of his semen, breeders lay off the young sire until several of his daughters have production records available from their first lactation. This takes 4-5 years. Breeders only select 10-20 per cent of such young sires to join the AI stud.

The genetic evaluation of both males and females is via the Individual Animal Model (IAM). This computer model integrates the performance information on the relatives of an individual. It includes details on its grandparents, parents, half brothers and sisters, offspring and, for cows, their own lactation records. The result is a Predicted Transmitting Ability (PTA); this is the expected value for a trait that a bull transmits to his offspring. Pedigree indices are the predictions of genetic merit for young animals based on half the sire PTA plus half the dam PTA.

Growing numbers of farmers are using multiple ovulation and embryo transfer (MOET) to speed up the rate of genetic progress. In the future its use is likely to increase. The production of many brothers and sisters from the same mating allows breeders to evaluate the transmitting ability of young sires more quickly. They do this by assessing the performance of a sire's sisters, instead of from their daughters as in the progeny test. MOET schemes usually involve setting up nucleus herds in which testing takes place. An added advantage over progeny testing is that breeders can measure and select for other traits, such as food intake and food conversion efficiency.

### NATURAL HISTORY AND BEHAVIOUR

Cattle were first domesticated over six thousand years ago and have been mainstays of agriculture in most countries. Their ruminant digestive system has greatly extended the population's food supply. This is because it can convert inedible fibrous foods into meat and milk suitable for human consumption.

Breeders have selected dairy cattle for their behaviour in the farming environment. Dairy cows are docile by nature and only change their behaviour in response to environmental or management changes. Farmers tend to cull cows from the herd that do not conform to the system. Cows will explore their surroundings when put into a different building or field. Dairy cattle do not normally display territorial activity.

Wild cattle and suckler beef cows produce one calf each year, and over 10 calves in a lifetime. Cows naturally hide their young calves rather than keep them following at foot, and return to them at infrequent intervals during the day for suckling. At night cows usually remain with their calves. A dam will suckle her calf 4-6 times a day. She will naturally wean her calf at 6-8 months of age. This compares with the dairy cow, which farmers separate from the calf at 24 hours, milk usually twice a day and cull after four lactations.

## Social structure

Dairy cattle quickly establish a social hierarchy within a group. It is mostly linear, although it is usual for some triangles of dominance to occur within. Quite often, the most dominant animals are not the most aggressive. Also, the cows leading the herd to and from the field for milking, and through the milking parlour, are not at the top of the dominance order. These cows are usually middle ranking. The most submissive animals, however, are often the last through the parlour and, in competitive feeding situations, are the last to feed.

It is unclear what the maximum group size should be for dairy cow management systems. Herds of over 150 individuals are now becoming increasingly common. However, for convenience of housing and management, such large herds are often split into groups of 50-100 during the winter months. It is important that the herd is stable, because any interchange of cows between groups will upset the social hierarchy. Submissive cows thrive better in much smaller herds when housed, whereas under grazing conditions they can cope in relatively larger groups.

Those parts of the body which a cow cannot lick, it often rubs against solid objects. One cow may groom another by licking its head and neck. The grooming animal is often slightly below the other in the social hierarchy.

The social hierarchy can be upset when cows are in oestrus or are sick. Submissive cows in oestrus can temporarily become dominant in encounters with higher ranking cows. Sick or lame animals often lose their position, particularly if they are unable to isolate themselves.

## Feed and elimination

Feeding is a major activity of the cow. Factors associated with the feed and with the animal determines intake. Animal factors include the genetic potential of the cow, the physiological state (pregnant, lactating, etc.) and behavioural factors. Feed intake can be expressed as:

intake = time spent eating (min/day) x rate of eating (g DM/min)

where,

time spent eating = number of feed bouts per day x mean bout length (minutes)

and,

rate of eating = number of bites per minute x mean bite size.

Table 2.4 shows the typical feeding behaviour of dairy cows. Sward conditions, particularly, modify grazing. The maximum number of grazing bites per day is about 40 thousand; this places a limit on herbage intake. With indoor feeding, restricting rations leads to faster rates of eating and to a greater variation in intake between cows.

Rumination occurs in many bouts of differing length. They are mainly when the cow is lying down, which is mostly after nightfall. Longer rumination times follow large intakes of fibrous feed. The DM content of the diet dictates the frequency of drinking and the amount an animal consumes each time.

A cow typically urinates 8-10 times a day and defaecates 10-15 times. The amount and type of feed the animal ingests determines the volume and number of these eliminations. Cattle often excrete 10-20kg of urine a day, and 30-50kg of faeces.

**Table 2.4**      **Typical feeding behaviour of lactating dairy cows.**

|  | Time spent feeding (min/day) | Rate of intake (g DM/min) | Total DM intake (kg/day) |
|---|---|---|---|
| Early season grazing | 500 | 36 | 18 |
| Late season grazing | 600 | 20 | 12 |
| Indoors ad libitum silage[a] | 220 | 50 | 11 |
| Silage as buffer to pasture[b] | 100 | 60 | 6 |
| Complete diet 60:40 ratio[c] | 300 | 70 | 21 |

[a]In addition to 9kg/day of concentrates.
[b]Offered after each milking for 45min.
[c]Ratio of concentrate DM to forage DM.

### Resting

Dairy cows generally lie down for 9-11 hours/day, during which most of the rumination occurs. The hours of darkness and weather conditions dictate the diurnal activities of lying, standing and grazing. Housed cattle have a more regular diurnal pattern of lying and feeding, which the milking time and hours of darkness prescribes. Cows usually lie in the same area of a building each time, although, with cubicle housing, not necessarily in the same cubicle. Young cattle spend a greater amount of time lying; three to six month-old animals spend about 14 hours/day, while younger calves lie for 20 hours/day.

### Reproductive behaviour

At oestrus the cow shows signs of excitability. She also increases grooming and occasional bellowing. The main behavioural signs are the attempts to mount other cows, which prompts them to mount her. If she stands to let others mount, it confirms that she is in oestrus.

A bull can detect pre-oestrus in cows. He noses the perineum and probably smells the urine. The bull sometimes exhibits an olfactory reflex, during this he extends his neck, contracts his nostrils and curls his upper lip. This is known as the 'Flehmen' posture. If the cow remains standing after nudging and courtship, the bull will mount and mate her. When mounted, a bull will normally pump his tail and at ejaculation will thrust forward. Afterwards, the bull often remains with the cow, and mates her again.

The pregnant cow generally shows no behavioural signs, but when parturition approaches she becomes restless. If in a field, she will often find a sheltered position away from the herd. Calving occurs most frequently at night, but there is some evidence that late-night feeding can delay parturition until the following day.

Most births occur without complications. In the first stage of labour, uterine contractions commence and continue for several hours and eventually the water bag and anterior feet of the calf appear. During a normal birth, the second stage, in which the calf is born, lasts about an hour. The third stage, involving the expulsion of the placenta, takes up to eight hours.

## REPRODUCTIVE BIOLOGY

Dairy cattle attain puberty when they reach about 40 per cent of mature weight. The exact age depends on the plane of nutrition; under good management conditions it will be at 10-12 months of age. There is a marked increase in follicular activity just before sexual maturity and, at puberty, normal oestrous cycles commence. These average 21 days in length, with the majority between 19 and 23 days. The cow is in oestrus for about 15 hours, although this can range from three to over 24 hours. The length is shorter in winter and for cows in poor body condition. Ovulation occurs 10-12 hours after the end of oestrus. The normal gestation length is 283 days, but this varies between animals, and depends on the sire of the foetus. Sires of continental beef breeds produce longer gestation periods. After parturition, involution (return to normal size and function) of the uterus continues for 4-6 weeks, but reproductive cyclical activity normally recommences within three weeks of calving.

### Management of breeding females
*Reproductive efficiency*
Dairy cows have their first calf at about two years of age. Farmers cull most by the time they reach their fourth lactation because they will usually be very thin, infertile and/or chronically lame.

The reproductive efficiency of a dairy herd is a major factor affecting profitability. A high annual milk production per cow is dependent on regular calving at 12 month intervals. Poor reproductive efficiency leads to longer calving indices and a reduction in milk sales per cow per annum. Producers therefore cull individuals for failing to conceive. On average, in the UK, only about 85 per cent of cows calve per annum. Of the 15 per cent loss, 8 per cent is due to cows failing to breed; these the farmer culls. The other 7 per cent is due to calving intervals of over 12 months (the average being 13 months).

Three management factors wholly account for the calving index of a herd. These are: time from parturition until the time when the decision is made to start serving; oestrous detection rate; and pregnancy (or conception) rate per service. Table 2.5 shows average and target values for these factors.

**Table 2.5** **Target and average values for reproductive performance.**

|  | Target | Average |
|---|---|---|
| *Calving to start serving (days)* | 60 | 70 |
| *Oestrus detection rate (%)* | 80 | 60 |
| *Pregnancy rate to all services (%)* | 65 | 55 |
| *Calving index (days)* | 365 | 395 |

In well-managed dairy herds, delay in the onset of oestrous cycles post partum is not a problem; most cows showing cyclical activity by 21 days, and almost 100 per cent by 42 days post partum. High yielding cows may resume cycling slightly late, but rarely more than 10-14 days. Abnormalities of the reproductive tract following difficult calvings (dystocia) cause delays in resumption of cyclical activity. Such abnormalities include retaining of the placenta, endometritis, or luteal or follicular cysts on the ovary. A veterinary surgeon should examine any cows exhibiting these symptoms. Underfeeding of cows can also lead to a delay in resuming normal oestrous cycles.

*Oestrous detection*
The oestrous detection rate (ODR) is the number of cows handlers observe in oestrus during a three week period, in relation to the number they expect to observe. Sixty per cent is the average ODR on farms. Over 90 per cent of those cows not obviously in oestrus are cycling, the remainder are acyclic. The problem is therefore mainly one of stockmanship. Duration and intensity of oestrous activity is variable

and in winter, when activity is less, stockpersons miss many oestrous periods.

The presence of a bull enhances oestrous activity. Siting a bull pen next to the housing area or collecting yard can improve detection rates. Fixing heat-mount detectors to the back, or applying paint to the tail-head of the cow, are aids to improving detection rate. When another animal mounts a cow in oestrus, the pressure on her back triggers off a colour change in the heat-mount detector. With tail paint the mounting action rubs off the colour.

One alternative is to synchronize the oestrous period of several animals, by controlling the length of the luteal phase with prostaglandin or progesterone. Prostaglandin F2$\alpha$ is an abortive agent; it is therefore vital not to administer it to females that might be pregnant. It is also dependent for its action on the cow exhibiting normal cyclical activity, so treatment of non-cycling females will not induce oestrus. Normally, the treatment involves two injections at 11 day intervals with inseminations at 72 and 96 hours after the second injection.

Farmers administer progesterone as an impregnated intravaginal coil, which they insert for 9-12 days. Following withdrawal, the cow is ready for insemination 48 and 60 hours later. This treatment, unlike prostaglandin, is also beneficial for females that have not resumed normal oestrous cycles, as it stimulates ovarian activity.

Unfortunately, synchronization techniques, while overcoming the need to detect oestrus, result quite often in poor conception rates. Their use, therefore, is not widespread. Instead, farmers use it for individual problem cases where the veterinary surgeon has confirmed that the cow has normal reproductive function, but oestrus has been impossible to detect.

*Conception*
Inseminating cows at the optimum time (mid to late oestrus) results in the spermatozoa fertilizing about 90 per cent of ova. The average pregnancy rate (diagnosed 6-8 weeks post-insemination), however, is only about 55 per cent. Embryo losses account for the difference, most of them occurring within 14 days, with the cow returning to oestrus after the normal 21 day period. About 10 per cent of embryos are lost after 14 days and most losses occur before

implantation is complete at about 35 days. These losses, noticeable by a delayed return to oestrus, are more common in older cows.

Many farms still use natural service by bulls. However, the risk of slowing genetic progress, and the inherent danger of handling bulls (in particular those of the dairy breeds), has persuaded most to rely mainly on artificial insemination (AI).

### Artificial insemination

Generally it is the AI stations that keep the semen collection bulls. After collection, AI staff examine the semen microscopically to find out the proportion of live spermatozoa. They then add a diluent before putting the mixture into 0.25cc straws and freezing them in liquid nitrogen. Each straw contains about 20 million spermatozoa. The station stores these in tanks of liquid nitrogen until they use them. Trained operators from breeding companies carry out most inseminations. The optimum time to AI cows is during mid to late oestrus. The insemination technique involves the operator inserting a gloved hand into the rectum and gripping the cervix through the rectum wall. They then insert a catheter with the straw of semen into the vagina, and pass it through the cervix. Finally they deposit the semen at the anterior end of the cervix. Careful training and hygiene are necessary for the technique to be effective.

To reduce costs, and as an attempt to increase pregnancy rates, some farmers train their own staff to carry out do-it-yourself (DIY) AI. Farmers store the semen in tanks on site, which the AI centre tops up with liquid nitrogen at intervals. Farmers can select the time of day they serve cows according to the stage of oestrus. With the trained operator, the once daily farm visit restricts the time of insemination. Nevertheless, this potential advantage of DIY has to be set against the relative inexperience of the on-farm operators.

Appropriate handling facilities, such as AI stalls, are necessary for inseminating cows in. It is important to only hold the cow in such stalls for a minimum period. Deprivation from food and water and from the rest of the herd could be stressful.

### Pregnancy diagnosis

Traditionally, a veterinary surgeon diagnoses pregnancy at 6-8 weeks post-insemination by palpating the reproductive tract per rectum. However, the use of ultrasound to scan the developing embryo per rectum now allows diagnosis as early as 28 days. The development of tests that measure the progesterone levels in milk allows even earlier diagnosis at 21-26 days post-insemination. At this time levels in the pregnant animal will be markedly higher than in the non-pregnant, cycling cow. These tests have the advantage that they only take a few minutes and that it is possible to carry them out on the farm.

### Recording

To attain a high level of reproductive efficiency it is essential to have a satisfactory system for recording breeding data. Dates of calving, oestrous periods, services, pregnancy diagnoses, and veterinary treatments are particularly important. Many dairy producers use circular and horizontal wall charts to record the reproductive progress of individual cows. Farmers of large herds are increasingly using computerized recording systems.

## MILK PRODUCTION

Breeders have selected the modern dairy cow to produce a large amount of milk relative to its size. When a dairy cow is at her peak she may produce 6-12 thousand litres over a lactation period of 10 months, with a two month break before she calves again. Maximum daily yields are over 40 litres. This is considerably greater than the suckler beef cow who secretes 8-10 litres/day. Dairy cows therefore need an ability and willingness to consume large amounts of feed daily and an efficient digestive system to provide the basic components for milk production.

### Mammary system

At birth, the mammary system is only partially developed, and it continues to grow during the rearing period. Between three and nine months of age there is a rapid multiplication of cells in the mammary gland. There is also a continuing proliferation of the duct system after puberty. The onset of pregnancy causes rapid development of the gland under the influence of the pregnancy hormones. During the final month the development of alveoli accelerates. This mammary growth continues after parturition until milk yield peaks six weeks after calving. The balance between production and loss of cells then swings in favour of loss. A cow's udder has four quarters, with the hind ones producing about 60 per cent of the milk. In older cows, the ligaments that hold the gland in place may stretch, producing the pendulous appearance of the udder.

Alveolar cells synthesize and secrete constituents of milk continuously. The cells discharge these into the alveolar lumen, and from there they pass into the ducts, sinuses and cisterns. The udder of a dairy cow, on a twice daily milking routine, may hold over 20 litres at any one time; this is ten times the maximum for a beef cow suckling her own calf (Webster, 1993). Milk secretion rate does not fall unless the interval between milkings is greater than 15 hours. The build-up of a protein inhibitor in the lumen of the alveolus influences the rate of milk secretion. Increasing the frequency of milking removes more of the inhibitor and, as a result, milk secretion per day rises. A large amount of blood flows through the udder (about 500 litres per litre of milk). Some constituents transfer directly from the blood to the milk.

The major precursors of the constituents of milk are free fatty acids, triglycerides, amino-acids, acetate and glucose. Figure 2.1 outlines these components. A cow produces milk fat from fatty acids that it both synthesizes in the udder and transfers from the blood. Cows manufacture fatty acids, which contain 4-10 carbon atoms, from acetate and $\beta$-hydroxybutyrate arising from the volatile fatty acids (VFAs) she produces in her rumen. Those with 18 or more carbon atoms pass directly from blood triglycerides within the udder. The intermediate length fatty acids derive from both sources. Milk fat contains a mixture of both saturated and unsaturated fatty acids.

The lactose in milk (milk sugar) comes mainly from blood glucose. Milk yield is especially dependent on the amount of water secreted in the udder, and this in turn controls the water-soluble constituents influencing osmosis, in particular lactose. Thus, milk yield relates closely to the amount of lactose the udder synthesizes.

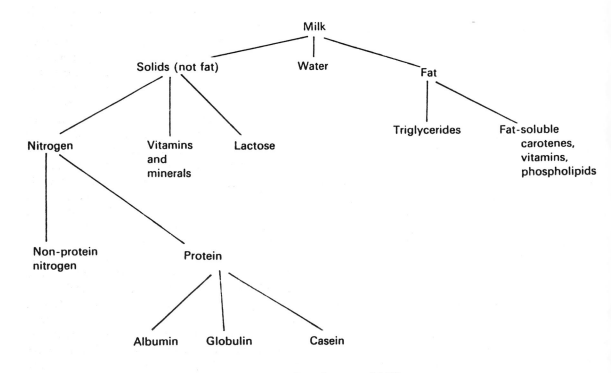

**Figure 2.1    Constituents of Milk**

Milk protein comes mainly from the amino-acids in the blood. Calcium, phosphorus and magnesium, besides potassium, sodium, chlorine, and trace elements, make up the ash content of milk. The udder absorbs vitamins from the blood, in particular vitamins A and the B complexes. Other important vitamins include C, D, E and K.

Figure 2.2 shows typical lactational trends in milk yield and composition. Nutritional factors, such as energy and protein intakes and diet formulation, and non-nutritional factors, such as genetic potential, age and disease, modify these trends.

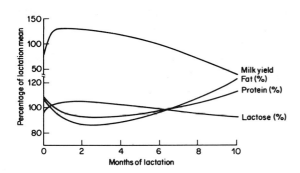

**Figure 2.2**      **Lactation curves of milk and its constituents.**

### Milking management

The efficiency of the milking process has become more important as herd sizes and the ratio of cows to stockperson have grown. Hygienic extraction of milk and transfer to the bulk tank, with little stress on cow and person, are the main objectives.

It is important not to over-feed 3-9 month old heifers when cells in their mammary gland are rapidly multiplying. Growth rates of over 0.75kg/day can lead to a reduction in secretory tissue production, and to a permanent depression in subsequent milking performance.

From the fifth month of pregnancy milk yield begins to decline more rapidly. This usually coincides with the eighth or ninth month of lactation. It is normal practice to dry-off the cow 6-8 weeks before parturition is due, the simplest method being simply to stop milking. Farmers may have to offer only straw and water for two or three days to cows giving large yields at drying off (over 15kg/day) to reduce milk secretion. During the dry period the cow should continue to lay down fat in preparation for the next lactation, and therefore it is important to feed the animal accordingly.

### Milk output

The performance of a dairy herd requires close monitoring and control. Within a quota system, month-by-month comparison of actual versus anticipated milk sales is an integral part of management. Computer programs normally produce this prediction. It uses the expected calving dates of cows to assess the mean lactation yields, with that calving pattern and number of cows, to produce the annual quota. Farmers base their winter feeding programme on the quantity and quality of forage available, and on lactation yields they require. Irrespective of the decision to use a flat-rate or a feeding-to-yield system, a ration formulation programme can be helpful to assess concentrate inputs.

It is therefore most important to record total daily milk sales and compare these with predicted sales. A graph in the dairy or farm office shows at a glance whether changes in feeding management are necessary. If daily sales are below expected levels, remedial action is necessary. The average response to an additional 1kg of concentrates is 1kg of milk, but farmers can obtain a greater or lesser response, depending on a variety of factors. Milk yield response per kg concentrate will increase when there is limited quantity of forage (less than 90 per cent of appetite), when it is of poor quality (less than 10MJ of ME/kg DM) or when concentrate feeding levels are already low (less than 0.25kg/kg milk). Extra concentrates may also increase liveweight and reduce forage consumption. In examining the economics of concentrate feeding, therefore, it is important to account for the effects on milk yield, milk composition, weight change and forage intake.

Except in higher yielding herds (over 6500kg milk/cow), there is little benefit in having a more even time interval between twice daily milking than 14 hours and 10 hours. Thrice daily milking increases yields by about 15 per cent at all stages of lactation compared with twice daily milking. However, when nutrition is not limiting the high yielding dairy cow, it is not the synthesis of milk that limits production but the infrequency of milking (Webster 1993).

*Milking machines*

Machines extract milk by applying a vacuum to the teat end within the teat-cup liner (Figure 2.3). The resulting pressure difference between the open liner and the teat sinus causes the teat sphincter to open, and the milk to flow out. A pulsation of alternate opening and closing of the liner prevents the damage a continuous vacuum would cause a teat. Milking machines work at a vacuum level of 45-50 kilopascals (kPa) ie about half an atmosphere. In the pulsation cycle, the liner is open for 50-75 per cent of the time, and there are 45-70 cycles per minute.

The teat cup liners are made of rubber or synthetic rubber. It is vital that this liner does not slip down the teat during milking. The milk passes from the liner into a claw-piece where the four liners meet, and this normally has an air-bleed that enhances the speed of milk removal. It then passes down the long tubes either into individual cow, glass recorder jars or directly to the milk line and then to the bulk tank.

*Milking parlours*

With the increase in herd size, most farmers milk their cows in parlours. Some farmers with small herds, however, still milk in cowsheds, with a line transferring the milk to the bulk tank. The most common milking parlour in operation is the static herringbone (Figure 2.4). Rates of throughput average about 30 cows per person hour for cowsheds, 55 cows per person hour for static herringbones and over 100 cows per person hour for rotaries.

The time the operator spends on the work routine per cow, and for cows to milk out, determines milking performance (cows milked per person hour). Work routine is the time a stockperson takes per cow to carry out all the operations associated with milking (eg cleaning teats and applying clusters). Waiting time depends on the work routine, and the length of time the cows take to milk out. The shorter the work routine, the more likely that waiting time will occur. Longer parlours (more cows per side) reduce the waiting time per cow. This is because the operator spends longer carrying out the work routine on one side, before returning to the other where milking is already in progress.

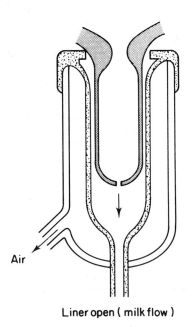

Liner open ( milk flow )

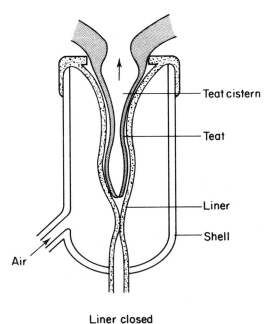

Teat cistern

Teat

Liner

Shell

Liner closed

**Figure 2.3**          **Extraction of milk from the teat during machine milking.**

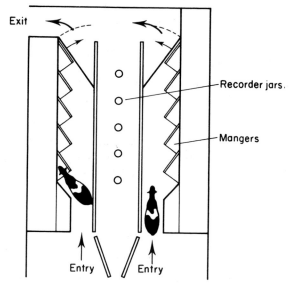

**Figure 2.4    A 10 stall, 5 milking unit herringbone parlour.**

The dairy buying the milk carries out a total bacterial count of milk (TBC). The cleanliness of the cows' teats at milking and hygiene in the milking plant mainly affects TBC. Washing teats with clean water (containing disinfectant), and drying them with individual paper towels, reduces the TBC level. In well-managed herds TBC is less than 15 000 bacteria/ml. Milking parlour plants are cleaned automatically at the end of each milking, normally by circulation cleaning, which takes about 15 minutes.

### Rearing dairy replacements

Most dairy farmers rear their own replacements rather than buying them in. The main reason is the job satisfaction from influencing genetic progress in the herd through the choice of sires. Also, being self-contained prevents the introduction of disease, and it is possible to train and prepare heifers before they enter the dairy herd.

Very often, however, rearing is inefficient. Most dairy farmers rear too many replacements, leading to high culling rates in the herd. In an efficient business the culling rate per year is only about 20 per cent. A farmer rearing heifers to calve at two years of age will need about 50 youngstock per 100 dairy cows. Nonetheless, many herds carry over 100 female youngstock per 100 cows. In most enterprises a dairy sire need serve no more than 60 per cent of cows. To enhance calf values a beef sire should cover the remainder (see Chapter 3). Where possible, farmers should serve the cows of high genetic index with the dairy bull. Reducing calving age to about two years, and having a replacement rate of less than 25 per cent, is financially more beneficial than conventional calving at 2.75 years at a replacement rate of over 25 per cent. The advantages stem from the need to have less land for youngstock, less working capital and associated interest charges, labour, mechanization and housing, and a higher rate of genetic progress.

Calving heifers at two years of age at target weights (Table 2.6) does not require intensive rearing, and much of the total diet is grass and forage (Table 2.7). There is clear evidence that intensive feeding, producing very high pre-puberty growth rates, can be detrimental to lifetime milk production. The period of rearing that is most critical is the six months before puberty. At this time growth rate should not exceed about 0.8kg/day, otherwise it may adversely affect mammary development and lead to permanent harmful effects on performance.

**Table 2.6    Target liveweights for heifers of different breeds.**

| | Target weight (kg) | | | |
|---|---|---|---|---|
| | Birth weight | Mating | Pre-calving | Mature |
| *Holstein-Friesian* | 43 | 360 | 550 | 660 |
| *Ayrshire* | 32 | 280 | 430 | 510 |
| *Guernsey* | 27 | 260 | 390 | 450 |
| *Jersey* | 24 | 220 | 350 | 400 |

*Newborn calves*

A clean dry pen is essential when calving takes place indoors. After birth, it is important to remove any

**Table 2.7    An example of a rearing system for autumn-born Holstein Friesian heifers.**

| Month | Liveweight (kg) | Total concentrates | Bulk feed |
|---|---|---|---|
| *September* | 43 | 15kg milk replacer + 60kg concentrate | Hay/straw |
| | 0.60 | | |
| *November* | 80 | 350kg concentrate | Silage |
| | 0.75 | | |
| *April* | 200 | 100kg barley | Grazing |
| | 0.70 | | |
| *October* | 330 | 250kg barley | Silage |
| | 0.60 | | |
| *April* | 430 | – | Grazing |
| | 0.80 | | |
| *September* | 550 | | |

mucus from the nose and mouth of the calf, and dress the navel with iodine solution or an antibiotic spray. A calf should suckle its dam within six hours of birth to ensure that it absorbs adequate immunoglobulins, which the colostrum contains. The ability of the calf's alimentary tract to absorb the immunoglobulins declines after about six hours, and is negligible by 24 hours after birth. Immunoglobulins are important in preventing *Escherichia coli* infections in the first few weeks of life, and the onset of other diseases.

If the stockperson does not see the calf suckle in the first six hours, then they should remove colostrum from the dam or another recently calved cow, and feed it by hand. It is important to feed a minimum of three litres with a teat and bottle, or by using a stomach tube to transfer it directly to the abomasum.

*Calf rearing systems*
The calf should remain with its dam for at least 24 hours. If, however, a cow continues suckling her calf a few times a day between milking, even for a couple of weeks, she and her calf will develop a strong emotional bond. Separating them before natural weaning takes place (at six months) is extremely distressing to both. As it is generally impractical to allow calves to suckle for so long, most farmers transfer calves to individual pens after 24 hours. There they either train them to drink from a bucket or feed them by teat. This way the distress to the dam

and calf is less (Webster 1993). It is important to continue feeding colostrum at a minimum of five litres/day in two or more feeds, on days two to four inclusive.

At five days of age, it is possible to change to a milk substitute. Table 2.8 shows the four most common methods of feeding (see also Chapter 3 pp 67-72). In all systems, clean water, fresh concentrates and a forage (hay or straw) should be available. The herdsperson can wean abruptly providing the calf is eating at least 650g/day of concentrates. After weaning, farmers should house calves in groups and offer forage, with an appropriate level of concentrates (1-3kg/day depending on forage quality). This will give a growth rate of 0.65-0.80kg/day.

**NUTRITION AND FEEDING**

The high genetic merit lactating dairy cow is faced by an intense and continuous demand for nutrients to sustain the great demand of the mammary gland to synthesize milk. Dairy cows have difficulties in meeting this demand for nutrients due to the quality and quantity of the food, gut capacity to process the food, and the time available for eating. Most of the feed eaten annually by the dairy cow is grazed grass and forage; the remaining part of the diet is concentrate feeds. This enables cows to receive high energy intakes. In total, high yielding cows have a daily dry matter (DM) intake of over 3.5 per cent of

**Table 2.8**     **Example of milk substitute feeding systems for calves following 5 days of colostrum feeding.**

|  | Once-daily bucket | Twice-daily bucket | Cold[a] *ad lib* | Warm[b] *ad lib* |
|---|---|---|---|---|
| *Age at weaning (days)* | 35.0 | 35.0 | 35.0 | 35.0 |
| *Amount milk powder (g)[c]* | 500.0 | 250.0 | 1000.0 | 1000.0 |
| *Amount water (litres)[c]* | 3.0 | 2.5 | 7.0 | 7.0 |
| *Milk powder/calf to weaning (kg)* | 15.0 | 15.0 | 35.0 | 40.0 |
| *Live-weight gain (kg/day)* | 0.5 | 0.5 | 0.7 | 0.9 |

[a]   Fed through a teat connected to a container of milk substitute

[b]   Fed through a teat connected to a dispensing machine of milk substitute

[c]   For once- and twice-daily bucket, amounts refer to amount per calf at each feed, and for *ad lib* systems, amounts refer to concentration of milk substitute on offer.

their bodyweight, this equates to over 5.5 tonnes DM per annum. Cows require energy, protein, minerals, vitamins and water for body maintenance and productive purposes, including lactation, reproduction and growth.

**Digestive system**

Cattle are ruminant animals and can digest large amounts of fibrous material. The animal grasps the food with its powerful rough tongue and its eight incisor teeth (in an adult) on the lower jaw and the dental pad on the upper. Molar teeth masticate the food with copious amounts of saliva. Cows secrete over 100 litres of saliva a day. It contains phosphate and bicarbonate, and is an important buffering agent for the rumen, maintaining the pH of the contents at about 6.5.

A unique aspect of digestion in the ruminant is rumination (chewing the cud). The animals regurgitate boluses of coarse food from the rumen into the mouth and chew it for about a minute. They then swallow the bolus and repeat the process. Cows spend about 6-8 hours/day ruminating; the actual length of time depends on the amount of fibrous material consumed. To maintain normal rumen function, some fibrous feed is necessary. This should not be less than 0.5kg DM/100kg bodyweight as grass or forage.

The rumen has a capacity of about 150 litres and is subdivided into four compartments (for an illustration of the digestive system, see Chapter 3, Figure 3.5). It acts as a large fermentation vat where the microorganisms digest the food. Protozoa and yeasts help the bacteria in the rumen; the balance of the various microbes varies with the type of diet. It takes 1-2 weeks for the population to adjust to a large change in diet, such as from high forage to low forage. Farmers must, therefore, make changes gradually or digestive upsets will result.

A cow hydrolyses carbohydrates, such as cellulose and starch, in its rumen to monosaccharides. It subsequently ferments these to the organic acids - acetic, propionic, butyric and some longer-chain acids. The animal absorbs these volatile fatty acids (VFAs) through the rumen wall where they pass into the bloodstream. High fibre diets lead to 60-70 per cent of the acids as acetic acid, 16-20 per cent as propionic acid and 7-12 per cent as butyric acid. Substituting forage with concentrates reduces acetic acid to about 50 per cent, with a balancing increase in the proportion of propionic acid. The proportion of VFAs a cow produces has important effects on its metabolic hormone balance, particularly of insulin and growth hormone; this has implications for milk production and body fat deposition. Diets producing much propionic acid reduce milk fat content and encourage the partition of energy to body fat at the expense of milk production.

An important function of the rumen microbes is the synthesis, from non-protein nitrogen, of microbial protein. Ruminants subsequently digest this in the stomach (abomasum) and small intestine, in a similar manner to a monogastric animal. Microbes also synthesize B-complex vitamins.

Fermentation produces large amounts of gas, particularly methane and carbon dioxide. Cattle expel these from the rumen by the regular reflex action of eructation (belching).

The reticulum is a sac at the anterior end of the rumen. It has a honeycomb appearance and it passes food boluses up the oesophagus during rumination, and digesta from the rumen to the omasum. In the pre-ruminant calf, a reflex action stimulated by sucking or drinking milk causes the closure of the muscular reticular groove. This gives direct access to the abomasum, and allows the milk to bypass the as yet poorly developed rumen.

The omasum removes a large amount of water and organic acids from the digesta. It has a capacity of about 15 litres and, like the rumen and reticulum, has no secretions. The abomasum is the true stomach secreting the normal gastric juices. It and the small intestine digests the food further, as in monogastric animals.

**Food intake**

A major factor affecting the performance of dairy cows is the voluntary food intake. Both animal and feed factors determine this.

Food intake changes as lactation progresses, and peak intake occurs around 15 weeks post-partum. This peak occurs earlier in, and is greater for, thin rather than fat cows. Intake increases by about 1kg DM/50kg bodyweight, and by approximately 1kg DM/4kg milk.

Feed factors concern the availability and type of feed. The most important feed factor affecting intake is the metabolizable energy (ME) concentration of the diet (MJ/kg DM). The microorganisms of the rumen slowly break down foods that have a low energy density and high fibre content (such as low-quality forages). Consequently, the rate of passage of material out of the rumen is slow and intake is limited. Increasing the energy concentration of the

diet, by substituting concentrates for forage, leads to a faster throughput of the rumen and a higher intake. At very high energy levels the main limitation to intake is not the physical capacity of the rumen and the rate of passage of digesta. Instead the metabolic control mechanism, which acts in the same way as in the monogastric animal, limits consumption. In this situation further concentrate feeding does not increase ME intake.

**Energy**

In the UK the energy requirement of the dairy cow is measured as metabolizable energy (ME). This is the amount of energy available for metabolism after subtracting the energy losses in faeces and urine, and as gas from the rumen (Figure 2.5).

Alderman and Cottrill (1993) list the ME requirements of dairy cattle, and the ME contents of feeds. Gross energy (GE) of concentrate feeds is greater than for forages, but in the total diet GE is normally about 18-19MJ/kg DM. The main variation between rations is the faecal energy loss (and therefore digestible energy content). Losses in the faeces range from under 15 per cent for some concentrates to over 60 per cent for some straws. Urinary energy losses are normally about 2-8 per cent of GE and gaseous losses from 6-10 per cent. It is therefore possible to estimate ME from digestible energy (DE) with some accuracy using a factor of 0.81.

**Protein**

Cows require protein for both maintenance and production purposes. Formerly, protein requirements were expressed as digestible crude protein (DCP), with maintenance requirements of 350g/day and production requirements of 55g/kg milk for a 600kg cow. This system has limitations and does not allow for how the cow digests and metabolizes dietary protein. The protein proportion of the diet is made up of true protein and non-protein nitrogen (NPN). Microorganisms in the rumen degrade part of the true protein of the diet and the NPN to ammonia. The protein which has not been degraded bypasses the rumen and flows into the abomasum and small intestine, where enzymes digest it. Microorganisms use much of the ammonia in the rumen for growth, and this microbial protein also passes along the

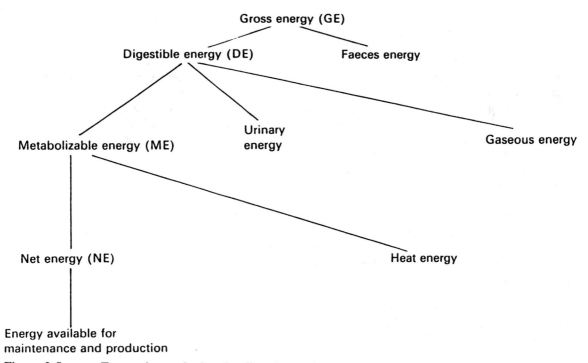

**Figure 2.5**    **Energy losses during the digestion and metabolism of a feed.**

alimentary canal for normal enzymatic digestion. The blood absorbs the ammonia that is surplus to the requirements of the microorganisms. It transports the ammonia to the liver where it is converted to urea. Saliva recycles some urea via the rumen, and the animal excretes the remainder in the urine.

The Metabolisable Protein (MP) System quantifies the animal requirements for protein, and the protein supply from the feed. Metabolisable protein is that which is available for metabolism following digestion and absorption, from the small intestine. It is made up of the digested microbial true protein and the digested undegraded protein.

Feed therefore has a rumen-degradable protein (RDP) content and an undegradable protein (UDP) content. For high yielding cows in early lactation, microbial protein production is inadequate to supply their protein requirements. Rations therefore also need to supply UDP. The choice of protein is thus very important in early lactation. The digestibility and amino-acid composition of the UDP are additional factors affecting the supply of protein to the tissues of the animal.

Matching protein supply to protein requirements is a complex process. This is because associative effects with other ingredients, and feeding level, which affects the rate of passage through the rumen, influences the degradability of feeds. A further factor in protein nutrition is the effect of the crude protein (CP) content of the diet on voluntary feed intake. Total DM intake rises as the CP content of the diet increases up to about 190g/kg DM. Alderman and Cottrill (1993) describe the Metabolisable Protein System in detail.

**Minerals and vitamins**
Both minerals and vitamins are necessary for the basic functioning of the body. Underwood (1981) discusses fully the requirements. Calcium and phosphorus account for about 70 per cent of the ash content of a cow's body, and they are closely interrelated. The requirements for calcium are about 15g/day for maintenance and 1.7g/kg milk, and for phosphorus 13g/day for maintenance and 1.5g/kg milk. It is not easy to mobilize magnesium from the bones where the body stores it. It is therefore essential to feed 11g/day for maintenance plus

0.7g/kg milk. In spring, and occasionally in autumn, grazed grass is low in magnesium; cows then require supplementation to prevent the onset of hypomagnesaemia.

## Water

Cows need to drink four or five times a day (more on dry forages than on silage). An adequate supply of clean water in troughs or bowls is therefore essential. Non-pregnant, non-lactating cows require about 4kg water/kg DM; this comes from the feed and drinking water. At low temperatures (0°C and below) this falls to 3.5kg/kg and at high temperatures (over 27°C) it increases to about 5.5kg/kg feed DM. Cows in late pregnancy require about 50 per cent more water, and lactating cows have an additional requirement of about 0.87kg/kg milk they produce. Thus, a 600kg cow yielding 30kg milk/day, on a diet of silage and concentrates, will drink about 70kg water/day.

## Practical winter feeding

The choice of strategy for feeding the dairy herd depends on the available resources of the farm, in particular the land, labour, capital and milk quota. Thus the farmer has to decide how many cows to keep at what milk yield level to meet the quota target. The choice of few cows at high yield or more cows at a lower yield depends on land and labour availability, cost, and the opportunity for other profitable enterprises.

### Forage

In the UK, silage, mainly from grass although in many areas from maize, has gradually replaced the traditional basal winter feeds of hay and roots. The amount of grass silage each hectare of grassland produces depends on the rainfall, soil type, fertilizer policy, and the frequency of cutting. Earlier and more frequent cutting leads to better quality (digestibility and ME), but a smaller quantity. Forage quality and the amount of concentrates determine the cow's requirement for forage for winter feeding, as both factors affect forage intake.

Digestibility, ME and the fermentation characteristics of silage influence its quality, and this in turn affects consumption. Silages that produce a butyric acid fermentation with a high pH and high ammonia nitrogen (over 100g/kg total nitrogen)

content, are not particularly acceptable, and feed intakes are low.

If the forage a farm produces is deficient either in quantity or quality, it will be necessary to purchase feed to make up for the deficiency. Substitutes, if silage is scarce, include hay, straw, wholecrop cereals and brewers' grains. If the available silage is of poor quality, the farmer can purchase more expensive concentrates to supplement it.

## Concentrates

Dairy cow concentrates can comprise a whole range of materials. These include 'straights', such as barley, wheat, sugar beet, maize gluten, rapeseed meal, soyabean meal and fishmeal; protein concentrates, which are mixtures of high protein straights fortified with minerals and vitamins; and supplements, such as small amounts of vitamins and minerals. The most common pelleted compounds contain least-cost formulations of straights plus supplements.

The system farmers use to allocate concentrates to individuals is not as important as the amount of concentrate the herd receives. This latter amount is a major factor affecting profitability. Traditionally, dairy producers offered concentrates in proportion to the milk yield of the cow (eg 0.4kg concentrates/kg milk). However, feeding according to yield gives no advantage in milk production compared with providing all the same daily ration. This is true for cows in early and mid lactation (weeks 1-24) and where the farmer feeds the same total quantity of concentrates to the herd. Consequently, many use a simple flat-rate system. They feed all cows in early and mid lactation identical levels of concentrates. For females later in lactation (post 24 weeks of lactation) it is possible to reduce the concentrate level at a rate of 1kg every four weeks. This will not adversely affect the persistency of lactation.

In larger herds, which use more straights than compound concentrates, farmers provide mixed diets of concentrate and forage (often termed total mixed rations or TMR). This system also places less emphasis on individual cow feeding. The normal practice is to offer mixed diets with a ME concentration of 11.5-12MJ/kg DM in early lactation, 11-11.5 in mid lactation and 10.5-11 in late lactation.

## Feeding systems

Farmers with herds of less than 120 cows normally provide silage from self-feeding systems. Cows graze the silage face, and a barrier or electrified wire controls the feeding and prevents waste. If the height of the silage is greater than 2m, it will be necessary to remove the top material mechanically and feed it in troughs. This is because cows have difficulty in pulling silage from the face due to its density.

When the silage is available for about 20 hours a day, a feed face width of 15-20cm/cow is adequate. In systems where farmers provide low amounts of concentrates (6kg/day or less), or where there is limited time available for feeding, it is vital to increase feed face width (e.g. up to 30cm/cow). If this is not possible, farmers should offer some forage in troughs. The barrier needs maintaining at a distance of 40-60cm from the feed face, otherwise it will restrict intake. If using an electrified wire the height above the ground should be about 80cm.

In trough, bunker or passageway feeding systems, farmers transport the silage from the silo with a fore-end loader on the tractor, or by a forage box. The feed face width per cow should be similar to that for self-feeding, providing there are no other feeds, such as concentrates or roots. When feeding other rations in addition, all cows need to have access simultaneously and a minimum width of 65cm/cow.

The storage of high DM silage (over 30 per cent DM) in concrete or metal towers, with automatic filling and emptying, is a high capital and maintenance cost system. Weighed amounts of forage, with added concentrates, travel directly to troughs without needing much labour. Few dairy units use such fully mechanized storage and feeding systems.

Farmers feed much of the concentrate ration in the milking parlour. It is possible to programme the dispensers for individuals. They transfer the exact amount of concentrates into the trough in front of each milking stall. The amount of concentrates that it is possible to feed at each milking depends on the rate of eating, and the milking parlour size and work routine. The maximum amount it is possible to feed is 3-4kg.

Out-of-parlour feeders work on the same principle but are in the housing area. Transponders on collars individually identify cows. When they enter the feeder, it recognizes them and delivers a programmed amount of concentrates. This type of dispenser normally divides the daily allocation between four six-hour periods so that the cow does not eat all at once. It is essential with such dispensers that the feeding animal has adequate protection from dominant individuals stealing their allowance. Such feeders therefore incorporate a stall with a self-closing rear gate. Where cows receive their rations in or next to the housing area, it is important to provide ample space. This is so that all animals, irrespective of their place in the dominance order, are able to receive their daily requirement of feed.

If cows do not need individual rationing, then it is possible to mix out-of-parlour concentrates with the silage. Farmers do this either by putting a layer of concentrates in or on the silage in the forage box, or by using a mixer wagon. The latter method results in a more homogeneous mixture. Providing forage and concentrates as mixtures has the cost advantage of allowing the use of a wider range of feeds, including straight concentrates and silage substitutes. Frequent feeding of concentrates, either mixed with the silage or through out-of-parlour feeders, is less stressful nutritionally on the rumen. Large amounts of concentrates (over 3kg/feed) depresses rumen pH. Potentially this causes problems such as cows going off their feed, reducing total intake, and lowering milk fat contents.

## Practical summer feeding

### Herbage

Grazed herbage is the cheapest food available, but its efficient use often presents problems. Management requires both the efficient production and use of herbage, while also maintaining cow production levels.

The performance of grazing dairy cows depends on a range of feed and environmental factors that influence the animals' willingness and ability to harvest the grass (Figure 2.6).

Under poor pasture conditions, the rate of intake (g DM/min) declines and, as the time spent grazing (min/day) increases only slightly, the total herbage consumption falls. This is common where grass availability and quality decline in late season. The rate of intake depends on the rate of biting (bites/min)

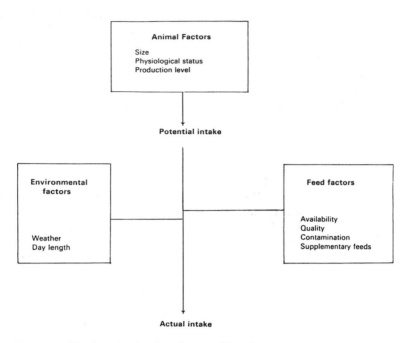

**Figure 2.6    Factors affecting the intake of grazed herbage.**

and bite size (g DM; see also p 20 ). Rate of biting varies to only a small extent and consequently the major factor determining intake is bite size. To maintain high herbage consumption, it is vital to manage pasture in a way that cows can take a large sized bite. Table 2.9 gives typical intakes and behaviour measurements over the grazing season.

**Table 2.9    Grazing behaviour and typical grass intakes of dairy cows.**

| Season | Intake[a] (kg DM) | Grazing time (min/day) | Rate of biting (bites/min) | Bite size (g DM) |
|--------|-------------------|------------------------|----------------------------|-------------------|
| *Early* | 18 | 510 | 64 | 0.55 |
| *Mid* | 14 | 570 | 65 | 0.38 |
| *Late* | 13 | 610 | 66 | 0.32 |

[a] Intake = grazing time x rate of biting x bite size.

### Supplements

Feeding concentrates to grazing cows is common practice, although the economics are often questionable. The mean response in milk yield is normally less than 1.0kg/kg extra concentrates, as the cow usually compensates by consuming less herbage. Under very good pasture conditions (at bite sizes of over 0.5g DM) the time an animal spends grazing falls by over 20min/kg concentrate DM. The best use of concentrates is therefore where sward conditions limit herbage intake through a reduction in bite size. Cows of high milk yield potential also require supplementation if the farmer wishes to reap the benefits.

When drought or overstocking lead to deficiencies in herbage availability, forages can be a useful supplement. Buffer feeding is where farmers offer silage, hay or a straw-based mix *ad libitum* for a short period daily. This system allows the cows to decide for themselves how much to eat. The level of consumption reflects the adequacy of the sward. It is possible to feed forage for 30-60 minutes after milking or, in more extreme circumstances, cows can be housed overnight and offered forage, and graze intensively during the day.

### Grazing systems

Rotational systems of grazing, such as strip-grazing or paddock grazing, control the amount of herbage available. Such systems use grassland efficiently, but involve the extra costs of fencing and watering compared with set-stocking.

Strip-grazing involves moving the electric fence daily or twice daily. The decision on how far to move the fence depends on the length of stubble remaining on the grazed area. Similarly, with paddock grazing, the length of the stubble indicates when it is necessary to move the cows to fresh pasture. Guidelines for the optimum stubble height are 6-8cm for dry cows and 7-10cm for lactating cows, the lower values reflecting spring and the higher late summer grazing. Rotation lengths are 14-21 days in spring increasing to 28-35 days in late summer.

Set-stocking (or continuous stocking), which allows the cows free access to the whole grazing area, produces a short, dense sward, providing the pasture is not understocked. If this occurs, a mosaic of grazed and ungrazed areas develops. Advantages of the system are the increased longevity of the sward due to the high tiller population density, and the reduced labour, fencing and water trough requirements. Optimum sward heights are 5-8cm for dry cows and 6-10cm for lactating cows, with lower values for spring, and higher values for late summer grazing.

Zero-grazing involves housing cows in the summer, and cutting fresh grass each day to feed indoors. While the system improves grassland use, it has the disadvantages of higher labour and mechanization costs. An alternative is storage feeding; this involves year-round indoor feeding of forage, which obviates the need for daily cutting of fresh grass. Both these systems face possible welfare problems because the cows live indoors all year. It is important to pay great attention to ventilation, cleanliness of the lying area with frequent scraping of concrete floors, and individual space allocation. If the farmers were to neglect these, the problems of lameness and mastitis will increase. For such systems to be acceptable, the cows should be allowed to use an outdoor exercise paddock for a minimum of three hours daily during the summer. This will also be beneficial for oestrous detection purposes.

*Stocking rate*
Grass is dependent on a variety of factors, including rainfall, soil type and fertilizer. Consequently, growth varies between geographical areas, between years and even between adjacent fields. The choice of stocking rate is therefore a more influential decision on performance than the actual grazing system.

High stocking rates increase both the amount of grass the animals graze, and productivity per hectare, but it does reduce individual cow performance. One method of overcoming this problem is to have high stocking rates to enhance use, but provide daily access to a buffer feed (forage or forage substitute). This allows the cows to top up any shortages from the pasture.

About 60 per cent of grass production occurs in the first two months of the grazing season. In eastern areas, where mid summer drought is a problem, the proportion of the crop produced at this time is greater. In the wetter western areas more suitable for grass production, the proportion is less. The stocking rate therefore should be high in early season (5.5-6.5 cows/ha), but will need reducing in mid and again in late season (down to 2.5-3.5 cows/ha).

## ENVIRONMENT AND HOUSING

In the UK, dairy cows often live inside for at least six months of the year. The overhead costs of housing and mechanization have risen with increasing herd sizes, and as systems have become less labour intensive. There is a need to ensure that this transformation in management is not detrimental to animal welfare.

Dairy cows have a ruminant digestion and metabolism system that generates heat and makes them very resistant to cold stress. Tissue insulation and the length of the coat enhances this hardiness. The heat production of a cow depends on its food intake, size and activity. High yielding cows have a large heat output and are more resistant to cold than, for example, dry cows on maintenance rations.

Dairy cattle have less tissue insulation than beef animals, mainly because they have thinner skin and less subcutaneous fat. The variation between individuals within breeds is quite high; this is due to differences in body condition. In cold environments cattle shed hair less readily and the coat becomes thicker and longer. However, high yielding dairy cows usually have thin coats in winter because of the high level of heat production from digestion and metabolism.

The range of thermal neutrality is variable. It depends on the animal's metabolic state and other environmental factors, such as wind and rain (see also Chapter 3, p 58 and Figure 3.2). The lower limit is the lower critical temperature, below which the animal has to increase heat production by shivering and other means. Exposure to wind and rain raises this critical temperature.

In the UK, heat stress is not usually a problem. If it does occur, it is likely to affect high yielding more than low yielding or dry cows. Cows normally respond by reducing their food intake. However, adaptation does occur and the air temperature at which consumption falls depends on the temperature individuals are accustomed to. Animals dissipate heat by sweating and panting. Also to some extent their coat absorbs and reflects solar radiation.

The milk yields of cows decline below an air temperature of about -5ºC, probably due to vasoconstriction and a consequent reduction in the blood flow to the udder. In European dairy breeds the yields and fat content of milk falls at temperatures over about 25ºC. Extremes of ambient temperature affects performance less than factors affecting body heat production (at temperatures within the zone of thermal neutrality).

The major justifications for housing cattle in winter are for the convenience of management and to avoid poaching damage to the land. Providing adequate shelter from the sun for cows outdoors in summer solves the problems of potential heat stress. In winter, shelter from the wind and adequate levels of nutrition are necessary to ensure cattle remain above their lower critical temperature.

## Housing

A housing system must satisfy the well-being of the cows and provide an economic workable environment for stockpersons. The housing area is the central component of the dairy unit complex. Other components associated with feeding, milking, handling and effluent disposal are integrated around it (Figure 2.7).

Under good management, loose housing systems allow greater cow comfort and more exercise, and enable the mechanization of feeding and cleaning.

The cubicle system of housing is now the most common. For dairy farms in non-arable areas it has the advantage of economy of bedding. However, capital costs are high, and it needs a satisfactory slurry removal, storage and spreading system. The most important aspect of cubicle division design is the prevention of injury to the cow. The lower rail should not catch the cow during lying and standing. There should also be ample lunging space in front (with head-to-head cubicles) or to the side (with wall-facing cubicles) for when the cow is rising. The dimensions of the cubicle are a most important feature - see Table 2.10. A headrail on the top cubicle division rail, or a brisket rail on the cubicle floor, will prevent dunging and urinating on the bed. They work by not allowing the cow to move too far forward in the cubicle.

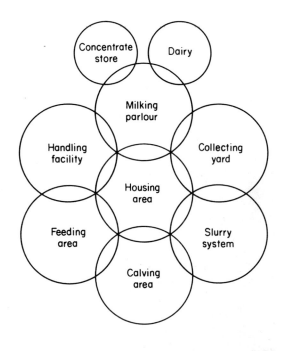

**Figure 2.7**     **Components of a dairy housing system.**

**Table 2.10**     **Recommended dimensions for cubicles for dairy cows.**

| Liveweight (kg) | Length (m) | Width (m) |
|---|---|---|
| 350-500 | 2.00 | 1.00 |
| 500-600 | 2.15 | 1.10 |
| 600-700 | 2.30 | 1.25 |

It is essential to use ample clean, dry bedding in cubicles to prevent injury, particularly to the hocks, to provide comfort and insulation, and to prevent mastitis. The most popular bedding materials are sawdust and wood shavings, with each cubicle requiring 1-1.5 kg/day. It is important that the sawdust and shavings are not damp, because they are likely to contain large concentrations of coliform bacteria, which can cause severe mastitis. Straw (usually chopped) is also a useful bedding material, with each cubicle requiring 1.5-2kg/day. Cow mats made of rubber or a similar material reduce the abrasive surface of the cubicle bed and provide insulation. However, they are expensive, and it is still necessary to have some bedding material to absorb any slurry a cow brings in on her feet.

Strawyards provide relatively low-cost housing in arable areas. The major advantages of this system is that the stockperson can handle manure in a solid form so that it is easy to store until spreading. This system needs 2.5-5kg straw/cow/day depending on whether all or only part of the house has bedding. In strawyard systems it is advantageous to have a concrete scraped area where the cows stand for feeding to prevent a dirty straw area developing. Also, the concrete is beneficial to hoof wear as it will help to prevent them from over growing, which is a problem on totally bedded systems.

The main types of dairy buildings are:
1. Framed (steel, concrete or timber) buildings used for cubicle or strawyard housing, with concrete block walls, space boarding (Yorkshire boarding) above the walls and roofs of corrugated asbestos cement or aluminium sheets.
2. Prefabricated timber kennel buildings containing cubicles whose uprights support the sheeted roof.

Framed buildings have a greater life-span and lower maintenance costs and are more adaptable, whereas the kennel type buildings have a lower initial financial investment and are warmer.

A British Standard Institute's (BSI 1981) *Code of Practice for the Design of Buildings and Structures for Agriculture* is available. The stockperson should give particular attention to space, lighting, ventilation and the provision of feed and water.

**Space allowance**
In cubicle systems, each cow should have her own cubicle. The loafing/feeding area should provide a minimum of $3m^2$/cow, including a minimum passageway width between cubicles of 2m and feeding passageway width of 3m. In strawyard systems, the bedded area should provide a minimum $5m^2$/cow plus a loafing/feeding area of at least $2m^2$/cow.

**Lighting**
Adequate lighting is necessary for the cattle to perform their normal routines and to allow the herdsperson to inspect them at regular intervals throughout the day and evening. In buildings with insufficient natural roof lighting it will be necessary to provide artificial supplementation during the daytime. Milk production may increase where artificial lighting extends day length to 16 hours when the natural day length is shorter than this. Cattle should have 'full lighting' for 16 hours, plus 'minimal lighting' for the remaining eight hours to allow them to move around.

**Ventilation**
There is a need for effective ventilation without draughts. Where it is possible, the design of the building should provide this naturally. This not only reduces costs compared with controlled ventilation, but is generally more effective, and obviates the need for emergency cover in case of a breakdown. Poor ventilation results in condensation, which aggravates the problem of damp bedding, and can increase the incidence of environmental mastitis. Where slurry is stored within the building in slatted tanks, personnel should take extreme caution in agitating the slurry during mixing or emptying. The gases slurry generates are fatal to humans and cattle, so additional ventilation, such as the opening of doors, may be necessary during these processes.

**Effluent disposal**

It is illegal to allow contaminated water to enter a water course. It is therefore important to contain slurry from cubicle housing systems, milking parlours and collecting yards, and silage effluent from silos. The legislation affecting the application of manure and waste to land is in several Acts of Parliament and regulations. The *Codes of Good Agricultural Practice for the Protection of Water* (MAFF 1998a) and *Air* (MAFF 1998b) outline the practices and legislative controls.

**Calving and isolation areas**

An area for calving cows and for the housing of sick or injured animals is a necessary requirement. Calving boxes are usually 12-25m², but this small size causes manoeuvring problems if it is necessary to help during calving. There is also the problem in a restricted area that a dam may trample or lay on her calf. The difficulties of cleaning out such boxes can lead to a build-up of disease organisms. This may cause endometritis or coliform mastitis in the newly calved cow, and enteric disease or navel ill in calves.

Larger straw pens, housing a number of cows together, are preferable and have benefits if the stocking density is low (over 10m²/cow). Such pens are easier to clean out mechanically, and therefore stockpersons are more likely to keep them clean and dry. Calving areas should have wall surfaces that are easy to clean and disinfect, and should contain ample feeding space and drinking water.

Sick animals require isolation pens that are at least 12m², with walls longer than 3m. It is important to site the pen away from healthy stock, where it has adequate ventilation and a separate foul drainage system. All surfaces should be easy to clean, and the pen should incorporate feeding and watering facilities. It is important to include a means of handling isolated animals; the most simple arrangement is a tying ring and a crush gate.

**Bull pens**

Farms with their own bull should have a specialized pen available. This should be at least 1m²/60kg, with ample ventilation. In addition there should be an open exercise area, twice the size of the housing area, with walls at least 1.5m high. A service pen 3.3m long by 1.2m wide should be next to it.

For safety reasons, it is important to incorporate escape routes via gaps, railings or refuge walls. The pen should have a warning notice on the door. Health and Safety Guidelines on bull handling are available from HM Agricultural Inspectorate, Health and Safety Executive.

**Welfare considerations**

It is important not to judge animal husbandry systems simply on whether they affect animal performance. Animals must be able to exhibit normal behaviour.

Problems of welfare are more likely to arise during the housing period, which in the UK is 4-8 months of the year. The *Codes of Recommendations for the Welfare of Livestock: Cattle* (MAFF 1983) outlines important requirements to ensure welfare. They include (amongst other things):

1. Adequate space for freedom of movement and to exhibit normal behaviour.
2. Daily exercise for tethered cows.
3. The provision of a clean dry lying area.
4. Ventilation that prevents build-up of gas and reduces smell.
5. Natural light during daytime and low level lighting over night.
6. A non-competitive feeding system.
7. Separate bedded areas for sick or calving cows.
8. Handling facilities for cows requiring treatment.
9. Emergency arrangements in case of fire or disruption of supplies and services.

A recent important report by the Farm Animal Welfare Council (FAWC 1997) draws attention to the welfare problems which may affect dairy cows and makes many recommendations to improve their care and management.

## HEALTH AND DISEASE

Herd health is important both for the financial loss resulting from ill health, and the suffering it may cause to the animals. Financial losses arise from deaths, culling, lower milk production and from the costs of treatment. It is rare for deaths in a dairy herd to average more than 3 per cent per year, the most common causes being hypomagnesaemia (staggers), hypocalcaemia (milk fever), and coliform mastitis. In calves, mortality rates often exceed 5 per cent if management is not of a high standard. The major problems are calf scour and pneumonia.

Disease prevention is more cost beneficial and causes less animal suffering than relying on curing of sick animals. Advice from professionals in preventive medicine, which producers then carry out, is an essential component of good farm management.

## Reproductive problems
### Dystocia
Difficult calvings (dystocia) arise in about 5 per cent of births in dairy cattle, although in first-calving heifers the incidence may be over 10 per cent. It is important to give prompt assistance or the calf may be stillborn. The sire of the calf affects both calf size and gestation length. There is considerable variation in sires both between and within breeds (see also Chapter 3, p 59 and Table 3.9). For maiden heifers, it is advisable to choose sires that produce a low incidence of dystocia. Heifers that are over fat at calving are more liable to dystocia. It is important to control the plane of nutrition in late pregnancy to prevent too much fat accumulating around the reproductive tract.

Cows will need examining internally if labour has been in progress for several hours without the water bag or the calf's feet appearing. It will also be necessary to conduct an internal examination if labour does not progress within two hours of the appearance of the amniotic sac. It is vital to conduct these examinations aseptically. Assistance with the calving is usually by attaching ropes to the legs and pulling during the normal contraction of the uterus. If this does not result in progress, or if the calf is in an abnormal position, it is essential to call veterinary assistance.

### Retention of the placenta
Any malfunction in the complex endocrine control of the third stage of labour can lead to retention of the placenta. This usually occurs following premature births (which are common with twins), and there is a higher incidence in herds with brucellosis infections. In many situations, a cow will expel a retained placenta naturally after 6-10 days. However, if there is any sign of endometritis, it is important to call the veterinary surgeon to remove it, and treat the infection.

### Endometritis
Endometritis is an inflammation of the uterus that occurs after calving, and leads to a discharge from the vulva. In severe cases, the cow becomes very ill, goes off her feed and loses condition, milk yield may fall and poor fertility may follow. Prompt treatment by the veterinary surgeon is necessary. To reduce the incidence of this infection, it is important to calve in hygienic conditions and to wash and disinfect hands and arms, and equipment when assisting a birth.

## Metabolic disorders
These disorders are a result of an imbalance between nutrient input and nutrient output. Selection of cows for ever higher yields may exacerbate welfare problems of metabolic origin, associated with the impossibility of resolving the conflict between metabolic hunger and digestive overload. It is therefore vital to have simultaneous improvements in nutrition.

### Hypocalcaemia
Hypocalcaemia or milk fever is the most common metabolic disorder. It usually occurs in cows in their third lactation or later, within 24 hours of calving. However, many cases may arise from the day before to three days after birth. Clinical signs are unsteadiness in walking, followed by the cow lying quietly and then being unable to rise. Continually falling blood calcium leads to a cessation of intestinal movement and muscle tremors. Eventually, if untreated, the cow rolls on to her side, and death can occur from the pressure of gas in the rumen. Subcutaneous administration of calcium borogluconate and, in severe cases, intravenous treatment, gives a rapid recovery. Preventive measures include feeding a low calcium diet in the dry period and introducing a high calcium one just before the calving date. Alternatively, it is possible to drench with or inject vitamin D, 6-10 days before calving, and feed a high forage diet indoors for 2-3 weeks before parturition.

### Hypomagnesaemia
Hypomagnesaemia, or low blood magnesium, occurs mainly in animals on spring grass, which has a low magnesium content, and occasionally on autumn grass. As the body has virtually no stores of

magnesium, the cow requires an adequate daily intake. The symptoms of this condition are a high excitability followed by falling over with spasms in the legs. Death due to heart failure can follow quickly. Farmers should administer magnesium sulphate solution subcutaneously immediately they recognize the condition, and call the veterinary surgeon. Intravenous therapy can be fatal and such treatment requires professional expertise. It is possible to prevent hypomagnesaemia by mixing 60g/day of calcined magnesite with the feed. It is also possible to supply the magnesium in drinking water using a metering device that delivers an exact amount into the water trough.

*Acetonaemia*

Acetonaemia, or ketosis, is a problem of high yielding cows in early lactation. The cow has a low appetite and becomes dull and lethargic. Milk yield falls, and there is a sweet smell from the breath and in the milk, due to the formation of ketone bodies. The liver produces these from acetate when, because of inadequate energy intake, there is insufficient propionate available from the rumen. Treatment of the disease can include intravenous injection of glucose, although this effect is only transient. Other clinical problems such as mastitis, lameness, or change of diet, or any problem that causes the cow to reduce its feed intake, can trigger the condition. It is usually possible to prevent the disease by feeding good quality forage *ad libitum*.

**Mineral deficiencies**

Mineral and trace element deficiencies occur rarely in dairy cattle on mixed diets. However, on all grass or all forage diets, deficiencies in some trace elements may occur in lactating cows in some geographical areas. The most common problems are from low levels of copper, cobalt or selenium in the grass or forage. Blood analyses to examine the energy and protein status of cows can also be used to determine their trace element (and major mineral) status. Farmers can arrange these through their veterinary surgeon.

**Disorders of the digestive tract**

The most common digestive disorders are those relating to the rumen. The type of diet and changes to it are the main causes.

*Bloat*

During bloat the rumen fails to expel, by belching, the gas formed during the normal fermentation process. Some diets produce a stable foam in the rumen that prevents the escape of gas. Other causes are the cessation of ruminal contractions, such as in cows with hypocalcaemia, and blockages of the oesophagus. Pressure that builds up in the rumen causes the animal to lie down on its side. Subsequently, death can result from heart failure, or from choking following inhalation of rumen contents. In severe cases, the veterinary surgeon uses a trochar and cannula to gain entry to the rumen through the left side of the animal. This allows the gas to escape. It is possible to prevent bloat with antifoaming agents such as peanut, linseed and paraffin oils, and synthetic detergents. Farmers can administer these as a drench, in the feed or in drinking water, or by applying it to the flanks for licking, or by spraying the pasture. Bloat occurs most commonly on legume pastures. Feeding long roughage will reduce the incidence of this condition.

*Scouring*

Diets containing a high protein level and/or little fibre, as with spring grass, can cause scouring (diarrhoea) in cows. This is not necessarily a digestive upset, but other causes might be more serious. They include liver fluke, Johne's Disease and, rarely, parasitic gastroenteritis. If such scouring occurs, it is important to call the veterinary surgeon.

*Acidosis*

Acidosis is a digestive disorder resulting from a lowering of the rumen pH from its normal level of 6-6.5. Consuming large amounts of starchy concentrate usually causes this condition. The lower pH reduces rumen motility, and feed intake falls or ceases altogether. At very low levels of pH (4.5 and below), caused by gorging on concentrates, the lactic acid concentration in the rumen reduces microbial activity, and leads to toxin production. This may cause liver damage and, sometimes, death. More typically, cows on high concentrate and low forage rations show a more chronic acidosis, which is characterized by low milk fat levels and laminitis.

Preventive measures include providing high quality forage *ad libitum*, feeding concentrates little and often (or mixed with the forage), ensuring that concentrates are not all of a starch type, and feeding sodium bicarbonate (as 1-2 per cent of the concentrate).

### Displacement of the abomasum

Around the time of calving, the abomasum occasionally becomes displaced, migrating under the rumen from the right side to the left. Subsequent gas formation in the abomasum causes the cow to go off her feed, and the milk yield declines. Acetonaemia may follow as a result. It is possible to return the abomasum to its normal position, either by surgery or by manipulation through the abdominal wall with the cow lying on its back. A high forage diet during the dry period reduces the incidence of the condition.

### Ingestion of foreign bodies

Cows with apparent indigestion may have eaten a foreign body such as a piece of wire. The symptoms are grunting, an arched back and forward kicking of the hind legs. Only the veterinary surgeon can deal with such cases, which may require surgery.

### Lameness

Lameness is a painful and crippling disease that practically all cows suffer at some time in their life. Dairy cows have about a 25 per cent incidence of lameness per year. Problems are usually associated with the winter months and shortly after calving. Cubicle housing with slurry systems, silage feeding, high planes of nutrition, and genetic factors are all possible predisposing causes. The problem of lameness is worse than 40 years ago; this is probably due to the rise in cubicle housing and silage feeding. Table 2.11 summarizes a number of factors predisposing dairy cows to lameness. The lameness represents the symptom of a variety of possible underlying problems, although foot disorders cause most cases. About 80-90 per cent of cases are associated with the outer claws of the hind feet. This is due to the extremely large udder distorting the stance of older cows. The udder displaces the stifle outwards and the hock inwards, throwing the weight onto the outer claw.

**Table 2.11    A summary of factors predisposing dairy cows to lameness.**

|  | Examples |
|---|---|
| Breeding | Hind leg conformation, foot colour |
| Nutrition | Direct: laminitis on starchy feeds |
|  | Indirect: wet, acid slurry on grass silage |
| Housing | Poor cubicle design. Hard, wet, slippery concrete |
| Management | Poor foot care. Twice daily milking |
| Behaviour | Prolonged standing of submissive cows |
| Stress | Physiological changes about the time of parturition |

*Source: Webster 1993.*

Most foot problems result from laminitis, which occurs in early lactation, and results in poor quality horn growth. Initially the disorder appears as tenderness in the feet; foot problems, such as solar ulcers, follow some weeks later as the poorly developed horn reaches the sole surface. Both high concentrate and high protein diets are predisposing causes. It is possible to reduce the incidence of the problem by not 'lead feeding' concentrates in early lactation, by ensuring the total diet has not more than 18 per cent CP in the total DM, regularly scraping (minimum twice daily) concrete passageways, and regularly trimming the hoof to prevent overgrowths. The toe length from hair-line to point of toe should be about 7cm. Glueing a wood block under the unaffected claw to take the load-bearing will alleviate severe problems.

In summer and winter, foul of the foot, caused by *Fusiformis necrophorus*, can reach a high incidence. The symptoms are a swelling above the hoof, severe lameness and a putrid smell between the claws. Footbathing in formalin, copper sulphate or zinc sulphate solutions will prevent the condition. It is possible to treat clinical cases with subcutaneous sulphonamide drugs, or by antibiotics, which the veterinary surgeon injects.

Lameness in dairy cattle leads to a loss of cow condition and a fall in milk yield, and may necessitate culling more animals. This economic loss, combined with the suffering, is the reason stockpersons have to give regular attention to this problem.

## Mastitis

Disease organisms that enter the udder via the teat orifice, cause mastitis (inflammation of the mammary gland). Many infections remain at a subclinical stage and clots do not appear in the milk, although it will adversely affect milk yield. Cows may eliminate such infections naturally, or they may become clinically affected some time later. The stockperson can detect the presence of clots (and of clinical disease) by taking a foremilk sample from each teat before milking. The precipitation of milk proteins by leucocytes and epithelial cells cause such clots to form. In severe cases the quarter may become swollen and, if the infection gets into the bloodstream, it will raise temperature, and death can occur.

Mastitis is one of the great health and welfare problems of dairy cattle. It may affect from 10 to 90 per cent of cows in a herd at any one time, although clinical cases are normally under 5 per cent. Depression in milk yield averages about 15 per cent; the milk is characterized by a lower fat and lactose content.

Cows contract most infections during the milking phase and during the time of peak yield. The teat cups transfer infection from diseased quarters to the others. Common organisms are *Streptococcus agalactiae, S. dysgalactiae, S. uberis* and *Staphylococcus aureus*. Good hygiene at milking will maintain low levels of these infections. Preventive measures include washing and drying of udders before attaching the cluster, and teat disinfection immediately after removing the cluster. This is by dipping or spraying the teats with an iodine, chlorhexidine or hypochlorite solution. A routine check on the milking plant, in particular the reserve vacuum level, also helps prevent mastitis. Infusing all quarters with antibiotic at drying off reduces generally the level of infection in the herd. During lactation, the stockperson should detect clinical cases of mastitis early by foremilking at each milking. Treatment is by intramammary infusion, or by injection of antibiotic or other preparations.

A cow may pick up other organisms between milkings. *Escherichia coli (E. coli)* infections generally originate in the housing area. They often arise during early lactation in herds with low cell counts in milk. Contamination of the teat orifice with dirt is the main cause. This type of infection is often severe, with no subclinical phase, and affected quarters, even if they respond successfully to treatment, often fail to produce milk subsequently. Very severe cases can result in fever and death. Prevention through keeping cows clean and dry should be a major objective.

Farmers must discard the milk for the period the manufacturer prescribes after giving intramuscular or intramammary antibiotic therapy for clinical mastitis during lactation. The dairy that receives the milk from the farm tests for antibiotic content, and will penalize farmers selling milk containing antibiotics. The dairy also tests for white cell numbers; it shows the level of mastitis in the herd. Average herd levels are about 300 000 cells/ml. In well-managed herds cell counts are below 150 000 cells/ml.

Summer mastitis in maiden heifers and dry cows occurs in July and August. *Corynebacterium pyogenes*, which flies transmit, are the main cause and infection invariably results in the loss of a quarter. Farmers control the problem by keeping flies away from the udder, by treating at intervals with fly spray or Stockholm tar, and by long acting intramammary dry-cow antibiotic therapy.

## BSE

Bovine spongiform encephalopathy (BSE) is a notifiable disease first recognised in the mid 1980s. The symptoms are a loss of coordination and ultimate recumbency. A post-mortem examination shows a characteristic spongy nature of the brain. It is a slow disease taking on average five years from infection to death. Some suggest that cattle acquired the disease from consuming, at the calf stage, concentrate feed containing meat and bone meal derived from sheep infected with scrapie (which is another slow spongiform disease). In 1988 the Government banned the inclusion of any material of animal origin (except fishmeal) in cattle feed. The incidence of the disease began to decline from its peak of about 1000 cases/week in the early 1990s.

## Health problems of youngstock

Most calf deaths occur in the first four weeks of life, mainly from septicaemia and scours (see also Chapter 3, p 77-78). Rectal temperature is a good indicator of the calf's general health. The stockperson should take the temperature of any calves that are reluctant to feed, which have sunken eyes, starey coat, hunched back or loose dung. A veterinary surgeon should examine immediately any animals that show a deviation of $\pm 1°C$ from the normal temperature of 38.5°C.

### Septicaemia
Septicaemia arises from calves absorbing *E. coli* bacteria through the gut wall into the bloodstream. This usually takes place soon after birth, in calves that do not receive adequate amounts of colostrum. These animals, therefore, have little or no passive immunity from immunoglobulins in the blood. The result is often sudden death.

### Scouring
Localized infections of the intestine cause scouring in calves (diarrhoea). The faeces of these animals is usually white and pasty in appearance, but sometimes dark and watery. Without treatment dehydration results and death can occur. Farmers should treat this condition by reducing the concentration of the milk substitute or milk, while also feeding an electrolyte solution. A veterinary surgeon should see those calves that do not respond to these treatments.

### Pneumonia
Following weaning, the major problem in housed calves is pneumonia, initially caused by viruses, but often with secondary bacterial infections. The predisposing factors are inadequate air space per calf resulting from low roofs, high stocking densities and too little ventilation. The building should not be draughty. In older calves that are ruminating, low air temperatures are unimportant. Naturally ventilated buildings, such as the monopitch design, generally result in a low incidence of pneumonia. Vaccines are available that give a degree of protection against some viruses causing pneumonia.

### Parasitic diseases
In summer, two nematode parasites common to calves in their first grazing season are lungworm (*Dictyocaulus viviparus*), causing parasitic bronchitis (husk), and stomach worms (*Ostertagia ostertagi*), causing parasitic gastroenteritis (see also Chapter 3, p 72, 73, 78).

Lungworm infection results in a husky cough, bronchitis and pneumonia, and, 3-4 weeks after infective larvae reach the lungs, severe weight loss. Calves pick up these larvae from the pasture. The severity of the disease depends on the amount of larvae the animal ingests, and high mortality rates can occur. The most dangerous period is from June to November. It is possible to treat the disease with anthelmintics. The best procedure, however, is to vaccinate calves with an oral vaccine containing irradiated larvae before turning them out to grass (two doses with a four week interval between).

The symptoms of stomach worm infection are diarrhoea and a rapid loss of weight of calves in their first grazing season, usually from July onwards. It causes clinical disease in many animals, but mortality is low. In the spring calves eat the larvae that overwinter in the grass sward. These develop into adult worms in the abomasum and, as with the lungworm, the eggs pass out in the faeces and develop on the sward into infective larvae. Vaccination against the disease is not possible, but grassland management in conjunction, where necessary, with anthelmintics, can control the disease.

One method of control is to treat the calves with anthelmintic in July to remove the adult worms, and then move the calves on to clean pasture. Farmers should have previously cut this pasture for silage or hay. Alternatively, cattle in their second or later grazing season, or sheep, should have grazed it. A second method is to integrate calves, older animals and grass cutting (for silage or hay) into a rotational system, to reduce the numbers of infective larvae on the sward (Figure 2.8), because clinical disease only occurs when there is a high concentration of these.

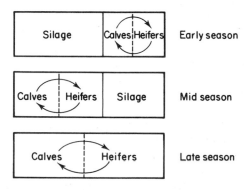

**Figure 2.8    Grazing system for dairy youngstock.**

A second type of stomach worm disease can occur in the later winter months. Larvae that calves have ingested in autumn and that become embedded in the abomasum wall cause this disease. The larvae develop into adult worms in late winter and produce similar symptoms to the summer disease, but with low morbidity and high mortality. It is possible to control this disease by grazing calves on clean aftermath in late summer and autumn.

## Notifiable diseases

The Ministry of Agriculture has the authority to control the movement of livestock and regulate imports when these measures are necessary for the control of certain diseases. In cattle, these include BSE, anthrax, brucellosis, enzootic bovine leucosis, foot-and-mouth disease, bovine tuberculosis, and warble fly infestation. By law the person in charge of an animal suspected of having one of these diseases must report it to the Ministry of Agriculture or to the police. Orders under the *Animal Health Act 1981* confer legal powers. The *Zoonosis Order 1975* requires that persons must also report zoonotic diseases, such as salmonellosis and brucellosis (which are transmissible from animal to man).

It is a legal requirement for owners to record movements of cattle to and from the farm, to identify animals from birth with an ear-tag, and to report all cases of abortion and sudden death.

## GENERAL CARE AND HANDLING
### Stockmanship

Good stockpersons often have 'confident introverted' personalities. As dairy herds become larger, stockpersons need to be more self-reliant and have the intelligence and drive to spend time observing and caring for their animals besides doing routine jobs.

Cattle are creatures of habit, both individually and as a herd. Regular observation of cows and youngstock alerts the stockperson to any abnormal behaviour. This helps to identify many situations, such as sick animals, cows about to calve, and oestrus. Abnormal behaviour, such as restlessness and bellowing in the herd, will also suggest where there are problems of feed availability or disturbance. A good stockperson observes such anomalous behaviour and knows when and how to react. It is essential to

consider equally the welfare of the animals and their productive performance.

The *Agriculture (Miscellaneous Provisions) Act 1968* makes it an offence to cause unnecessary pain or distress to animals. Failure to observe a provision of the *Codes of Recommendation for the Welfare of Livestock: Cattle* (MAFF 1983) can be used in proceedings against a person. *The Welfare of Livestock Regulations 1994* (as amended in 1998) specify the general conditions under which farmers should keep livestock, with additional conditions for intensive systems and for the housing management of calves.

### Condition scoring

Body condition of the cow is another indicator to how satisfactory a feeding method is. Handlers estimate condition of dairy cattle based on fat cover around the tail-head. In beef cattle it is more common to assess the fat cover over the spinous processes in the loin area. The scale of scoring runs from zero (extremely thin) to five (very fat). Most dairy cattle are in the range of 1.5-3.5. Target condition scores at calving are 3-3.5 for the traditional British Friesian type and 2.5-3 for the Holstein.

After calving, dairy cows will mobilize body fat, as they are in negative energy balance. Those of high genetic potential will mobilize more fat than animals of low potential, and those on low levels of nutrition will mobilize more fat. The lowest condition score will coincide approximately with peak milk yield and the lowest protein content. After peak milk yield, the intake of ME will exceed requirements for maintenance and milk production. Consequently, the body condition score and milk protein content will increase.

The body condition of dry cows does not often improve, because of the high energy requirements of the foetus. It is beneficial, therefore, for the condition at day 250 of lactation to be within 0.5 of the target score at calving. If the mean group or herd score is not within 0.5 it will be necessary to raise feeding levels.

Condition scoring can have advantages over the weighing of cows. Weight fluctuates widely each day, due to milk extraction, drinking, feeding, urinating and defaecating. Farmers can overcome this disadvantage by more frequent weighing, such as with an automatic weighing platform at the exit of the milking parlour. Results are linked to a computer to give mean weights, and weight changes over a period. However, even this sophisticated method does not define nutritional status accurately, as heifers and second parity cows in particular, are still growing and mobilizing or laying down fat. In early lactation, such animals are often mobilizing fat (reducing in condition score), but also gaining weight. Thus, a simple monthly assessment of condition scores in a herd can provide a very useful assessment of nutritional status.

### Identification (marking)

Calves, sent away to market, must have a MAFF approved ear tag which gives the official holding (farm) number and the unique number for that animals, put into each of their ears. Calves remaining on a farm must be so marked within 20 days of birth. Once such a calf is tagged an application must be made to the British Cattle Movement Service (BCMS) for a so-called cattle passport. The BCMS will record the details of the animal in its computer-run Cattle Tracing System (CTS) and will send the passport document to the farmer. A farmer who is sending a cow or calf away from his premises must inform the BCMS and must make sure that the passport accompanies the animal to its new destination. The CTS will register cattle and their movement from birth to death. It will hold details of all cattle born in or imported into Great Britain.

### Disbudding etc

Farmers should disbud calves at 1-3 weeks of age using a gas or electrically heated cauterizing iron. A local anaesthetic is necessary for this operation (see Chapter 3, p 79). At this time, farmers remove supernumerary teats of heifer calves with a sharp pair of sterile scissors and apply iodine to the area.

### Restraining devices

A crush is an essential piece of equipment for controlling cattle during treatment. The best place to site the crush is next to the milking parlour. Handling systems should incorporate a gathering pen, a forcing funnel, a race incorporating a footbath, the crush, and a shedding gate into two collecting yards (Figure 2.9). Most veterinary treatments involving dosing,

injecting, blood sampling or rectal palpation can use such a system. To treat lame cows, the crush must restrain the animal satisfactorily, and should incorporate a yoke, a bellyband and a means of securing the leg.

If using a standard crush for AI, it is beneficial for the floor level to be at the same height for the cow and the inseminator. An alternative is to have separate AI stalls about 2.5m long by 0.6m wide. Self-locking yokes in a feeding passage are a useful facility for some veterinary treatments, for AI and for routine stock-tasks such as tail cleaning. Also, locking the cows in the yokes daily is a useful aid to oestrous detection. It allows the handler to examine the vulva and tail area of each cow in detail.

### Transport

The movement of animals from one environment to another creates stress, and any transportation of animals has to take full account of their welfare. Besides the psychological stress, cattle can suffer minor injuries if travelling in a lorry or trailer where there is insufficient bedding material or inadequate restraint. Good stockmanship can overcome many of these difficulties.

The statutory requirements for the welfare of transported animals are now mainly laid down in *The Welfare of Animals (Transport) Order 1997*. This Order gives detailed structural requirements regarding vehicles transporting animals, and for the welfare of the animal during loading, carriage and unloading.

Hauliers with 'basic' trucks must provide food and water at intervals not exceeding eight hours, followed by a 24 hours rest period. It is possible to extend journey times when vehicles meet certain higher standards of construction. For example, adult cattle in specialist vehicles can travel for 14 hours with a one hour rest for watering and feeding, followed by 14 hours travel, then a 24 hours rest. The driver must have an agreed Route Plan for destinations outside the UK. Hauliers need an Animal Transport Certificate for journeys within the UK over 50km. Transport documents are not required if cattle are being moved within agricultural land in a small vehicle owned by the owner or occupier of that land - the internal length of the vehicle space available to the animal being not more than 3.7 metres.

available place of slaughter and even then only if the transportation does not cause additional suffering. For guidance on the complex matter of the transport of fit and unfit animals see MAFF 1993 and 1998c.

There is a great responsibility on stockpersons, attendants at markets and hauliers to ensure that animals have a comfortable journey without suffering. The transport of sick or injured animals requires veterinary certification. For severe injuries or sickness, the veterinary surgeon may arrange for slaughter of the animal on the farm, before transportation takes place.

## THE WAY AHEAD

The economic pressures on dairy farming will ensure that the number of herds in the UK continues to reduce and herd sizes increase. It is unlikely, however, that there will be any swing back to labour intensive systems. Technological developments will continue to produce labour saving husbandry methods. This need not be detrimental because, with good management, the removal of drudgery from the daily work routine should allow more time for the stockperson to oversee the animals.

New breeding developments, including multiple ovulation and embryo transfer, will speed up genetic progress in milk yield potential. However, good management will be essential if farmers are to take advantage of this potential in individual cows, and maintain it over several lactations.

Breeding policies so far have assumed that selection for milk fat plus protein yield will produce the best genotype for all farm situations. It is likely in future, however, that farmers will demand more diverse genotypes eg for high concentrate input intensive systems, compared with low input grassland-based enterprises.

Feeding systems will need to provide ad libitum access to feed for much of the lactation cycle. Thus, buffer feeding of forage or forage substitutes in the summer, and free access to good quality forage in the winter, should ensure that undernourished cows, and associated fertility and metabolic problems, are less prevalent in the future. The development of early maturing maize varieties will lead to the spread of maize growing in the UK, and the increasing incorporation of maize silage into winter diets.

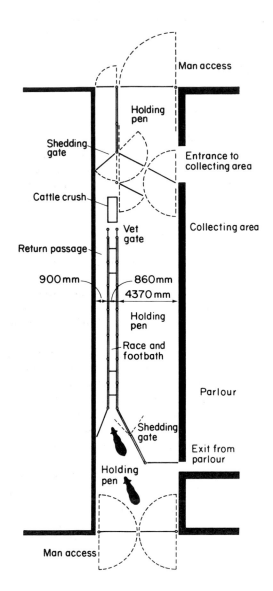

**Figure 2.9     An example of a handling facility for dairy cows.**

The welfare of unfit ie sick or injured animals requires special attention. An unfit animal may only be transported if it is being taken for veterinary treatment or if it is being taken to the nearest

Farmers are likely to use mixed diets of forages, by-products and straight concentrates more.

Environmental pressures on the spreading of slurry may encourage a swing back to straw-bedded systems of housing dairy cows in arable areas. The possible development of automatic milking cluster attachment has wide ranging implications for management. This could allow the elimination of the milking parlour, because cows could at their leisure enter stalls to be milked automatically. More frequent milking would increase production and place less stress on the udder. The cow may be not as prone to mastitis and lameness, due to less distension of the udder. However, an automatic milking machine would only improve overall welfare if farmers matched it with better management. An important benefit would be to allow the stockperson to spend more time observing and tending his/her animals and less time on routine laborious work. The poor cost/benefit ratio, however, may limit the extent of this development.

Pressure from consumers of animal products is likely to ensure that these are free from artificial additives. This may prevent the use of stimulants to enhance milk production, even when their presence in milk is negligible or undetectable. Interest in farm animal welfare by the public is also increasing. This awareness should lead to a continuing improvement not only in the standard of management of dairy cattle, but also in the systems of production.

As dairy cow potential increases through selection, and as economic pressures on farming mount, the need to have highly productive, profitable, long living cows will be a priority. Producers can only achieve this if they give welfare a high profile.

Consumers will increasingly demand to know how and where animal products are produced in order to ensure that only those products from animals farmed, transported and slaughtered under welfare-friendly systems enter the food chain.

## REFERENCES AND FURTHER READING

**Alderman G and Cottrill B R** 1993 *Energy and Protein Requirements of Ruminants*. CAB International: Wallingford, UK

**Blowey R W** 1988 *A Veterinary Book for Dairy Farmers, 2nd edition*. Farming Press: Ipswich, UK

**British Standards Institute** 1981 *British Standard Code for the Design of Buildings and Structures for Agriculture, BS 5502, Section 2.2*. BSI: London, UK

**Bramley A J, Dodd F H, Mein G A and Bramley J A** 1992 *Machine Milking and Lactation*. Vermont Insight Books: Berkshire, UK

**Farm Animal Welfare Council** 1997 *Report on the Welfare of Dairy Cattle*. PB 3426. MAFF Publications: London, UK

**Federation of Milk Marketing Boards** 1995 *Dairy Facts and Figures*. Federation of Milk Marketing Boards: Thames Ditton, UK

**Fraser A F and Broom D M** 1990 *Farm Animal Behaviour and Welfare, 3rd edition*. Ballière Tindall: London, UK

**McDonald P, Edwards R A, Greenhalgh J F D and Morgan C A** 1995 *Animal Nutrition, 5th edition*. Longman: Harlow, UK

**Ministry of Agriculture Fisheries and Food** 1983 *Codes of Recommendations for the Welfare of Livestock: Cattle* (reprinted 1991) PB0074 MAFF Publications: London, UK

**Ministry of Agriculture Fisheries and Food** 1993 *Guidance on the Transport of Casualty Farm Animals* (reprinted with updating 1998). MAFF Publications: London, UK

**Ministry of Agriculture Fisheries and Food** 1998a *Code of Good Agricultural Practice for the Protection of Water*. PB0587 MAFF Publications: London, UK

**Ministry of Agriculture Fisheries and Food** 1998b *Code of Good Agricultural Practice for the Protection of Air*. PB0618 MAFF Publications: London, UK

**Ministry of Agriculture Fisheries and Food** 1998c *Guidance on The Welfare of Animals (Transport) Order 1997*. PB 3766. MAFF Publications: London, UK

**Webster J** 1993 *Understanding the Dairy Cow, 2nd edition*. Blackwell Scientific Publications: London, UK

**Underwood E J N** 1981 *The Mineral Nutrition of Livestock*. CAB International: Wallingford, UK

# 3 Beef cattle and veal calves

## A J F Webster

### THE UK INDUSTRY

In the UK the British beef industry is closely integrated with the dairy industry.

### Trends in the UK industry

In 1994 there were approximately 2.8 million dairy cows and 1.8 million beef cows (see Table 3.1 and footnote on this page). Over the last nine years dairy cow numbers have dropped by approximately 400,000, but this has been matched by a similar increase in the number of beef suckler cows. Approximately half the calves born to dairy cows will be sired by a bull from a beef breed (eg Charolais), reared for prime beef and killed at 1–2 years of age. Intensive beef producers rear some of the male dairy-type calves. The dairy industry currently exports the remainder to continental Europe where farmers there rear them for white veal using methods that are illegal in the UK. Table 3.2 reveals that the export trade in calves grew from 153 thousand in 1985 to 520 thousand in 1994.

The increase in the numbers of beef cows relative to dairy cows (Table 3.1) is due partly to the imposition of quotas on milk production and to the

**Table 3.1    Statistics relating to the UK beef industry in 1985 and 1994.**

|  | Year | | |
|---|---|---|---|
|  | **1985** | **1994** | **1994:1985 ratio** |
| *Livestock numbers (thousands)* |  |  |  |
| dairy cows | 3131 | 2786 | 0.89 |
| beef (suckler) cows | 1339 | 1760 | 1.31 |
| total cattle and calves | 12847 | 11709 | 0.91 |
| *Meat supplies (tonnes carcase wt, thousands)* |  |  |  |
| UK production | 1115 | 919 | 0.82 |
| imports | 203 | 186 | 0.92 |
| exports | 189 | 223 | 1.18 |
| UK consumption | 1129 | 957 | 0.85 |
| UK production: consumption (%) | 99 | 96 | 0.97 |

Source: MAFF (1986, 1995) (see footnote)

*Footnote:* 'The economic and other statistics given in this chapter were current at the time it was written. Publication of this, the fourth edition of the UFAW Handbook, coincides with the most severe economic depression which the modern beef industry has experienced, largely due to the ban on exports of British beef following the establishment of a link between BSE (bovine spongiform encephalopathy) and new variant CJD (Creutzfeldt-Jakob Disease) in humans. I have not given current (1999) economic statistics because they are both abnormal and unsustainable. Either the beef industry will recover, in which case the 1994 figures, will be relevant, after adjustments for inflation, or there will not be a beef industry to describe.'

**Table 3.2**      **Sources of UK production of beef and veal, 1985 and 1994**.

| | Numbers (thousands) | | % total | |
|---|---|---|---|---|
| | 1985 | 1994 | 1985 | 1994 |
| 'Prime beef', UK-bred young cattle | | | | |
|   from the dairy herd | 2278 | 1487 | 52 | 38 |
|   from the beef herd | 1031 | 1149 | 23 | 29 |
|   imported Irish store cattle | 171 | 34 | 3 | 0.9 |
|   total prime beef | 3480 | 2560 | 80 | 65 |
| 'Cow beef' | | | | |
|   dairy cows | 345 | 372 | 8 | 9.4 |
|   beef cows | 252 | 270 | 6 | 6.9 |
| Calf disposals | | | | 0.5 |
|   slaughtered in UK | 99 | 20 | 2 | 14 |
|   exported | 153 | 520 | 4 | |

Adapted from Beef Yearbook (MLC: 1995) (see footnote on page 49)

increase in the average yield of milk per cow. Table 3.2 also reveals that the UK beef production and consumption fell by 18 per cent and 15 per cent respectively between 1985 and 1994. Production and consumption are almost in balance and there has been a small rise in beef export.

**The product**

Conformation and fatness determine the value of a beef carcass. Conformation describes the shape of the carcass in terms of muscle to bone ratio and the proportion of meat in the expensive cuts. The value of a cut of beef depends on attributes of quality such as taste, tenderness and juiciness. Age at slaughter, position on the carcass (eg rump, shoulder etc) and fat concentration determines these qualities. Generally, meat from cows contains more mature connective tissue (gristle) than meat from prime beef carcasses, and is therefore tougher. Prime cuts such as fillet, sirloin and rump fetch higher prices because they contain little or no gristle or visible strips of inter-muscular fat.

The decision when to slaughter a prime beef animal depends on its conformation and fatness. In practice, this occurs when the fat concentration in the carcass is 15–20 per cent. A butcher will trim off about half this fat before the beef reaches the consumer. This appears wasteful but it is because quality beef requires a sufficient quantity of intra-muscular fat (ie marbling). This only occurs when the animal has deposited substantial amounts of 'waste' fat in the subcutaneous tissues and within the abdomen (ie kidney, knob and channel fat or KKCF). However, one objective of selective breeding in all meat animals is to reduce the deposition of unsaleable subcutaneous and intra-abdominal fat while preserving a tasty amount of marbling fat.

The standard procedure for carcass classification within the European Union is as follows. There are eight grades of conformation ranging from: E (excellent), U+,-U,R,0+,-O,P+, -P (very poor), and seven grades of fatness: 1(leanest), 2,3,4L,4H,5L,5H (fattest). Table 3.3 illustrates the distribution of young castrate males (steers or bullocks) and young bulls within the conformation and fatness classes. In general young bulls are leaner than steers at slaughter and a greater proportion are grade E or U (24.2% in comparison with 10.1%). Heifers are usually fatter than steers at the same conformation class. Cull cows, especially from the dairy breeds, have poorer conformation than steers but carry more intra-abominal KKCF fat at slaughter.

**STRUCTURE OF THE INDUSTRY**

**Beef from the dairy herd**

Farmers producing prime beef from calves born to dairy cows remove them shortly after birth and rear them artificially. This involves providing them with

**Table 3.3**      **Carcase classification and percentage distribution of steers and young bulls in the classification grid for Great Britain 1993.**

| | Steers, fat class | | | | | | | Young bulls, fat class | | | | | | |
|---|---|---|---|---|---|---|---|---|---|---|---|---|---|---|
| | increasing fitness | | | | | | | increasing fitness | | | | | | |
| | 1 | 2 | 3 | 4L | 4H | 5L | 5H | 1 | 2 | 3 | 4L | 4H | 5L | 5H |
| *E* | - | - | - | - | - | - | - | - | 0.2 | 0.4 | 0.3 | - | - | - |
| *U+* | - | - | 0.3 | 0.6 | 0.3 | - | - | - | 0.6 | 2.2 | 2.0 | 0.3 | - | - |
| *-U* | - | - | 1.4 | 4.8 | 22.5 | 0.2 | - | 0.2 | 2.0 | 7.6 | 7.1 | 1.2 | - | - |
| *R* | - | 0.6 | 7.0 | 22.5 | 10.2 | 0.8 | - | 0.3 | 4.3 | 15.2 | 12.6 | 2.0 | 0.2 | - |
| *O+* | - | 0.7 | 6.8 | 17.2 | 7.0 | 0.8 | - | 0.3 | 3.3 | 9.9 | 7.0 | 1.0 | - | - |
| *-O* | - | 0.5 | 3.3 | 6.4 | 2.5 | 0.3 | - | 0.3 | 2.6 | 7.6 | 3.8 | 0.4 | - | - |
| *P+* | - | 0.2 | 0.8 | 0.8 | 0.2 | - | - | - | 1.4 | 1.9 | 0.4 | - | - | - |
| *-P* | - | - | 0.2 | - | - | - | - | 0.2 | 0.4 | - | - | - | - | - |

Blank spaces indicate an incidence of less than 0.2%

a liquid milk-replacer diet until they are eating sufficient digestible dry food to support maintenance and growth. This usually occurs at about five weeks of age. Phenotype, the cost and availability of food (grass, forage crops and cereals), and season of birth determines the type of rearing system to which the calves move later.

*Semi-intensive beef systems*
The expression 'semi-intensive' describes systems in which young cattle live out at grass for one or two summers. Table 3.4 outlines two options, where farmers either finish cattle in yards during their second winter, or after a second summer at grass at 20–24 months of age. Beef producers normally castrate and rear male calves as steers.

*Finishing in yards*
This system is the most common for calves born from September to December. Despite financial incentives for summer calving, it is still the peak calving period for dairy cows. These calves can grow at 0.7–0.9kg/day over the first winter so that they weigh 200-220kg when the spring comes and they go out to grass. At this stage of growth, their ruminant digestive system is sufficiently mature to enable them to thrive on a diet of grass alone (without milk or cereals). Such calves achieve weight gains close to 1kg/day and reach 340-360kg by October or November when they return to the yard. If they are to finish before the following spring, they will then

need to receive a highly digestible diet *ad libitum*. This is primarily a mixture of forage (grass or maize silage) with a cereal-based concentrate ration supplement if necessary. An early maturing Hereford x Friesian steer may finish by January at a slaughter weight of 420kg (Table 3.4). Heifers will finish even younger and lighter. A Charolais x Friesian steer will grow faster but finish a little later at a much greater weight. A Holstein type male calf may achieve a similar slaughter weight to the Charolais x Friesian, but will take much longer to finish.

*Summer finishing*
Traditionally farmers used this system either for calves that were born in mid to late winter. This is because such animals are sufficiently mature to obtain the maximum benefit from grass during their first summer. Farmers also summer-finish calves from big, late-maturing bulls, like the Charolais, that are difficult to fatten in yards over their second winter. The EU has now created a further incentive to turn beef cattle out for a second summer at grass by paying the subsidy for prime beef cattle in two stages, at 10 and 20 months of age. If there is no intention of finishing cattle out of yards during their second winter, producers will put them through a 'store' period. During this time the animals receive only silage and supplements of essential minerals and vitamins. They may continue to gain weight at 0.5–0.8kg/day but lose body fat, so that they end the winter bigger but leaner than when they entered it.

**Table 3.4**     **Semi-intensive beef production systems for calves from the dairy herd.**

| | Finishing in yards | | | Finishing at grass |
|---|---|---|---|---|
| | **He x Fr** | **Ch x Fr** | **Holstein** | **Ch x Fr** |
| *Live bodyweight (kg)* | | | | |
| *weaning (5wk)* | 65 | 70 | 65 | 70 |
| *turnout (year one)* | 200 | 220 | 220 | 200 |
| *yarding* | 340 | 360 | 340 | 360 |
| *turnout (year two)* | - | - | - | 480 |
| *slaughter* | 420 | 520 | 510 | 600 |
| *Approx age at slaughter (months)* | 12-14 | 14-18 | 18-24 | 20-24 |
| *Carcase wt (kg)* | 205 | 277 | 260 | 318 |
| *Killing out (%)* | 51.0 | 53.2 | 51.3 | 53.0 |
| *Saleable meat:bone ratio* | 3.8 | 4.0 | 3.4 | 4.0 |

He - Hereford; Fr - Friesian; Ch - Charolais. (Cattle finished in yards are not turned out to grass in year two).

When such lean and hungry animals return to pasture for their second summer, their appetite for good grass is initially high because they sense that they are thin (or underweight for age). In these circumstances they can achieve growth rates well in excess of 1kg/day on low-cost summer pasture. However, cattle in good to fat condition initially, moving to spring grass, lose considerable weight that they may not make up for several weeks. It is therefore important to ensure that cattle on such a rearing system are not too fat at the end of winter. It is equally important not to turn out cattle that are nearly finished at the end of winter on to grass; they will need feeding on in yards until they are ready for slaughter.

*Intensive systems for beef and veal*

Intensive calf-rearing systems are those that involve rearing calves from the dairy herd in confinement from birth to slaughter. It is most common to rear male dairy-type (Holstein/Friesian) calves that have a poorer conformation, so are cheaper to buy than Beef x Dairy calves, in intensive systems. The strategy in these systems is to offer the highest quality feed available at an economic price and finish the calves as quickly as possible. This is because rearing calves entirely indoors or in yards is capital-expensive. Inevitably this reduces both the age and weight, and therefore the price, of the animal at slaughter. For a Friesian male calf worth comparatively very little (ie £120) at birth, this can be the best economic option.

For a good Charolais x Friesian calf, worth double, it pays to rear it semi-intensively to a greater slaughter weight even though it takes longer.

The three main feeding options for intensive beef are based on grass silage plus concentrates, maize silage plus (less) concentrates or 'barley beef' (Table 3.5). In the barley beef system, the main feed is cereal with small amounts of protein, minerals and vitamins plus a little straw to encourage rumination. Since barley is more energy-rich than grass or maize silage - this system finishes calves most quickly. It is, however, less profitable than silage-based systems, except where the barley is home grown and more likely to provoke digestive disorders.

Intensive beef production systems require confining animals throughout their lives at a high stocking density. This increases the risk of infectious diseases, such as pneumonia, relative to systems that permit the animals to spend half the year at pasture. These problems are particularly acute for calves in the first three months of life. In the early days of intensive beef systems, calves moved to the rearing accommodation and were exposed to carriers of infection directly after weaning at 5–6 weeks of age. This created serious losses from respiratory disease. In recent years specialist contract-rearing units have developed. They provide 12 week old calves at a weight of 110–130kg and, ideally, an acquired immunity to the major respiratory pathogens for intensive beef units.

Table 3.5 outlines two more options for rearing male Holstein/Friesian calves, namely 'white' and 'pink' veal. Farmers produce 'white' veal by rearing calves entirely or almost entirely on a liquid milk-replacer diet, deficient in iron. This restricts the synthesis of the oxygen-carrying pigments myoglobin and haemoglobin. A deficiency of myoglobin produces the white flesh, which the veal trade (allegedly) requires. A deficiency of haemoglobin produces anaemia. In the UK, legislation bans the methods of white veal production that continental Europe uses. However, the UK exports approximately half a million calves per annum who are subsequently destined for such production (Table 3.2). An alternative option is to produce pink veal; this involves rearing calves to 300kg on a mixture of milk replacer and cereals. Irrespective of the legal position, neither white nor pink veal production appear to be an economic option for British farmers relative to intensive systems of beef production (Table 3.5).

*Beef from the suckler herd*
Calves born to beef cows stay with their mothers until weaning in late summer or autumn. Producers sell them into yards for winter feeding before they are finished out of yards in the spring or at pasture the following summer. Once again, phenotype determines this choice. Most beef farmers house their cattle over winter, not to protect them from the cold (since their cold tolerance is phenomenal – see Table 3.11), but for convenience of feeding and to protect the pastures from poaching. Calving is usually concentrated either in early spring, normally before turnout, or at pasture

in the autumn. When calves are born in the spring, the peak period for lactation in the cows coincides with the peak period for production of grass. This reduces the cost of feeding concentrates to cow and calf (Table 3.6). It is common practice to house autumn-born calves with their mothers and provide them with access to a special creep area. Such an area provides a place where they may rest and receive concentrate feed without competition from the adult cows. Farmers then turn them out to pasture in spring and wean calves in mid summer at 9–10 months of age and weights of 250-350kg (depending on breed).

The Meat and Livestock Commission (MLC) annually publishes summaries of the physical and economic performance of recorded beef herds. Table 3.6 compares the average performance of lowland herds calving in spring with hill herds calving mainly in the autumn, and the average and top third of upland herds, calving in the spring. Differences in physical performance between cattle on the hills, uplands and lowlands are small; this shows how tough cattle are. The main difference in profit margin per cow is due to the size of the cow subsidy. Differences in gross profit margins between spring and autumn calving herds are also relatively small and vary annually. The additional cost of providing concentrate feed to autumn calving herds is matched by their heavier weight (and therefore better price) at sale. The most obvious difference between the average and the top third upland spring calving herds is that the latter farmers feed concentrates directly to the calves. This is more efficient than feeding it to the calves' mothers.

**Table 3.5** **Intensive systems for beef and veal production for Holstein/Friesian bull calves from the dairy herd (values from MLC 1995).**

|  | Grass silage concentrates | Maize silage concentrates | Barley beef | White veal | Pink veal |
|---|---|---|---|---|---|
| *Daily liveweight gain (kg)* | 1.00 | 1.15 | 1.30 | 1.60 | 1.40 |
| *Slaughter weight (kg)* | 510 | 490 | 460 | 220 | 300 |
| *Slaughter age (months)* | 16 | 14 | 11 | 4 | 7 |
| *Concentrates feed (tonnes)* | 1.3 | 0.8 | 1.8 | - | 0.8 |
| *Milk powder (tonnes)* | - | - | - | 1.2 | 0.4 |
| *Target gross margin (£/head)* | 160 | 140 | 115 | 30 | 70 |
| *Equivalent sale price (pence/kg carcase)* | 209 | 204 | 214 | 380 | 240 |

**Table 3.6**     **Physical and economic performance of suckler herds.**

|  | Lowland herds | | Upland, spring | | Hill cattle |
|---|---|---|---|---|---|
|  | Spring | Autumn | Average | Top third | Autumn |
| *Live calves reared (per 100 cows)* | 90 | 90 | 92 | 93 | 89 |
| *Age of calf at sale/transfer (days)* | 229 | 370 | 268 | 283 | 320 |
| *Weight of calf at sale/transfer (kg)* | 263 | 337 | 276 | 300 | 296 |
| *Daily gain (kg)* | 1.15 | 0.91 | 1.03 | 1.06 | 0.8 |
| *Feed concentrates (kg)* | | | | | |
| *to cow* | 115 | 185 | 138 | 50 | 179 |
| *to calf* | 4 | 113 | 70 | 126 | 141 |
| *Cow subsidy (£)* | 57 | 55 | 111 | 118 | 113 |
| *Gross profit margin per cow (£)* | 282 | 329 | 369 | 441 | 380 |

(Source MLC Beef Yearbook 1994)

## BREEDING AND GENETICS
### The Holstein/Friesian dams
Most UK beef comes from the dairy herd, therefore most beef calves have Holstein/Friesian mothers. Until recently the Holstein and the British Friesian were two different breeds in the UK but the two breed societies have now amalgamated. There remains however a profound difference in conformation. The typical British Friesian has a superior muscle to bone ratio and proportion of meat in the expensive cuts relative to the long legged, bony American Holstein. There is a steady, and probably unchangeable, trend towards the extreme Holstein type of dairy cow that gives more milk but has a considerably inferior beef shape. One of the reasons for the UK increasing the number of calves it exports into white veal units in continental Europe is the rise in the number of poor quality Holstein-type male calves. These calves are unsatisfactory for beef production, but are so cheap that it is profitable to rear them quickly for veal.

### The beef breeds
The most desirable traits for a beef bull are: rapid growth rate, especially lean tissue; excellent carcase conformation, without excess subcutaneous and KKC fat; and good appetite, especially for grass and forages to achieve early finishing. These traits are somewhat incompatible and the relative importance of each depends on the nutrition and environment within the rearing system. Table 3.7 shows the variation in growth rate and fatness of different breeds of pedigree bulls (ie weights and backfat thickness at 400 days of age). There is no such thing as a perfect breed, different types suit different circumstances.

*Hereford*
The Hereford was, for many years, the most popular breed for beef production from the dairy herd. It is a relatively small animal, with an average 400-day weight of approximately 470kg and a mature weight of 900kg. The breed has an excellent conformation and finishes quickly due to a large appetite for forages relative to its energy requirement for maintenance. The bulls carry a low risk of dystocia (Table 3.8). The gene for a white face is dominant and therefore the beef producer seeing a white-faced calf at market knows that it was sired by a Hereford bull. The bulls are relatively gentle natured. For all these reasons they remain a popular breeding animal for dairy farmers to keep and run with their heifers. They also remain excellent sires of crossbred suckler cows. However, Charolais and Simmental have largely over taken their role as a first-choice sire of prime beef calves. The beef trade has for many years favoured larger carcasses, although the largest carcasses are now about as big as men and current machinery can reasonably handle. Also the quality of

**Table 3.7**     **Performance of principal beef breeds (pedigree recordings).**

| | No recorded | Bulls 400-day weights | | | Backfat (mm) | |
|---|---|---|---|---|---|---|
| | | **Range** | **Mean** | **1975 to 1990** | **Range** | **Mean** |
| *Aberdeen Angus* | 1134 | 434-670 | 471 | +86 | 1.6-8.5 | 4.6 |
| *Belgian Blue* | 508 | 511-655 | 519 | - | - | - |
| *Charolais* | 3315 | 453-831 | 617 | +49 | 1.5-4.1 | 2.6 |
| *Hereford* | 1683 | 361-635 | 469 | +35 | 1.3-9.3 | 4.2 |
| *Limousin* | 3404 | 410-669 | 525 | +94 | 1.0-5.3 | 2.6 |
| *Simmental* | 3569 | 445-810 | 600 | +73 | 1.5-5.2 | 3.3 |
| *South Devon* | 1228 | 424-730 | 546 | +21 | 1.5-5.5 | 3.1 |

**Table 3.8**     **Carcase composition of calves from sire breeds crossed with Friesian/Holstein cows.**

| | Sire breeds | | | |
|---|---|---|---|---|
| | **Angus** | **Charolais** | **Hereford** | **Limousin** |
| *Weight at slaughter (kg)* | 393.0 | 494.0 | 410.0 | 454.0 |
| *Feed conversion ratio (g gain/kg of feed)* | 86.0 | 82.0 | 88.0 | 85.0 |
| *Killing out (%)* | 52.5 | 54.8 | 52.3 | 54.7 |
| *Saleable meat in carcase (%)* | 72.5 | 72.7 | 71.9 | 73.3 |
| *Saleable meat in expensive cuts (%)* | 44.1 | 44.8 | 44.1 | 45.4 |
| *Fat trim in carcase (%)* | 9.6 | 9.0 | 9.7 | 9.2 |

(Source: Southgate in More O'Ferrall, 1982)

feed for beef cattle has generally improved. The facility of the Hereford to finish early on poor forage has become something of a disadvantage. When forage quality is good, Herefords fatten too soon. Table 3.7 reveals that there has been relatively little progress in selecting Hereford cattle for better growth rate and many of them are still too fat.

*Aberdeen Angus*
These cattle are black and polled and both these characters are dominant. Breeders developed the Angus to produce high-quality beef, with conspicuous marbling, on good land but in the short growing season of north-east Scotland. The traditional Angus is a small, early finishing animal, suitable as a mate for heifers but fattens too quickly. There has recently been a move to increase size in this breed, mainly through importation of bulls from North America. Table 3.7 illustrates that average 400 day weights for Angus bulls are now higher than Hereford and the top

of the range is as heavy as any breed other than Charolais and Simmental.

*Limousin*
The Limousin is a red-coloured breed originating from central France. It is, on average, intermediate in weight between the British beef breeds and the very large Charolais and Simmental. It is lean and has excellent carcass conformation with a high killing-out percentage, and a high proportion of meat in the expensive cuts (Table 3.8). This makes it very popular with butchers. Calving problems are lower than for Charolais and Simmental; this makes it a suitable bull for mating with dairy cows to produce beef. Selection for leanness and a high killing-out percentage (which implies a small gut) inevitably tends to select against appetite, especially for highly fibrous forages. Limousin-crosses are therefore difficult to finish without expensive concentrate feeds. Intensive systems, which house animals in yards and slaughter them by 18 months of age, are most

suitable for Limousins. Limousin x Friesian male calves in these systems are usually entire and can be difficult to handle.

*Charolais and Simmental*

It is possible to consider these popular large breeds together with other, less numerous large breeds, such as the Blonde d'Aquitaine and South Devon. Pure-bred Charolais and Blonde d'Aquitaine have rather similar cream-coloured coats. The Simmental is similar in colour to the Hereford. At best, these bulls and their calves grow fast and have the ability to reach a large size and excellent carcass conformation without getting too fat. Since the prime factor determining the price of a beef animal is its slaughter weight, these breeds, especially the Charolais, have become increasingly popular as sires for both dairy cows and suckler beef cows. The dairy farmer who sells male calves at 10 days of age can get much more money for a Charolais-cross that may finish at 580kg, than for a Hereford-cross that may finish at 450kg. However, the dairy farmer cannot afford obstetric problems since they compromise the prime source of income, namely subsequent lactation performance. The suckler beef industry can only generate income from the sale of calves (and ultimately the cow herself) so, inevitably therefore, they often use very large bulls and carry the greater

risk of dystocia (Table 3.9). To maximize slaughter weights, calves from these large bulls, whether out of dairy or beef cows, are best suited to 20–24 month systems. Most calves have the capacity to finish well, (if slowly) off grass. Table 3.7 shows that the Charolais and Simmental breeds are continuing to increase in size and growth rate to 400 days. The rate of improvement in these traits has been less impressive in the native red South Devon breed.

*Belgian Blue*

Breeders have developed the modern Belgian Blue from a breed similar in appearance to the traditional blue-roan beef Shorthorn. They have selected for 'double muscling' (ie heavy muscling), especially in the hind quarters (Figure 3.1). This leads to a very high muscle to bone ratio but abnormalities of skeletal development, especially a reduction in pelvic dimensions. Cows carrying the double muscling trait in the homozygous form are seldom able to give birth normally and will need a Caesarian section. The homozygous double-muscled animal also does not thrive, partly because of a poor appetite and partly because of a predisposition to infectious diseases, especially pneumonia. In the natural state, this trait is a lethal recessive. However, F1 hybrid calves from a double-muscled bull do grow rapidly (although they are difficult to finish) and have a good carcass

**Table 3.9** **Effects of sire breed on calving difficulties and calf mortality in suckler cows, dairy cows and heifers (from Allen, 1990).**

| Breed of sire | Suckler cows | | Holstein/Friesian | | | |
| | | | Cows | | Heifers | |
| | Dystocia (%) | Mortality (%) | Dystocia (%) | Mortality (%) | Dystocia (%) | Mortality (%) |
|---|---|---|---|---|---|---|
| *Aberdeen Angus* | 3.1 | 1.3 | - | - | 2.3 | 5.8 |
| *Hereford* | 3.8 | 1.6 | 1.2 | 2.9 | 3.1 | 6.2 |
| *Friesian* | - | - | 2.5 | 5.0 | 7.4 | 7.9 |
| *Limousin* | 7.2 | 4.4 | 3.2 | 6.1 | 8.1 | 9.7 |
| *Charolais* | 9.6 | 4.8 | 4.2 | 5.2 | - | - |
| *Simmental* | 9.3 | 4.2 | 3.1 | 5.6 | - | - |
| *South Devon* | 7.4 | 4.1 | 2.4 | 5.6 | - | - |
| *Belgian Blue* | - | - | 4.3 | 5.4 | - | - |
| *Average* | 6.7 | 3.4 | 3.0 | 5.1 | 5.2 | 7.4 |

quality. The incidence of dystocia when using a Belgian Blue bull on a Holstein/Friesian cow is no worse than for a Charolais bull. There are no special welfare problems in the crossbred calves. However, the breeder who elects to use Belgian Blue semen must accept the welfare problems involved in producing the pure bred double-muscled bulls.

**Figure 3.1      A Belgian Blue Bull (courtesy Genus Breeding).**

### The suckler cow breeds

Most suckler cows in the UK are not pure-bred but first crosses between two pure breeds. This is partly because genetic traits relating to maternal behaviour and calf survival are more likely to improve by heterosis (hybrid vigour) rather than by selection within breeds. It is also partly because dairy farms are regularly producing and selling suitable half-bred heifers for the suckler herd, such as Hereford x Friesian and Limousin x Friesian. Such animals, that dairy farmers have reared artificially, with little or no contact with their own mothers, manage the responsibilities of motherhood very well.

Over the last 200 years the Hereford has become the most popular breed of suckler cow in the world outside the hottest areas of the tropics. The ability of the breed to thrive and fatten on poor quality grasses

and forage may now be a disadvantage in a bull intended as a sire of calves destined for slaughter as prime beef. It is, however, an excellent trait for a bull siring hardy beef cows able to subsist and regularly produce calves, at least cost, on range in the USA or on the hills of the UK. Other traditional hardy British beef breeds include the Aberdeen-Angus, Beef Shorthorn, Galloway and Welsh Black. Cross-breeding the black Angus or Galloway with the Shorthorn produces the Blue-Grey, a small, very hardy cow able to outwinter with little supplementary feeding on the Scottish Highlands and Islands and regularly produce calves for ten years or more.

**Table 3.10      A comparison of Blue-Grey and Hereford x Friesian suckler cows.**

|  | Blue-Grey | Hereford x Friesian |
|---|---|---|
| *Average cow weights (kg)* |  |  |
| mature | 450 | 475 |
| at first mating | 280 | 325 |
| *Lactation yield (kg)* | 1600 | 1940 |
| *Calf 200-day weights (kg)* |  |  |
| lowland | 203 | 217 |
| hill | 190 | 198 |
| *(lowland minus hill)* | 13 | 19 |

(Adapted from Allen and Kilkenny, 1984).

Table 3.10 compares some aspects of the maternal performance of Blue-Grey and Hereford x Friesian cows on lowland and hill farms. The Hereford x Friesian is a little larger and milkier and, on lowland pastures, produces a calf about 14kg heavier at 200 days of age. This advantage declines to 8kg on the hills. These figures indicate a small genotype and environment interaction. The extent of the advantage of the Hereford x Friesian (measured only in calf 200 day weights) diminishes as the environment becomes harsher. The Hereford x Friesian is hardy enough for most UK environments, even if they live out over winter. When housing cows over winter, selection for hardiness becomes almost irrelevant. However, this is not entirely true since the hardier cow is the one that will thrive on the poorest, cheapest forage. There has been, to my knowledge,

...rehensive study of nutrition/genotype interactions in Hereford x Friesian or Limousin x Holstein suckler cows. On first principles, the latter is likely to require considerably more feed for maintenance, especially if they live out for the winter.

## ENVIRONMENT AND HOUSING
### Temperature

Cattle are able to tolerate a wide range of thermal environments with little or no loss of thermal comfort or stress to their welfare, and at little cost to production. This is because they are a large, well-insulated species with an excellent ability to regulate body temperature. Figure 3.2 illustrates the heat exchanges of two extreme types of cattle, a Friesian cow (*Bos taurus*) at peak lactation and a West African Fulani cow (*Bos indicus*) at maintenance. Homeothermy is achieved by a balance between metabolic heat production (Hp) and heat loss, partly by the pathways of convection, conduction and radiation (Hn) which occur at a rate proportional to the temperature gradient from the core of the animal (eg rectal temperature) to the air, and partly by evaporation of moisture from the skin (ie sweat) and from the respiratory tract. Insulation against Hn depends on the thickness of the hair coat, the skin and the subcutaneous layer of fat. The Hereford, with a thick winter coat, therefore has better insulation than the dairy Holstein/Friesian.

Metabolic heat production (Hp) per unit of metabolic size (kj/kgW$^{0.75}$per day) rises with increasing food intake. All cattle are able to regulate He very effectively by active sweating and controlled thermal panting. Thermal panting involves increasing the rate, but reducing the depth, of respiration so as to improve evaporation of moisture from the membranes inside the nasal cavity.

The thermoneutral zone for cattle is wide, mainly because of their ability to regulate heat loss by evaporation. Below the lower critical temperature an animal must increase metabolic heat production by shivering to maintain homeothermy. Above the upper critical temperature it must reduce metabolic heat production; it usually achieves this by reducing food intake. Both extremes constitute thermal discomfort and both reduce productivity. The lower critical temperature for the Friesian is below -20˚C (in still air), although the exposed udder of the dairy cow is

likely to experience local chilling at air temperatures below 0˚C. Suckler cows are as cold tolerant as dairy cows. They are likely to have less to eat (so metabolic heat production is less) but their thicker coat provides better insulation. This is not just genetic. Poorly fed, or sick animals that live out over winter are the last to shed their coats in spring because they produce little metabolic heat.

The more productive the animal, the more metabolic heat it produces, so the greater its tolerance to cold but the greater its susceptibility to heat. The animal most susceptible to cold is the one that has little food.

**Table 3.11** **Lower critical temperatures for beef cattle housed in still-air conditions (0.2m/s) and in a draught (2 m/s).**

| | Lower critical temperature (˚C) | |
|---|---|---|
| | Still air | Draughty |
| *Newborn calf* | +10 | +18 |
| *Weaned calf, 6 weeks old* | 0 | +10 |
| *Veal calf, 12 weeks old* | -12 | 0 |
| *Yearling beef cattle* | | |
| *rapid growth* | -30 | -15 |
| *store condition* | -15 | -5 |
| *Beef cow, maintenance* | -15 | -8 |

(From Webster, in Swan and Broster, 1981).

Outside thermal comfort does not depend on air temperature alone but also on other features of the environment, namely sun, wind, rain and snow. Table 3.11 gives approximate values for the lower critical temperatures of different classes of beef cattle either in still air conditions (0.2m/sec) or in a moderate draught (2m/sec). At birth, cold begins to affect the calf in a well-bedded box and out of draughts at about +10˚C, but the stress is slight until air temperature falls below -10˚C. Cold stress, while not ideal, is less of a killer than over-stocked, under-ventilated and polluted calf houses. By the time of weaning, under UK conditions, low temperatures alone are unlikely to stress the calf. Well-grown veal calves, by virtue of their very high intakes of ME from a high-fat, liquid diet, are extremely tolerant to cold but sensitive to heat. Yearling cattle and cows

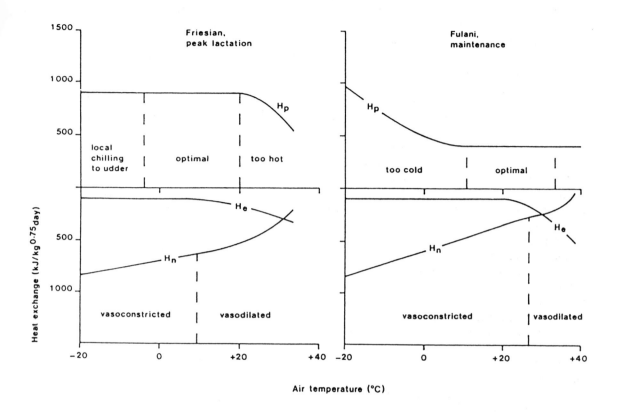

**Figure 3.2**     **Heat exchanges of two extreme types of cattle, a Friesian cow (*Bos taurus*) at peak lactation and a West African Fulani cow (*Bos indicus*) at maintenance.**

are unlikely to be cold stressed unless rain or snow thoroughly soaks them. A cow standing in driving sleet at an air temperature of 20°C is colder than if she were in dry air at –40°C.

**Bedding area**

The goal of a reasonably clean, dry, comfortable bed is easier to define than achieve. Cattle are undoubtedly very comfortable when lying in deep, clean straw. However, straw is probably too expensive unless a farm produces its own, and

prohibitively so in some areas of good beef country like the Orkney Isles. The main problem is that cattle urinate and defaecate indiscriminately. This makes it difficult to maintain the bedding reasonably dry and clean even when the feed is dry (eg barley and straw) and practically impossible when the main feed is low dry matter grass silage. It is possible to house adult, non-lactating beef cows and growing cattle over six months of age on slatted concrete floors. These are far from ideal as a mattress to lie on but they can be kept clean and dry if the gaps between the slats are

40 mm wide and the cattle are stocked densely enough to tread the dung between the slats. However, young calves below 6 months of age should not be kept on slatted floors because 40 mm gaps are too wide for their feet and the stocking density necessary to ensure that the slats stay clean increases the risk of provoking pneumonia in these young animals. It follows that while pregnant beef cows may be housed on slatted floors, it is no place for a beef cow to rear her calf, still less give birth.

Beef cows and fattening heifers can be accommodated in cubicles. They are obviously unsuitable for males, who urinate into the middle of the cubicle bed. The dimensions of the cubicle should be determined by the size of the largest animal that they are likely to accommodate. Each animal should be able to lie down entirely within the cubicle without interference from its neighbours (eg being trampled upon), change position (stand up and lie down) without difficulty, yet, so far as possible, urinate and defaecate in the dunging passage. A cubicle for a Hereford x Friesian suckler cow should be at least 2 by 1m. Fine tuning of the standing position can be achieved by adjusting the neck rail or brisket board at the front of the cubicle.

### Space allowances

These are listed in Table 3.12. When cows or fattening cattle are given restricted amounts of concentrate feed, it is essential that all the animals can feed at the same time. Recommended trough spaces for cows and growing cattle are 0.6–0.8m, and 0.4–0.6m respectively, depending on age and breed. A group of 10 yearling cattle on slats with 0.4m trough space per head and a floor area of 1.6m$^2$ per head, would require a square pen 4m long by 4m deep. However, the recommended minimum air space is 12m$^3$ per head. For pens 4 by 4m either side of a 4m wide feeding passage, the average height of the building should be 5m. Examples of housing designs accommodating these basic environmental requirements are given later.

### BEHAVIOUR
### Social structure

Cattle in the wild form large, stable herds whose size appears to be limited only by the availability of pasture. Being part of a herd contributes to a sense of

security since it reduces the risk of capture for a large animal that cannot hide. Cattle perceive a predator as a threat but not as a source of real alarm if they can maintain a satisfactory flight zone and can see (or think they can see) an escape route. It is possible to build these principles into handling systems for wild range cattle (see Figure 3.3 and Grandin 1993).

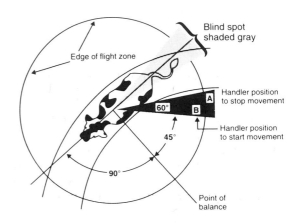

**Figure 3.3**    **Flight zone diagram showing the most effective handler positions for moving an animal forward.**

**Table 3.12**    **Space requirements for beef cattle.**

|  | Calves | | Fattening cattle over 1 year | Beef cows |
|---|---|---|---|---|
|  | 0-6 weeks | 6-12 weeks | | |
| *Floor space (m²)* | | | | |
| *straw bed* | 2 | 2 | 3-4 | 4 |
| *slatted floor* | - | - | 1.4-2.2 | 3.0-3.7 |
| *Air space (m³)* | 6 | 10 | 12 | 12 |

Removing an individual from a herd will cause it distress. One exception is the cow about to calve, who will isolate herself from the herd to give birth. Having licked her calf clean and suckled it she will then leave it to lie hidden and go back to the herd. She will only return to feed her calf 4–6 times during the first few days until it is strong enough to run with the herd.

On open range and given plenty of space, cattle form stable sub-groups. A dominance hierarchy is established, whereby each cow knows its place.

## Reproductive biology

The cow normally has one calf per year, after a pregnancy of nine months. There are small genetic differences in the duration of pregnancy that are largely due to the genotype of the calf. A Holstein/Friesian cow carrying a calf to a bull of her own breed has a gestation period of 281 days. If (to use an extreme example) the sire had been a Limousin bull, gestation would take 287 days. Table 3.9 gives values for the incidence of dystocia (calving difficulties) and calf mortality in suckler cows and dairy cows (Holstein/Friesians) artificially inseminated to bulls from different breeds. The traditional British beef breeds, Hereford and Aberdeen-Angus, carry a lower incidence of dystocia, which is one reason why farmers traditionally use them to inseminate heifers. Their relatively small size and equable temperament is another reason why they are suitable to have running with heifers.

The average incidence of dystocia in Holstein/Friesian cows is less than in suckler cows although, for calf mortality within the first two days of life, the reverse is true. The incidence of dystocia is, on average, higher for the cows mated with very large beef bulls, Charolais and Simmental. However, the risk of dystocia is due at least as much to the pelvic dimensions of the cow, as to the size of the bull. Cows of typical beef type usually have smaller pelvic dimensions and a greater incidence of dystocia although, left to themselves, they are more likely to ensure that their calf stays alive (Table 3.9). Breeders have selected Belgian Blues for very heavy muscling, to the extent that many pure-bred cows are unable to calve normally; such cows require repeated, premeditated Caesarian section and this constitutes a major welfare problem. However, the incidence of dystocia in Holstein/Friesian cows mated to Belgian Blue bulls is no greater than for the Charolais.

Artificial insemination (AI) studs, for dairy and beef cows, keep records of the performance and progeny tests of their individual bulls. These include records of calving difficulty. When selecting a bull for AI it is possible to take this into account, to avoid, for example, mating a bull who carries a high risk of dystocia to a cow with a history of calving difficulties.

Cattle have no discrete breeding season. A non-pregnant sexually mature female will show oestrus at intervals of about 21 days throughout the year. When beef cows are on pasture or open range, natural service by a bull running with the herd is practically obligatory. When cows are confined, it is possible to induce and synchronize oestrus and ovulation by the use of hormones. There are two approaches, both of which rely on controlling the end of dioestrus, which initiates the period of rapid follicular development. This involves either injection in dioestrus of prostoglandin to destroy the corpus luteum, or insertion and timed withdrawal of progesterone-releasing intravaginal devices (PRIDs).

The probability of fertilisation following synchronisation of oestrus followed by AI is unlikely to exceed 65 per cent, so most beef farmers need a 'sweeper' bull to serve those cows that do not conceive to AI. However, AI does make it possible to use bulls of much higher genetic merit than one is likely to find on the average commercial beef farm.

## PRODUCTION

The age and weight at which a producer finishes a young beef animal depends on its breed and sex (which together determine its phenotype) and its plane of nutrition. Cattle are 'early maturing' if it is possible to finish them at a relatively young age and light weight. The Hereford is an earlier maturing breed than the Charolais; the heifer also matures earlier than the bull. The animal (within any phenotype) that eats the most digestible nutrients over and above its requirement for maintenance will not only gain weight most rapidly but also deposit more fat relative to lean tissue. In other words, the higher the plane of nutrition, the greater the proportion of fat to lean meat in the body gains and the lower the finishing weight and age of the animal.

Figure 3.4 illustrates the effect of plane of nutrition on optimal slaughter weight for an early-maturing animal, like a Hereford x Friesian steer, and a late-maturing animal, like a Charolais x

Friesian bull. The appetite of a growing ruminant is driven by its potential for growth, which determines its metabolic hunger for nutrients, and restricted by the constraints of rumen fill. The rate at which an animal can ferment and break down plant fibre determines rumen fill. Given a highly digestible diet containing a high ratio of cereal starch to plant fibre, the Hereford x Friesian steer will consume enough energy and protein to finish at slaughter weight of 400kg within 10–12 months. On a diet consisting almost entirely of fresh grass and silage, it may be 460kg before it is ready for slaughter. It may also take twice as long to achieve that weight and satisfactory degree of finish. The larger, late-maturing Charolais x Friesian bull, on a highly digestible diet, achieves the right degree of finish at a weight of 500kg in perhaps 12–14 months. On a low-quality diet consisting mainly of plant fibre it may be unable to consume enough food to achieve a minimum fat concentration of 15 per cent. This chapter will consider optimal strategies for matching feeds to breeds in more detail later.

## NUTRITION AND FEEDING
### Digestion
When a cow grazes at pasture, all plant material enters the large paunch, or reticulo-rumen (Figure 3.5) where it is mixed, diluted with saliva and subjected to microbial fermentation. The end-products of microbial fermentation of plant carbohydrates (sugars, starch, cellulose and hemicellulose) are absorbed, largely across the rumen wall, as the volatile fatty acids - acetic, propionic and butyric acid - which form the major source of dietary energy to the ruminant. The digesta that leave the reticulo-rumen and pass into the abomasum (which corresponds to the true stomach in man) contain large amounts of microbial protein. These are subjected to further digestion and form the ruminant's principal supply of amino acids.

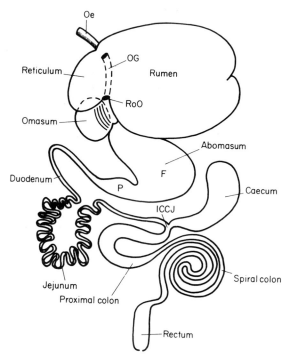

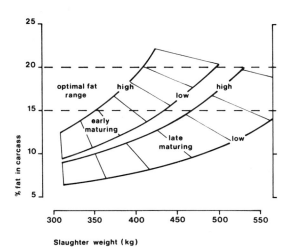

**Figure 3.4**   **Effect of type (early versus later maturing) and plane of nutrition (high versus low) on the relationships between slaughter weight (kg) and percentage fat in the carcase.**

**Figure 3.5**   **The digestive tract of the adult ruminant (from Webster 1993), Oe Oesophagus; OG oesophageal groove; RoO reticulo-omasal orifice; F fundus of abomasum; P pylorus; ICCJ ileo-caeco-colic junction.**

The adult ruminant derives most of its nutrients from reactions in the rumen. The small intestine (duodenum, jejunum) is the major source for absorption of amino acids and minerals. The large intestine (caecum, colon, rectum) acts as a second fermentation chamber, but normally contributes less than 8 per cent to nutrient uptake. (See also Chapter 2 p 29-30).

**Metabolizable energy**

By far the greatest quantity of nutrients absorbed from the gut is used as a source of energy (ie Metabolizable Energy, ME) to fuel the work of maintenance, activity and production. ME requirement is expressed in megajoules per day (MJ/day) and the ME concentration in the feed is expressed in MJ ME/kg dry matter (DM), which abbreviates to M/D. The DM intake of a ruminant is stimulated by its requirements for nutrients, especially energy, but restricted by the rate it can digest the food it eats, principally the amount of unfermented matter it can carry within the rumen. A growing calf or lactating cow will have a greater hunger for nutrients per unit of body weight than an adult, non-pregnant, non-lactating animal. However the less digestible the food (ie the lower the M/D) the more difficult it becomes to meet its energy requirements within the constraints of gut fill. Except for unweaned calves, farmers generally feed beef animals in yards a small rationed allowance of relatively expensive cereal-based concentrate feeds with a high M/D, and allow them to eat forage to appetite. At a prejudged *ad lib* intake the mixture of forage and concentrate achieves the M/D an animal needs to match ME intake to requirement for maintenance, growth or lactation.

The ME requirement for maintenance (MEm) of any mammal depends on its metabolic body size, or bodyweight ($kgW^{0.75}$). For cattle, MEm is $0.5MJ/kgW^{0.75}$. For example, a 460kg Blue-Grey cow has a MEm of 50MJ/day. At peak lactation in a beef cow, ME requirement is approximately twice maintenance. Table 3.13 gives the ME allowances growing cattle need to meet a range of target weight gains. The blanks in Table 3.13 indicate liveweight gains that an animal cannot achieve because, at any given M/D, ME requirement exceeds predicted DM intake. A diet with an M/D of 11MJ/kg can sustain a gain of no more than 0.5kg/day in a 100kg calf but 1.0kg/day at 300 or 500kg. A diet with an M/D of 12 should be able to sustain a weight gain of 1.5kg/day in an animal genetically capable of growing that fast.

Table 3.14 (on page 65) gives energy and protein values for some common feeds for beef cattle using the conventions of the new UK Metabolizable Protein system (see Alderman 1993, McDonald *et al* 1995). ME and fermentable metabolizable energy (FME) (the amount of feed energy that can be fermented to volatile fatty acids by the rumen micro-organisms) defines feed energy. In this system dietary protein is described by crude protein (CP), effective rumen-degradable protein (ERDP) and digestible protein that escapes degradation in the rumen and is available for acid digestion (digestible undegradable protein or DUP). Table 3.14 gives values for CP and ECDP only.

**Table 3.13**  Metabolizable energy (ME) concentrations required for growth in cattle.

| Liveweight (kg) | M/D (MJ/kg DM) | Daily liveweight gain (kg) | | |
|---|---|---|---|---|
| | | 0.5 | 1.0 | 1.5 |
| *100* | 11 | 26 | 35 | - |
| | 12 | 25 | 33 | - |
| *300* | 11 | 54 | 69 | - |
| | 12 | 52 | 66 | 66 |
| *500* | 11 | 76 | 97 | - |
| | 12 | 74 | 93 | 119 |

The main principle underlying the development of the Metabolizable Protein system is to ensure a balanced diet for the ruminant animal and the rumen itself. FME and ERDP describe the capacity of the feed to provide fermentable energy and degradable N for the rumen microbes. When the ratio of ERDP to FME is 9-10gERDP/MJ FME beef cattle achieve optimal fermentation and microbial protein synthesis. Table 3.14 (on p 65) indicates, for example, that ryegrass silage contains a relative excess of ERDP but meadow hay, not enough.

The ME concentration of pasture grass and the best silages appear to be able to sustain weight gains of 1kg/day or more. Where this does not occur, the problem is unlikely to be attributable to a simple energy deficiency. Cereals, with an M/D close to 13 MJ/kg can provide enough ME to sustain maximum weight gains. However, they are deficient in protein, especially ERDP, and other essential minerals and vitamins. A diet based almost entirely on cereals will also not provide enough long fibre to stimulate rumination and can lead to problems such as bloat and rumen acidosis (see Chapter 2, p 40). (See also Chapter 2, pages 29-31).

### Practical feeding

The choice of genotype for a system depends on the quality of feed available and the purpose of the animal. Figure 3.6 provides a schematic illustration of the effect of the relationship between M/D and the efficiency of feed utilization for growth.

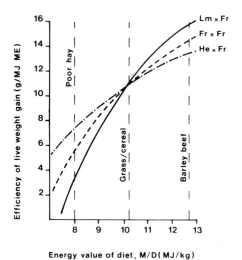

**Figure 3.6**     **A schematic relationship between diet quality (M/D) and food conversion efficiency (g gain/MJ ME) for three types of beef cattle.**

The efficiency of feed utilization for growth is defined by gram liveweight gain per MJ ME for three types of beef cattle fed to appetite.

A Holstein/Friesian calf on a poor quality diet of hay (M/D = 8) achieves only 4g gain per MJ ME but this rises to 12g/MJ ME on a barley-beef ration (M/D 12.8). It is able to consume more energy from

the barley-beef ration relative to its needs for maintenance and therefore converts food more efficiently. However, food consumption is not necessarily more cheap since barley is more expensive per unit of ME. Relative to the dairy-type Holstein/Friesian, the Hereford x Friesian is more efficient at converting feeds with a low M/D, because it has a relatively low ME requirement for maintenance. When giving diets with an M/D of 10.5 or above, the difference between the two types is small. The Hereford x Friesian, however, does tend to become too fat too soon on rations containing a high proportion of cereals.

The lean Limousin x Friesian cross has a relatively high potential for lean tissue growth but a relatively low gut capacity for indigestible fibre. It is therefore less efficient on poorer quality diets, but becomes progressively more able to express its genetic potential for lean tissue growth with improving feed quality. The Limousin x Friesian is not necessarily better than the Hereford x Friesian; they are best suited to different diets. As the quality of conserved forages for ruminants rises, so these diets progressively favour the larger and leaner continental bulls as sires of calves the producer intends to slaughter for prime beef. The suckler cow, however, may have to thrive on poor quality forage or dormant winter pasture. In this case it would pay to have a Hereford or Aberdeen-Angus father.

### Minerals and vitamins

Diets based wholly on fresh or conserved grass are likely to provide sufficient calcium, but phosphorus and magnesium may be marginal. In certain cases the diet may be deficient in copper, cobalt or selenium and all these deficiencies may stunt growth. It is possible to feed exact amounts of minerals by incorporating them into the concentrate ration. It is also possible to offer them free-choice in mineral licks. However, intake is erratic and does not relate to requirement, and selenium (and possibly copper) is toxic in excess. Some cattle may therefore consume too little and some too much. One of the most popular ways of delivering copper, cobalt and selenium to beef cattle is within a bolus that lodges in the rumen and releases these minerals at a steady rate; this is usually effective, but sometimes the bolus is lost.

**Table 3.14** **Energy and protein values of some common feeds for beef cattle.**

| | Dry matter (g/kg) | Composition of dry matter | | | | |
|---|---|---|---|---|---|---|
| | | ME (MJ/kg) | FME (MJ/kg) | CP (g/kg) | ERDP (g/kg) | ERDP:FME |
| *Pasture grass* | 200 | 12.2 | 11.4 | 175 | 130 | 11.4 |
| *Ryegrass silage,* excellent moderate poor | 250-300 200-350 150-250 | 11.4 10.0 9.0 | 9.4 9.0 7.0 | 170 150 120 | 140 100 90 | 14.9 11.1 12.8 |
| *Maize silage* | 250-300 | 11.2 | 9.8 | 100 | 80 | 7.2 |
| *Meadow hay* | 850 | 9.5 | 8.6 | 90 | 68 | 7.9 |
| *Barley straw* | 850 | 6,5 | 5.8 | 40 | 23 | 4.0 |
| *Barley* | 900 | 12.8 | 12.2 | 110 | 85 | 6.6 |
| *Soyabean meal* | 900 | 13.3 | 12.7 | 500 | 350 | 27.5 |
| *Maize gluten* | 900 | 12.7 | 11.5 | 200 | 140 | 12.1 |
| *Turnips* | 90 | 11.2 | 10.4 | 120 | 92 | 8.8 |

Cattle have no dietary requirement for vitamin C and are able to acquire B vitamins and vitamin K as an end-product of microbial synthesis in the rumen. Green grass and silage are good sources of carotenes, the precursors of vitamin A. Sun-cured hay is a good source of vitamin D, and whole cereals good sources of vitamin E. Generally, adult cattle are unlikely to suffer from deficiencies of the fat-soluble vitamins A,D, and E, but body reserves of these drop during the winter. Calves are born with almost no fat-soluble vitamins in their body and normally acquire them by drinking colostrum. The colostrum of beef and dairy cows in late winter may contain very low concentrations of fat-soluble vitamins. This gives the calves a poor start and makes them particularly vulnerable to infectious diseases of epithelial surfaces, eg diarrhoea and pneumonia.

**Water**

The *Codes of Recommendations for the Welfare of Livestock: Cattle* (MAFF 1983) state that 'cattle should have access to sufficient fresh, clean water at least twice daily'. This is necessary to maintain the water content of the body tissues while sustaining inevitable and regulatory losses. They also require water for milk production and to sustain the continuous culture of microbes in the rumen. Published figures for the water intake of cattle range between 50-150 g/kg bodyweight per day. It is greater during lactation and hot weather and less when the food is very wet (eg fresh grass or grass silage). The first response of cattle to a water shortage is usually to reduce food intake. Ideally, cattle should have access to clean water almost continually. If only allowed access to water twice daily they should be free to drink for as long as they want on each occasion.

Natural sources of water can also be a source of essential (or even toxic) minerals. It is also possible to dispense essential minerals, such as magnesium, in water troughs. However magnesium is rather unpalatable and this strategy will not succeed if the cattle have access to a more attractive water source such as a mountain stream.

**Feedstuffs**

*Grass*

This is almost the perfect food for beef cattle. The most nutritious part of the plant is the young leaf, so its digestibility (and M/D) is greatest before the emergence of the seed heads. Grazing management should aim to ensure that the grass is long enough (at least 7cm) to ensure that the cattle can consume enough during daylight hours, but not so long that it becomes stemmy and fibrous. Well managed grassland provides sufficient ME and protein for maintenance and lactation in beef cows and for maintenance and growth in calves receiving milk from their mothers. It also has the right balance of ERDP and FME. It cannot however sustain economically acceptable weight gains in calves from the dairy herd that are receiving neither concentrates nor milk from their mothers before they reach 200kg in the case of British breeds or 300kg for the larger continental breeds. This is because of the constraints of gut fill in the immature rumen. Cattle that receive all or nearly all their food from fresh and conserved grass may, in certain areas, suffer debilitating deficiencies of copper, cobalt, selenium or phosphorus. Such animals may be more susceptible to an acutely fatal attack of hypomagnesaemia or grass staggers. Increasing grass crop yields by application of inorganic fertilisers (N,P and K) inevitably reduces concentrations in grass of the trace elements (Cu, Co and Se). The claim of organic farmers that their cattle are far less prone to mineral deficiency diseases has a basis in that they do not reduce the concentrations of trace elements because they grow less grass/ha.

*Conserved forages*

It is possible to sun- or air-dry grass crops and conserve them as hay or compact them into an airtight clamp or big, covered bale, and conserve them anaerobically as silage. Well-made grass silage has a greater nutritive value than hay because the crop is cut at an earlier, leafier stage (Table 3.14). When making grass silage for dairy cows, it is important to cut the grass very young to maximize M/D. For beef cows, it is usual to leave the crop a little longer to achieve a greater yield but lower M/D (ca 10 MJ/kg). The anaerobic fermentation of grass sugars (ideally by lactobacilli) in the silage clamp, increases the acidity of the crop and inhibits further microbial

breakdown. In effect, the crop is pickled. One problem with this process is that the ME now consists almost entirely of slowly fermentable cell walls plus prefermented organic acids, which the cow can be metabolise but not the rumen microorganisms. Grass silage diets therefore ferment slowly and contain an excess of ERDP to FME. Appetite for grass silage, when it is fed on its own can be disappointing, because slow fermentation leads to greater rumen fill and absorption into the blood of ammonia resulting from degradation of excess ERDP. To improve both fermentation rate and appetite it pays to mix in a more rapidly fermentable feed with a relative excess of FME. Such ingredients include cereals, a root crop (turnips or fodder beet) or maize silage.

Well-made hay is less nutritious than silage but extremely palatable. Soft meadow hay made from a range of grass species is particularly suitable for encouraging young calves to develop an appetite for roughages. Barley straw has little nutritional value (Table 3.14) but is a good dietary supplement for cattle eating large quantities of cereals. The long, hard, relatively indigestible fibre dilutes the quickly fermentable starchy cereals and provides the necessary stimulation for increased rumination and therefore more salivation. Both these factors reduce the risk of ruminal acidosis.

It is possible to improve the nutritive value of straw by treating with alkalis – sodium hydroxide or ammonia. These cause the cell walls to open up and release more fermentable cellulose from its bonds to unfermentable lignin. Alkali-treated straws have an M/D close to that of hay. Ammonia treatment also increases the supply of ERDP to the rumen microbes. In any year, economics strictly determine the case for alkali treatment of straw; if the grass crop (hay or silage) is poor (in a wet summer) or in short supply (in a dry summer) then it pays to treat the straw crop.

*Maize silage*

In many areas of the southern UK it is now possible to grow maize and harvest the whole crop for silage in the autumn when the cobs are at a partially ripe, 'cheesy' stage. Yields of ME per hectare can be very impressive. The M/D is excellent but the ERDP to FME ratio is slightly less than optimum. Properly supplemented, maize silage can sustain growth rates in excess of 1.0 kg/day, not least because DM intakes are usually better for maize silage than when feeding grass silage alone. Feeding a mixture of 75 parts maize and 25 parts grass silage from a forage wagon is ideal.

### Cereals

The main nutrient in barley, oats and wheat is starch, which makes these cereals highly concentrated forms of ME for cattle. They are marginal or slightly deficient in protein, both as a source of ERDP for the rumen micro-organisms and as a source of amino-acids for growing or lactating cattle. When they constitute the major part of the ration (eg in a barley-beef system) they may promote very rapid growth, but this will cause early maturing breeds (eg offspring of Hereford and Angus bulls) to fatten too soon. All cereals may predispose to ruminal acidosis through too-rapid fermentation, unless diluted with long fibre. Oats is the safest cereal in this respect, being the most fibrous, but it is more expensive per unit of ME than barley or wheat.

### Root crops

The most popular root crops are fodder beet (in the south) and swedes or turnips (in the north). They have a DM content of less than 200g/kg but are rich in sugars and very palatable. They are deficient in CP and minerals. A small quantity of a root crop such as fodder beet (eg 5kg/day), can make an excellent supplement to grass silage. In the north and east of England and Scotland it is possible to grow roots as the main energy feed for wintering beef cattle. However, such a diet will need supplementing with protein, minerals and fat-soluble vitamins.

### By-products and crop residues

Most pelleted compound rations for cattle contain large proportions of by-products or residues from crops grown primarily for direct human consumption. The most common by-products that farmers feed their cattle are the protein-rich residues from plants such as soyabean, rapeseed or groundnuts. These are grown primarily as sources of vegetable oil for human consumption. Others include sugarbeet pulp, a highly palatable source of digestible fibre, and maize gluten, the crop residue after extraction of most of the starch from maize (Table 3.14). Both sugarbeet and maize gluten are reasonably well balanced in the ratio of

ERDP to FME and it is possible to feed them directly to cattle as 'straights'. Food manufacturers may also incorporate small quantities of fishmeal into compound feeds as an excellent source of amino-acids from DUP, minerals and vitamins. Some years ago manufacturers used meat and bone meal from abattoirs. However, when the processors changed their rendering procedure to leave more residual oil, it led to the epidemic of bovine spongiform encephalopathy (BSE). It is now illegal to include feed of animal origin (except fishmeal) in diets for cattle.

## MANAGEMENT OF ARTIFICIALLY REARED CALVES

This section deals with the artificial rearing of calves born to dairy cows for the first six months of life, whether destined as dairy replacements or for beef.

### Birth to weaning

Unless the cow is to calve out-of-doors, she should be confined in a spotlessly clean and disinfected calving box and well bedded down with fresh straw to minimise the risk of infection to cow or calf at the time of parturition. Assuming a normal delivery, the first tasks for the stockperson are to ensure that the calf can breathe properly, then disinfect the navel with iodine or with an antibiotic spray. The next essential is to ensure that the calf drinks an adequate amount of the cow's first milk (colostrum) in the first 12 hours. Colostrum provides not only food but also maternal antibodies to protect the young calf against the common infections that it is likely to encounter in early life. The abomasal secretions are not acid in the newborn calf and during the first 12 hours the epithelium of the intestine permits the passage of protein antibodies of the immunoglobulin G (IgG) class into the blood stream where they can protect against systemic infections such as neonatal septicaemia caused by the ubiquitous *Escherichia coli*.

Colostrum and (to a lesser extent) milk also contain maternal antibodies of the IgA class which cannot be absorbed from the gut but which confer protection from infection at the first point of contact with infectious agents such as Rotavirus, and again *E. coli*, namely the mucosal surface of the gut wall. This is one reason why the incidence of infectious diarrhoea tends to be less in calves given their mothers' milk, rather than fed artifically on a milk replacer diet. Suckler cows can be vaccinated to hyperimmunise them against Rotavirus and *E. coli*, so that their colostrum and milk will transmit more protective IgG and IgA to their suckled calves.

The teats of mature Holstein/Friesian cows may hang far below the abdominal wall and be hard for the young calf to locate. If the calf is left with its mother, the stockperson should ensure that it is feeding satisfactorily. Many good stockpeople like to remove the calf from its mother as soon as possible, before the maternal bond has developed, so as to minimise distress for both cow and calf. If so, the cow should be milked out and the calf offered three to four feeds of colostrum during the first day of life. For calves born to Holstein/Friesian cows, each of these first feeds should be of 1.5 litres.

Having started the calf, it is a matter of choice whether to rear it up to weaning in an individual pen on restricted amounts of milk replacer fed from a bucket twice- (or, later once-) daily (see Figure 3.7(a) on page 69), or to rear it in a group with free or controlled access to milk replacer sucked through a teat (Figure 3.7(b). The feeding options are as follows. (The amounts refer, once again, to the typical calf born to a Holstein/Friesian cow).

1. *Bucket feeding: twice-daily*
   *Week 1*: 2 x 1.5 litres/day at 125 g powder/litre = 375 g powder/day fed at blood heat (40°C).
   *During week 2*: increase intake to 2.0-2.5 litres/feed at 125 g powder/litre = 500-625 g powder/day.
   Introduce a dry calf starter ration (based on cereals, etc) and roughage in the form of hay (or possibly barley straw). Wean when intake of calf starter = 1 kg/day, which occurs at approximately 5 weeks of age. By this time each calf will have consumed approximately 20 kg of milk powder.

2. *Bucket feeding: once-daily*
   Weeks 1 and 2: feed twice daily as above.
   *Week 3*: feed 2.0-2.5 litres/day in a single feed containing 450-500g powder; offer starter feed and hay as before. It is *essential* to provide water for these calves. Intake of milk replacer prior to weaning will be approximately 16kg.

3. *Teat feeding: warm milk from dispensers.*
   These are machines which dispense freshly mixed milk replacer at blood heat (Fig 3.6b). Until recently, they simply provided milk on demand, so that calves drank *ad libitum.*. Most calves thrive on this system but intakes of milk powder to five weeks of age are likely to exceed 30 kg per calf, which substantially increases feed costs relative to bucket feeding. Moreover, calves that have had *ad libitum* access to milk powder may be eating much less than 1kg/day of calf starter ration at this time.
   This is likely to set them back post-weaning and may predispose to infections such as pneumonia. One strategy is to restrict time of access to the teat for the last week before weaning or make it more difficult for the calves to drink by progressively constricting the milk supply line. Once again, fresh water must be available at this time. It is now possible to buy milk dispensers which recognise individual calves wearing transponders around their necks or in their ears, dispense controlled rations (eg 400g/day of milk powder) and provide records of any calf which fails to drink its full ration. This provides an excellent early warning of incipient infection.

4. *Teat feeding: cold (acidified) milk.*
   Calves can be reared on cold milk provided they drink it slowly through a teat. The daily ration is mixed in advance and stored in plastic pails. The feeding system is completed very cheaply using a teat, a tube and a non-return valve. Milk powder for cold teat-feeding is usually acidified, mainly to prevent the milk souring in storage. Mild acid powders are based on conventional milk replacer powders containing skim milk and added animal fats, and pH is reduced to about 5.7. Strong acid powders (pH 4.2) are sometimes called milkless powders, since they do not contain casein (the main milk protein), as this coagulates into curds and whey at this pH. The protein source in strong acid powders is whey protein, (albumin and globulin) plus some vegetable proteins, specially treated to improve digestibility and reduce the risk of food allergies. In cold weather, the milk can be warmed to about 15°C using, for example, an aquarium heater with a thermostat.

**Figure 3.7**  **Calf rearing systems**
(a) **individual pens with bucket feeding**
(b) **group rearing on automatic teat system**

## The bought-in calf

The artificial rearing of calves on their farm of birth can usually be accomplished without mishap. Unfortunately, the majority of beef calves from the dairy herd are moved off their farm of origin at about 1 week of age into specialist rearing units. This may involve trips through two or more markets, and transport, sometimes for the full length of the country. Such animals are deprived of normal food, water and physical comfort, and are confused, exhausted and exposed to a wide range of infectious organisms, of which the most important are the *Salmonella* bacteria. By the time they reach their rearing unit, they are likely to be infected, dehydrated and stressed, and need special care if they are to survive.

On arrival, bought-in calves should be rested in comfort in deep straw. In very cold weather, they may benefit from a little supplementary heating for the first 2-3 days. Water should be available, but they should be offered no milk replacer for at least 2h after arrival. If they arrive in the evening, they can be left until the following morning. Some people feed 1.5 litres of milk powder at 125g powder/litre for the first feed. I prefer to give two 1.5-litre feeds of a proprietary glucose/electrolyte solution (or one tablespoonful of glucose and one teaspoonful of common salt in 1.5 litres of water) to rehydrate the calves and provide a minimal supply of energy but keep the gut empty of nutrients until the calves have been able to eliminate most of the enteric bacteria acquired in transit. Thereafter, feeding can be as described above. An injection of fat-soluble vitamins A, D and E is also advisable, especially for calves purchased in the late winter. A preferable alternative to this remedial action is for specialist calf rearers to negotiate contracts with dealers who can guarantee to supply calves from their farm of origin to the rearing unit on the same day and with the minimum of disturbance and mixing.

## Feeding after weaning

After weaning, calves are fed a concentrate ration plus hay, straw or silage *ad libitum*. The amount of concentrate fed and its protein concentration are governed by how fast the calf is expected to grow. If it is to be turned out to grass in the spring and reared for slaughter at 18-24 months, it should get no more than 3 kg/day of concentrate with a protein concentration of 180 g/kg plus hay or silage *ad libitum*. Calves in contract-rearing units, which need to achieve as much weight gain as possible by 12 weeks, and barley-beef calves, will probably be given unrestricted access to concentrates and consume more than 4kg/day. In this case the protein concentration in the post-weaning concentrate diet should be no more than 160g/kg, and it may pay to feed roughage in the form of straw rather than hay or silage. Since calves are likely to eat less barley straw than hay or silage it follows that they will tend to eat more concentrates when straw is the only source of roughage. If the calves are to be reared intensively on a high concentrate ration and slaughtered at 12 to 16 months of age, it pays to adapt them to a high concentrate ration as quickly as possible. If the mainstay of their subsequent diet is to be grass or maize silage then, equally, it makes sense to introduce these feeds as soon as possible.

## Housing young calves

The greatest threats to the health and welfare of the young calf are infections which may cause septicaemia, enteritis (leading to diarrhoea or scours) and pneumonia. The organisms responsible for these conditions are widespread and young calves, especially those which are moved through markets, are practically certain to be exposed to some degree of infection. It does appear, however, that both the incidence and severity of these infections are determined both by the numbers of pathogenic organisms and by pollutant substances presented to vulnerable epithelial surfaces of the gut and lungs. The prime specification for a calf house is, therefore, that it should be as hygienic as possible, not only on the surfaces of the walls, floor and feeding utensils, but also in the air itself. It should be power-washed with hot water or steam and disinfected before the first calves of the season arrive, and this process should be repeated between successive batches. Whenever possible, the rearer of bought-in calves should practice an 'all-in, all-out' policy, and avoid introducing new baby calves into an already infected building. Hygiene and humidity are also controlled by ensuring effective and appropriate drainage under the individual group pens.

Air hygiene is determined mainly by air space per calf (see Table 3.12) and, to a much lesser extent, by ventilation. This is because the animals are the prime source of pathogenic organisms, but ventilation only removes a small proportion of these organisms from the air in the building; most die *in situ* (for further explanation, see Webster, 1985). Unweaned calves require 6m³ air space per calf to ensure reasonable air hygiene, and post-weaning calves 10m³ (Table 3.12). Under UK conditions, effective air movement (a minimum of four air changes per hour) can be achieved by natural ventilation through strategically placed inlets and outlets. One approach is to place space boarding with 100 mm sections and 15 mm gaps below the eaves to provide 0.05m² open area per calf and an open ridge providing 0.04m² opening per calf (see Figure 3.10 on page 76). In still air, air enters the building below the eaves, is heated by the presence of the animals and exits through the open ridge. In windy conditions, eaves and roof openings can act as inlets or outlets, but the 15 mm gaps restrict the speed of the air entering the building and the pen covers prevent cold air descending to create cold draughts within the calf pens. An alternative way to prevent draughts troubling very young calves is to place straw across the backs of the pens (Figure 3.7a) which can be progressively added to the bedding as the calves grow bigger and stronger.

Whether reared in groups or individual pens prior to weaning, calves are normally grouped after weaning and reared in follow-on pens. The simultaneous stresses of mixing and weaning can increase the risk of disease at this time, especially pneumonia. One advantage of rearing batches of calves in groups and feeding them from a teat is that they can be kept in the same group and in the same accommodation from the time of arrival until turnout. Calves are initially restrained in the back of the pen by straw bales, which are taken down at weaning and used as bedding (Figure 3.8). The small increase in feed costs prior to weaning may be more than offset by reductions in the cost of housing and of disease.

### Rearing calves for veal

Most calves reared for veal in Europe are confined for life in individual wooden crates and fed only a liquid milk replacer diet, deficient in iron, to ensure white meat. In this system, which is now illegal within the UK, the calves are fed milk replacer in the same way as any bucket-fed calf, but given no access to solid food. The intake and concentration of the milk powder are increased, so that by 12 weeks they will consume 3 kg/day of powder at a concentration of 150 g/litre water at 40°C. Traditionally Holstein/Friesian calves are slaughtered after they have consumed about 200kg of powder and reached about 200kg live-weight. This occurs after 14-16 weeks on the unit. In recent years, as the cost of calves entering the unit has increased, it has become economically necessary to feed calves to greater weights so that some bull calves may be slaughtered for white veal at weights in excess of 300kg, at a time when they are becoming sexually active. The average iron concentration of the milk powder is 25mg/kg or less. This system abuses all five freedoms.

- The calves are malnourished, lacking dry, digestible food to encourage normal rumen development and sufficient dietary iron to prevent clinical anaemia
- They are enclosed in individual crates on slats. These crates make it difficult for the calf to stand when young, and impossible for them to adopt normal resting and sleeping postures when over 150 kg, nor groom themselves all over
- The calves are denied any exercise or direct social contact
- The incidence of both enteric and respiratory diseases is unacceptably high.

### Alternative husbandry systems for veal calves

Several alternative husbandry systems for veal calves have been explored in an attempt to discover one that is more humane and also economically competitive with the conventional crates. In the 1980s there was a vogue to rear calves in groups on straw and allow them to suck milk replacer *ad libitum* from an automatic dispenser. In the UK this became known as the Quantock system after its original proponents. The main drawback of this system is that the calves drink more milk than is necessary for optimal growth, or is indeed good for them, and this reduces food conversion efficiency (Table 3.15). In brief, 1 tonne of milk replacer will rear five Friesian bull calves in crates but only four by the Quantock system. Moreover, the system still does not permit normal development of the digestive tract.

Trials at the University of Bristol (Table 3.15 and Webster *et al* 1986) have examined a variety of alternatives. Of these the most effective has proved to be ACCESS (A Computer-Controlled Eating Stall System) in which calves equipped with transponders that are recognized by the computer enter either of two computer-controlled feeding stalls, to receive controlled amounts of milk replacer or digestible dry food contributing respectively 90 and 10 per cent of nutrient intake. With this system we were able to achieve food conversion efficiencies comparable to those in crates (Table 3.15). Moreover, the health of the calves was greatly improved by eliminating the worst abnormalities of feeding and housing.

*The Welfare of Livestock Regulation 1994* as now modified by *The Welfare of Livestock (Amendment)* *Regulations 1998* make it illegal, after the 1st January 2004, to keep calves over 8 weeks of age singly – they must be group housed. This rule only applied to accommodation which was in use for calf rearing before 1st January 1998. New calf houses do not have this 5 year exemption. Many aspects of calf husbandry are covered by the Regulations. For example, calves tethering is no longer permitted except in respect of group housed calves for a short period during liquid feeding; all calves must be fed twice daily and those which are housed must be inspected twice daily; standards are set of the provision of dietary iron and fibre; individual stalls and pens housing calves must allow tactile as well as visual contact; the size of the accommodation is specified etc etc. See the Regulations for details.

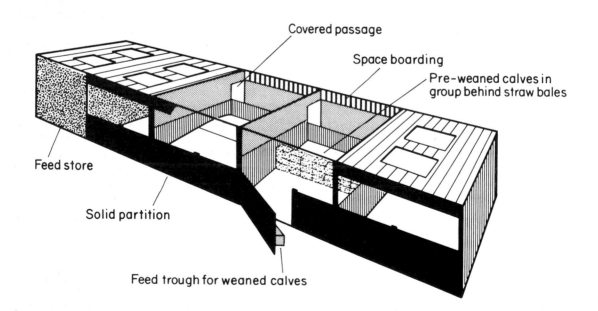

**Figure 3.8**     **Simple monopitch accommodation for rearing bought-in calves in groups. Calves stay in the same bay from arrival to turnout (from Webster 1985).**

**Table 3.15**  **A comparison of the performance, health and welfare of veal calves in different rearing systems (from Webster 1985).**

|  | Crates | Quantock | Access |
|---|---|---|---|
| *Growth* | uniform | some poor doers | uniform |
| *Food conversion ratio* | 1.5-1.6 | 1.6-1.8 | 1.5-1.6 |
| *Physical comfort* | bad | good | good |
| *Infectious diseases* | | | |
| enteric, % | 19 | 4 | <5 |
| respiratory, % | 9 | 18 | <5 |
| *Behaviour* | | | |
| oral | abnormal | some problems | normal |
| locomotor and social | impossible | normal | normal |

## MANAGEMENT OF GROWING BEEF CATTLE
### Summer grazing
The pastures grazed by young cattle during the summer need to be managed to guarantee that the grass is sufficiently well grown to ensure maximum consumption but still sufficiently leafy to ensure optimum nutritive value and as free as possible from roundworms that can infest the gut and lungs. The most efficient way to utilize high-quality spring grass is to advance an electric fence across the pasture so as to graze a new strip every day. This encourages the animals to eat all the fresh grass exposed, and defaecate on areas grazed previously. This practice is more often associated with intensive dairy herds than with more extensive beef systems. The next most efficient approach is paddock grazing, whereby the herd is put on to a relatively small area of fresh grass at a high stocking density so that it grazes it down in, say, 3 weeks, before moving on to new pastures. The paddock is left to recover for about 6 weeks and then grazed once again by the young cattle or, preferably, older animals (or possibly sheep) who will not be affected by any build-up of parasitic roundworms.

However well a pasture is managed, the quality of the grass declines during the summer, due in part to natural seasonal changes in the pattern of growth and in part to the accumulation of dung pats which encourage the growth of rich, green patches of grass which the cattle will not eat.

The two most important species of parasitic roundworms for grazing cattle are *Ostertagia*, which causes gastritis, and *Dictyocaulus*, which causes pneumonia. Both species overwinter on pasture and, therefore, can affect calves shortly after turnout. If, as is likely, the pasture was grazed by cattle the previous year, it pays to delay turning out young calves until early May, by which time most of the overwintered *Ostertagia* larvae will have died. If calves do become infested and excrete eggs on to the pasture, there is a second build-up of larvae on the pasture by the middle of July. Ostertagiasis in cattle can be controlled effectively by pasture management, but many beef farmers consider it necessary to dose young calves with anthelmintic drugs in their first summer at grass. Cows and yearling cattle who have previously been exposed to a low level of parasitic challenge develop an effective immunity.

The lungworm, *Dictyocaulus*, is a much more dangerous parasite which causes husk or hoose, a severe, incapacitating parasitic pneumonia. Affected animals have difficulty in breathing, develop a characteristic deep, husking cough and lose condition extremely fast. The most elegant way to control husk is to vaccinate the calves with two oral doses of irradiated (and thus attenuated) larvae before turnout. It is however often cheaper to control the condition with appropriate anthelmintics.

### Winter housing
When cattle are brought into an indoor yard system, having been out at grass all summer, they are well adapted to a fibrous diet and reasonably immune to the viral and bacterial organisms that cause infectious pneumonia. Management, therefore, presents few

problems. The objective is to feed the cattle so as to achieve pre-set targets for weight gain at the least possible cost. The most useful aid to management is, therefore, a good set of scales and an efficient arrangement for moving cattle in and out of their pens and over this weighbridge at intervals of about 4 weeks. Figure 3.9 illustrates such a system. Cattle may be driven fairly easily in groups into a collecting pen large enough to contain all the cattle from one pen in the beef rearing unit. They may then be driven down a race with solid sides and a curved approach to a weighbridge and a crush, where individuals may be restrained for routine treatment or preventive medicine. The high, solid sides to the race prevent the cattle from being distracted or startled by the presence of handlers or other alarming objects in front of them, and the curve in the race encourages them to follow their leaders until it is too late to try to escape.

Within the building itself, the cattle may be housed entirely on slats. Alternatively, and preferably, each pen may be divided into a well-strawed bedding area and a concrete area behind the feed fence at either side of the central feeding passage. Each pen is separated from the next by a gate which can be swung across daily to enclose the cattle in the bedding area and permit the concrete standing area behind the feed fence to be scraped down using a tractor.

The building should be designed to permit the maximum amount of air movement without incurring draughts. Once again this is best achieved by installing space boarding to a depth of at least 1 and preferably 2 m below the eaves, and an open ridge, with flashing upstands to either side (Figure 3.9) over the central feed passage. The air, which has been warmed by the cattle, and which leaves the building by this open ridge, will prevent the entry of rain or snow in all but the most severe conditions. In addition, over the winter, the ridge will probably let out about 100 times more moisture (from the cattle) than it will let in.

## MANAGEMENT OF THE SUCKLER HERD

I shall consider here only those suckler herds in which the animals are housed over winter.

### Autumn calving

The main advantage of autumn calving is that the cows can give birth to their calves when they are well dispersed at pasture, which is hygienic. Moreover, the weather is usually mild and the cows will have eaten plenty of good grass, which ensures that they should have plenty of high-quality colostrum and, subsequently, milk to give the calves a good start. One possible drawback is that calves by big bulls from cows in good condition may be very large at birth, and it is not exactly easy to give assistance to a half-wild beef cow having difficulty giving birth on open range.

If the cows calve between August and October they will require some concentrate feed, such as rolled barley, starting at about 0.8 kg/day and rising to 1.8 kg/day by November. This barley should be supplemented with magnesium to reduce the risk of deaths from acute hypomagnesaemia (see p 80). The bulls should be introduced to the cows in November to try to get as many cows as possible in calf before they are brought into their winter accommodation. If beef cows are wintered in a deep straw yard (Figure 3.11), mating indoors presents few problems. However, if farmers handle them in cubicles or on slats, both cows and the bull can suffer injury due to slipping or falling on slimy concrete floors. To avoid any risk to the bull, cows can be observed for signs of oestrus and brought to the bull pen for service. Irrespective of the type of accommodation for the cows it is essential to provide a creep area for the calves (Figure 3.11).

The amount of concentrate that is fed to the cows (0.5-1.5 kg/day) is determined by the quality of the forage. Cows that enter the winter in good condition (condition score 3-4, Figure 3.12) may reasonably be expected to lose 80kg over the winter and go out to grass the following spring at an average condition score of 2. This may appear harsh but is quite natural. Cattle evolved in environments in which the availability of food was seasonal, and are well-adapted to changes in body weight of up to 15 per cent.

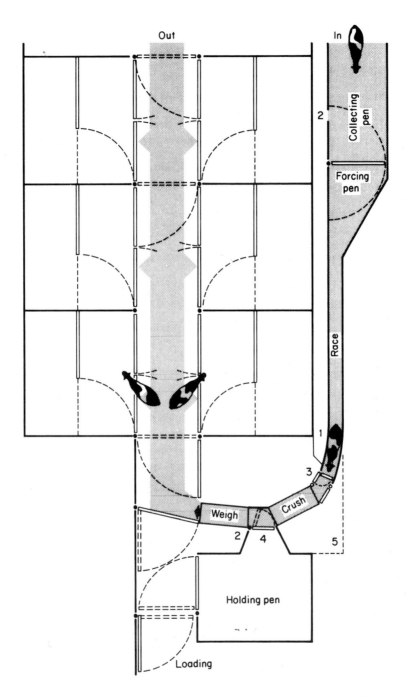

**Figure 3.9**     **Housing and handling facilities for in-wintered beef cattle (modified from MAFF/1984b Bulletin 2495) Crown Copyright. 1, raised walkway; 2, squeeze gap; 3, gate-access to rear of crush; 4, diversion gate; 5, line of roof over work area.**

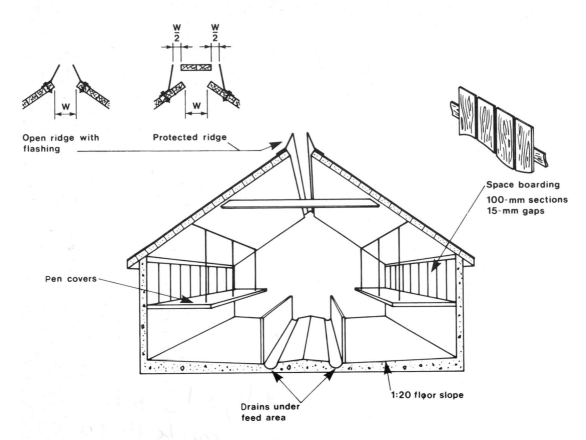

Open ridge with flashing

Protected ridge

Space boarding
100-mm sections
15-mm gaps

Pen covers

Drains under feed area

1:20 floor slope

**Figure 3.10      Specifications for a naturally ventilated calf house (adapted from Mitchell, 1978).**

The most effective means of controlling these two conditions is excellent hygiene at calving. There should be sufficient calving pens to allow each cow to deliver her calf on clean straw, stay with her calf for 2 days before returning to the herd, and enough time to permit the box to be cleaned out, disinfected and re-bedded with fresh straw before the next cow enters. This is expensive, but so are dead calves. Vaccines against *E. coli* and Rotavirus can also be given to cows to boost the transfer of passive immunity against these organisms in colostrum and milk. As indicated earlier, IgG in colostrum is absorbed into the blood stream and provides protection against septicaemia; IgA in milk confers local protection on the mucosal surface of the gut. These vaccines are effective but no substitute for good husbandry and hygiene, not least because *E. coli* septicaemia may be acquired through the navel

before the calf has had its first meal.

Turnout to grass takes place as soon as possible. The main risk for the cows at this time is hypomagnesaemia, and magnesium supplements are essential. If the cows are not getting a concentrate ration, these minerals can be incorporated into a block, added to the water supply or delivered from an intraruminal bolus. The bulls are introduced to the cows in June, when the weather is good and the cows are on a high plane of nutrition and gaining condition. In these circumstances, fertility should be high. The calves are weaned in October and usually moved directly into winter housing. As indicated above, simultaneous weaning and movement carries a high risk of provoking stress-related diseases like shipping fever.

Figure 3.11     A cattle shed with calf creep in the foreground.

months old. One effective procedure is to bring cows and calves into yards and separate them so that they can see each other but make no physical contact. Calves are fed hay for 2-3 days, then turned out once more. Cows are given a straw-only ration for 10 days to dry up their milk quickly, and a long-acting antibiotic is introduced into each quarter of their udders to minimize the risk of summer mastitis when they are returned to grass. There is no doubt that calves which are weaned and returned to grass for a period before being sold and/or housed for winter feeding are less likely to suffer problems such as shipping fever (pneumonic pasteurellosis) than those which experience the stresses of weaning, transport, mixing, winter housing and change of feed all at once.

### Spring calving

The cows in a spring-calving herd will have probably had their calves removed in October and then been kept out at grass for as long as possible, entering their winter accommodation in November in Orkney, or as late as Christmas following a mild autumn on the moorlands of south-west England. These cows will also be expected to lose about 80-100 kg over the winter and calve down at an average condition score slightly greater than 2. Once again, the amount of concentrate required (if any) will depend on the quality of the silage. It is, however, important to ensure that all beef cows have adequate access to a mineral supplement, which can be sprinkled over the silage to ensure even distribution.

Ideally, spring-calving herds should be turned out to grass before calving but, because of the fickle nature of the spring and the shortness of the summer in classic beef country, this is usually impracticable. This means indoor calving, which carries severe risks of infection.

The main problems of infectious disease for the newborn calf in a suckler herd are septicaemia, usually caused by *E coli*, and enteritis caused by viruses such as Rotavirus. Both organisms are inevitable inhabitants of winter accommodation for adult cattle. Calves born into such an environment can be infected with *E. coli* at birth and the resultant septicaemia may (at worst) kill them within 2-3 days. Rotavirus and other enteroviruses usually induce diarrhoea beginning in the second week of life, which

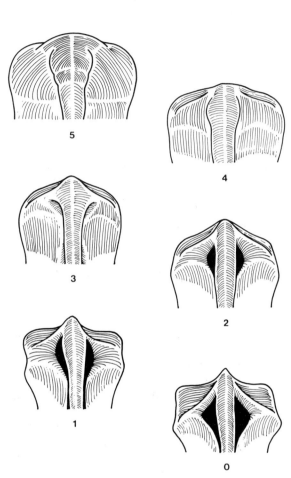

**Figure 3.12**   **Condition score of cows assessed for fat cover over tailhead and loin area. 5, grossly fat; 4, fat; 3, good; 2, moderate; 1, poor; 0, very poor (from Webster 1993).**

From turnout to the middle of summer, cows and calves run together on plentiful grass to ensure not only that the calves grow fast but also that the cows regain the condition lost over the winter. Weaning occurs in mid-summer when the calves are about 10

can also be fatal or severely stunt the calf's development.

## HEALTH AND DISEASE
### Preventive medicine
*Septicaemia and gastroenteritis*
To minimize the risk of septicaemia (*E. coli*) and viral gastroenteritis (Rotavirus etc), calving accommodation should be clean and calves should receive ample colostrum. In suckler herds calving indoors, it may be necessary to vaccinate the cows against *E. coli* and Rotavirus to boost antibody concentrations in the colostrum. Calves with diarrhoea rapidly become dehydrated and sodium-deficient. When treating such calves it is even more important to replace lost water and electrolytes (up to 5 litres/day) than to attempt to treat the primary infection with antibiotics.

*Salmonellosis*
Infection with (especially ) *Salmonella typhimurium* is a very common cause of enteritis and toxaemia in calves that have travelled through markets. Many strains of this organism are resistant to a wide range of antibiotics. The most effective preventive medicine appears to be to restrict food intake but otherwise minimize environmental stresses for the first few days after the calves arrive. Vaccination against salmonellosis *after* arrival can be dangerous.

*Pneumonia*
Infectious pneumonia of housed calves occurs most commonly at an age of 6-12 weeks. There is no vaccine that confers effective protection against all the infectious organisms that can damage the lung. The most effective approach to the control of the incidence and severity of pneumonia is to ensure good air hygiene through proper attention to stocking rate and ventilation, to avoid mixing calves as far as possible during their first winter and to minimize the stresses at and around weaning.

*Ringworm*
This is a common condition of housed calves in their first winter. The fungi are difficult to remove from porous surfaces like walls and woodwork and the disease is expensive to treat. However, it does not appear to cause the calves any great distress and

normally disappears when they are turned out to grass. Most cases are therefore left to clear up spontaneously.

*Lice and mange*
These conditions are caused by external parasites and can cause hair loss, considerable irritation and loss of body condition in housed cattle at any age. They are easily controlled by local or systemic treatment with insecticides.

*Parasitic bronchitis*
Parasitic bronchitis or husk affects calves in their first summer at pasture. It is effectively controlled either by oral vaccination with attenuated *Dictyocaulus* larvae before turnout or by regular dosing with appropriate anthelmintic drugs.

*Ostertagiasis* (stomach worm infestation)
This can also be controlled by regular dosing, or by pasture management.

*Warble fly*
The warble fly (*Hypoderma* spp.) lays its eggs on the hairs of the legs and abdomen of cattle. The larvae hatch out, burrow through the skin and migrate, eventually, to the back of the cattle from whence they emerge as maggots 25-30 mm in length. This painful and economically costly condition was made a notifiable disease and has largely been eradicated by treating cattle with a systemic insecticide between 15 September and 30 November to kill the larvae after they have penetrated the skin but before they have grown and migrated to the stage where they can do damage. Annual treatment is no longer compulsory but should be reinstated if there is a recurrence of the problem.

*Shipping fever*
This disease, also known as fibrinous pneumonia and associated with *Pasteurella haemolytica*, commonly affects calves from the suckler herd when they are brought off open range and concentrated into buildings or open feedlots. The calves are normally carrying the organism in the upper respiratory tract on arrival. Pneumonia develops in response to stresses of mixing, disturbance, exhaustion or improper introduction to concentrate rations. Once

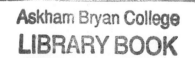

again, the disease is better controlled by attention to management than by vaccination.

*Hypomagnesaemia in cows*
Magnesium has a controlling, inhibitory effect on nerve and muscle activity. Cattle, especially adult cows at pasture, can teeter on the brink of magnesium deficiency for weeks before a sudden stress such as a rainstorm can push them over the brink into an acute crisis. The animal proceeds through nervousness to madness and may die in convulsions within 1 h of the first clinical signs. Because of the high risk of fatalities from hypomagnesaemia, it is important to make doubly sure that susceptible animals receive ample magnesium in the diet. Cows cannot be relied on to control their intake of high magnesium mineral licks, and these should be supplemented by adding magnesium to the concentrate ration or the water supply, or by administration of intraruminal slow-release boluses. Since none of these techniques is 100 per cent certain it pays to adopt a 'belt-and-braces' approach and use two of them.

For details of other diseases, and metabolic disorders of adult stock, see Chapter 2, pp 24-28.

## ROUTINE PROCEDURES

This section describes, very briefly, only the routine procedures in husbandry and preventive medicine common to most beef farms. The skills that enable the stockman or veterinary surgeon to carry out painful procedures like castration and dehorning with speed, safety and humanity are not to be learnt solely from textbooks but by direct, practical instruction from trained operators. The castration and dehorning of calves is controlled by the *Protection of Animals (Anaesthetic) Acts 1954 and 1964* (as amended in 1982) which deem the following to be operations performed without care and humanity unless carried out under anaesthetic:

1. The castration of a bull by a device that constricts the flow of blood to the scrotum (a rubber ring) unless it is applied within the first week of life
2. The castration of a bull by any means after the age of 2 months (after this age it must be carried out by a veterinary surgeon)
3. The dehorning of adult cattle or disbudding of calves except by chemical cauterization in the

first week of life.

For details of the complex regulations governing these husbandry and preventive medicine procedures see MAFF 1995.

**Castration**
The most usual time for castration, unless a rubber ring is used (see above) is when the calves are about 4-7 weeks of age. An anaesthetic is not compulsory at this age and it is possible that the operator who performs castration rapidly without anaesthetic causes less distress *at the time* than one who restrains the calves, injects local anaesthetic, waits for it to take effect and then restrains them again for surgery. The two methods used at this time are surgical removal or bloodless castration, ie crushing the spermatic cord, blood and nerve supply to the testes. A third method, used by a few, is to apply a rubber ring to castrate male calves at about 1 day of age. All forms of castration have been shown to be painful for several hours and the animals may continue to show some local pain for 2-3 weeks. It is now believed that pain is as severe in the first day of life as at one month of age, which runs contra to a conventional stockman's belief that it is kinder to castrate as soon as possible. Moreover it appears that surgical removal of the testes (traditionally favoured by many veterinary surgeons) is more painful than crushing the spermatic cord with a Burdizzo. The 'least worst' method appears to be crushing with a Burdizzo followed by the application of a rubber ring - an anaesthetic must be employed if this is done to animals over 1 week of age.

Everybody who rears male calves for beef should ask himself the question, "Is castration necessary?" If the cattle can reach slaughter weight in 18 months or less, the answer is almost certainly no. Entire male cattle grow lean tissue faster than steers or heifers, and the production advantages of bulls have increased in importance since the ban on the use of anabolic steroids as growth promoters for beef cattle.

**Dehorning**
The most common method is to remove the horn bud under local anaesthetic using a hot iron when the calf is at least 3 weeks old and the horn tip can be clearly felt (Figure 3.13). An alternative approach is to destroy the horn tip with hot air. Both techniques

**Figure 3.13     Restraint position for administration of anaesthetic and disbudding of calves.**

should only be attempted after personal instruction from a competent operator.

Dehorning and castration are commonly done at the same time. This is acceptable practice but it is not a good idea to combine these two stresses with that of weaning because, all together, they may predispose to stress-related diseases such as pneumonia.

## EMERGENCY SLAUGHTER

The most humane procedure for slaughtering casualty cattle is for either a veterinary surgeon or a licensed slaughterman to kill the animal on the spot using an approved procedure such as shooting it with a captive bolt and then cutting its throat. If, however, the animal can be transported to an abattoir, it is more likely to be killed and butchered in such a way as to render the meat fit for human consumption. The farmer is forced, therefore, to test the strength of his compassion against a cash loss that may amount to hundreds of pounds. The welfare of the animal is now largely covered by *The Welfare of Animals (Transport) Order 1997*. The general rule is that unfit animals may only be transported if the intended

journey is not likely to cause them unnecessary additional suffering.

Transport of weak, sick or injured cattle to a slaughterhouse will clearly cause unnecessary suffering if the animal is in severe pain which is exacerbated by movement; a broken leg is an obvious example. A calf that is too weak to move could reasonably be transported to a slaughterhouse if it is carefully carried on to (and off) the lorry. The larger animal, such as the downer cow that collapses and develops paralysis post-calving, presents a more difficult problem. She may be reasonably comfortable where she lies and equally so if on deep bedding in a lorry. Any procedure that involved dragging the live recumbent cow herself over the ground to enter and exit the lorry would constitute unnecessary suffering and the owner should arrange for the animal to be killed on the spot.

The whole issue of the moving of casualty animals is complex. For practical advice see *Guidance on the Transport of Casualty Farm Animals* (MAFF 1993).

## REFERENCES AND FURTHER READING

**Alderman G** 1993 *Energy and Protein Requirements of Ruminants*. CAB International: Wallingford, UK

**Allen D** 1990 *Planned Beef Production and Marketing*. Blackwell Scientific Publications: Oxford, UK

**Allen D and Kilkenny B** 1984 *Planned Beef Production, 2nd edition*. Granada Technical Books: London, UK

**Blowey R W** 1985 *A Veterinary Book for Dairy Farmers*. Farming Press: Ipswich, UK

**Farm Animal Welfare Council** 1986 *Report on the Welfare of Livestock at Markets*. RB 265. HMSO: London, UK

**Grandin T** (ed) 1993 *Livestock Handling and Transport*. CAB International: Wallingford, UK

**McDonald P, Edwards R A, Greenhalgh J F D and Morgan C A** 1995 *Animal Nutrition, 5th edition*. Longman: Harlow, UK

**Meat and Livestock Commission** 1994 *Beef Yearbook*. MLC: Milton Keynes, UK

**Metz J H M and Groenestein C M** 1991 *New Trends in Veal Calf Production*. EAAP Publication no.52, 1991. Pudic: Wageningen, Germany

**Ministry of Agriculture Fisheries and Food** 1983 *Code of Recommendations for the Welfare of Livestock: Cattle* (reprinted 1991). L 701. HMSO: London, UK

**Ministry of Agriculture Fisheries and Food** 1984a *Energy Allowances and Feeding Systems for Ruminants*, RB 433. HMSO: London, UK

**Ministry of Agriculture Fisheries and Food** 1984b *Cattle Handling*, B 2495. HMSO: London, UK

**Ministry of Agriculture Fisheries and Food** 1986 *Annual Reviews of Agriculture*. HMSO: London, UK

**Ministry of Agriculture Fisheries and Food** 1993 *Guidance on the Transport of Casualty Farm Animals (reprinted with updating 1998)*. PB 1381. MAFF Publications: London, UK

**Ministry of Agriculture Fisheries and Food** 1995 *Agriculture in the United Kingdom 1994*. HMSO: London, UK

**Ministry of Agriculture Fisheries and Food** 1995 *Summary of the Law Relating to Farm Animal Welfare (reprinted 1998)* PB 2531. MAFF Publications: London, UK

**Mitchell C D** 1978 *Calf Housing Handbook*. Scottish Farm Buildings Investigation Unit: Aberdeen, UK

**More O'Ferrall G J** (ed) 1982 *Beef Production from Different Dairy Breeds and Dairy Crosses*. Martinus Nijhoff: The Hague, The Netherlands

**Swan H and Broster W H** 1981 (eds) *Principles of Cattle Production*. Butterworths: London, UK

**Phillips C J C** (ed) 1989 *New Techniques in Cattle Production*. Butterworths: London, UK

**Wathes C M and Charles D R** (eds) 1994 *Livestock Housing*. CAB International: Wallingford, UK

**Webster A J F** 1985 *Calf Husbandry, Health and Welfare, 2nd edition*. Collins: London, UK

**Webster A J F** 1993 *Understanding the Dairy Cow, 2nd edition*. Blackwell Scientific: London, UK

**Webster A J F** 1995 *Animal Welfare: A Cool Eye Towards Eden*. Blackwell Scientific Publications: Oxford, UK

**Webster A J F, Saville C and Welchman D de B** 1986 *Alternative Husbandry Systems for Veal Calves*. University of Bristol Press: Bristol, UK

# 4  Sheep

## H Ll Williams

## THE UK INDUSTRY

### Development of the industry

During the 1980s sheep numbers in the UK increased from 15 million ewes to 20 million ewes (FAWC 1994). This is attributed to subsidy support, a decline in New Zealand imports and a growing export market to Europe. Due to the introduction of a sheep quota scheme in the early 1990s further expansion was curtailed.

In 1994 there were 88,000 flocks of sheep in the UK (MAFF 1995) with the average size of these flocks being 221 ewes; 28.5 per cent of breeding ewes were in flocks of over 500. It is forecast that by 2000 the number of sheep holdings will have decreased to 79,000 and the average flock size will be 526 ewes (MLC 1998).

An export trade in live lambs developed during the 1980's, mainly because of the growing demand in France for fresh home-killed meat. In 1993 the UK exported 1.4 million lambs; by 1996 this had reduced for a variety of reasons to about 0.65m and has continued to decline.

### Structure of the industry

The UK sheep industry is very diverse, largely because farmers undertake production in very contrasting environments. These range from the lowland cultivated pastoral and arable areas to the more severe upland and hill areas dominated by uncultivated rough grazings. In areas of rough grazing the combination of severe climate, difficult topography, low soil fertility and inaccessibility, allows little choice of animal enterprises. Such grazing represents one-third of the total agricultural land in the UK. In many areas only sheep production is appropriate. In lowland areas where there is more competition between species and between animals and crops, sheep may play only a minor role.

The industry is geared to supplying meat, mainly from finished lamb. Lamb meat production has been increasing because of the rise in the breeding ewe population since the early 1980s. Consequently, the level of self sufficiency is now approximately 116 per cent. Wool contributes less than five per cent of gross income in all systems of sheep production.

One unique feature of the sheep industry is the well-established integration between regions, largely according to altitude (eg hill and lowland). It uses land resources efficiently and exploits the wide range of genetic potential that exists within the 70 or so breeds in the UK. It is possible to allocate all breeds and resulting crossbreds to a certain category of flock depending on function (see Table 4.1). The National Sheep Association (1994) published a description of the main breeds and named crossbreds. Approximately 50 per cent of the then national flock of 20 million breeding ewes was in upland and hill flocks; these act as a reservoir of breeding stock. Each late summer upland and hill farms sell many of their sheep, at regional sales, to more favourable lowland farms. The type of animal they sell are: pure-bred ewes that have produced three crops of lambs on the hills, young cross-bred ewes (usually two-tooth), ewe lambs and store lambs (mostly wether lambs). Lowland farmers can buy such breeding ewes at an optimal stage of production and choose genotypes most suitable for local and national market requirements. Crossbred ewes from the uplands account for over 80 per cent of all lowland ewes. Crossbreeding provides scope for combining desirable characteristics of body size, reproductive potential and milkiness. It also allows rapid response to changes in market demand and systems of management.

**Table 4.1** **Classification of commercial sheep flocks in the UK.**

| Category | Main characteristics and selection objectives | Breeds[a] and their mature liveweight (kg)[b] | Main function |
|---|---|---|---|
| 1. Hill flocks | Hardiness; ability to rear 1 lamb; satisfactory quality and weight of fleece | Blackface (70) Welsh Mountain (45) Cheviot (64) Swaledale (64) | To make use of land, usually above 300m, that would otherwise be unproductive. |
| 2. Flocks producing sires of cross-bred ewes | Prolificacy; milking capacity; growth rate | Border Leicester (100) Blue-faced Leicester (96) Teeswater (95) Wensleydale (90) | To sire crossbred breeding ewes for lowland flocks as a result of crossing with draft ewes from hill flocks. |
| 3. Flocks producing fat lamb sires | Growth rate; carcase quality | Oxford Down (110) Suffolk (90) Texel (87) Dorset Down (77) Southdown (60) | To sire fat lambs, mainly in crossbred lowland flocks. Sometimes referred to as terminal sires. |
| 4. Self-contained flocks (upland/lowland) | Prolificacy; milking capacity; growth rate; carcase quality; weight and quality of wool | Clun Forest (73) Romney Marsh (75) Devon Longwool (95) Dorset Horn (82) | To produce their own replacements, surplus breeding stock and fat lambs. The Dorset Horn is used for out-of-season fat lamb production. |
| 5. Crossbred lowland flocks | Prolificacy; milking capacity | Mule (73)[c] Greyface (70)[c] Welsh half-bred (58)[c] Scottish half-bred (77)[c] | To produce lambs for slaughter. These are usually sired by terminal sires. |

[a] The most numerous breeds in each category
[b] Average of male and female sheep
[c] Average weight of female sheep

## SELECTION AND IMPROVEMENT

In the past, the sheep industry used visual appraisal and breed characteristics, such as colour, size, fleece and presence or absence of horns, to help it select potential breeding stock. Although producers maintained the purity of breeds, they made little or no progress in improving genetic potential, particularly in respect of criteria with direct economic value. The industry has recently established a sound foundation to selection by: better definition of its objectives, identifying more accurately superior animals and introducing schemes that aim to use superior genotypes advantageously. The Meat and Livestock Commission's (MLC) Sheep Yearbook for 1998 provides comprehensive descriptions of recent developments in this field. The main approaches to breed improvement include within-flock, across-flock and group breeding schemes. In all these schemes, the foundation to progress has been to adopt meaningful recording of measurable criteria. These criteria include: lamb weights, litter size and ewe bodyweight, and scanning of live animals for muscle and fat depth. Computer processing of field records involving new techniques of data analysis contributes greatly to the speed of progress.

Within-flock schemes rely entirely on data from official records, such as those from the MLC's SIGNET Flock Plan Service. Flock owners select the number of traits that are important in the particular situation and according to their objectives. Progress is usually relatively slow, largely due to the narrow variation within flocks of average size.

The across-flock scheme, involving the use of a team of reference sires using artificial insemination (AI) within many flocks, compares the performance data of their offspring within each flock. It allows flock owners to genetically compare sires on contrasting flocks. Breeders who supply sires for use on crossbred lowland flocks, where there is a need to produce well-muscled lambs without excessive fat, are the main users of such programmes.

The group breeding schemes require participants to pool superior animals to form a jointly owned nucleus flock with the main aim of providing replacement rams for the members.

## NATURAL HISTORY AND BEHAVIOUR

Sheep were domesticated about 10 000 years ago, probably in South-West Asia. They were one of the earliest species of domestic animal, largely because of their behaviour and size. The fact that sheep are ruminants, and therefore did not directly compete with man for food, also contributed to their adoption. It was relatively easy for migratory early man to move flocks with him, and sheep provided milk, fibre, skin and meat. Sheep are still an agriculturally important species, particularly in adverse environments, and inhabit a wide range of latitudes and altitudes. They are less well adapted to moist tropical conditions than to other climatic areas.

### Social organization

There is more opportunity for freely ranging sheep to display innate patterns of social behaviour than those in restrictive intensive systems. The basic behavioural characteristics of sheep include a strong flight reaction, marked flocking/follower behaviour, and the importance of vision in social organization (Kilgour & Dalton 1984; Lynch *et al* 1992).

On unfenced hills or pastures, sheep will graze only within their home range, and subgroups may occupy sub-ranges. Little is really understood about the social interactions that govern flock behaviour. Apart from reproductive behaviour, research has paid little attention to the communication between members of a flock. Leadership is probably not as precisely defined as in other species and is subject to frequent change. Sub-groups of less than four sheep tend not to behave as a group and are difficult to manage.

Visual contact is important to sheep. A sheep that cannot clearly see its neighbour will raise its head frequently to try and establish contact; in isolation, sheep will bleat loudly, frequently showing considerable distress. They alert other members of the flock by adopting a very upright posture, holding their head high and walking with tense steps and also sometimes by snorting. These steps are often audible to other members of the flock. During a disturbance a grazing flock will usually regroup on higher ground, usually not very far away from the cause. Where a predator or dog is pursuing, flight continues, often uncontrollably, and cases of ewes hurtling into ditches and streams do occur. Deaths and serious injuries resulting from attacks by stray dogs from urban areas is a growing problem.

### Resting and feeding

Sheep alternate between periods of grazing and rest. They graze in bouts of 20 to 90 minutes followed by 45 to 90 minutes when they lie down and ruminate or rest (Lynch *et al* 1992). Total daily grazing time depends on the availability of food, but is usually around eight hours and rarely more than ten. During rest, short periods of sleep occur, amounting to 3½-4 hours/day. Sheep spend most of the rest phase ruminating. Sheep have fixed eating patterns and, in some situations, are reluctant to accept new food if their diet changes abruptly, such as when flockmasters give hill sheep supplements during very bad weather. It is therefore important to train all sheep, during their first autumn, to accept cubed concentrate diets.

Trough-fed ewes exhibit dominance, so distributing troughs over a wide area helps to reduce this. Lowland sheep quickly become conditioned to the signals of feeding time, such as the sound or sight of a bucket, or the appearance of the shepherd. Stockpersons can use this to their advantage to move ewes between locations or into pens, or even into vehicles.

### Aggression

Ewes rarely show active aggression towards humans

and will accept interference around lambing time. On the other hand they can be very aggressive towards dogs at this time; flockmasters sometimes use the initial phase of this response to heighten the ewe's maternal instincts during fostering. Housed ewes, especially of horned breeds, tend to fight more than when at pasture. Stockpersons should watch rams carefully during the breeding season. Some rams are aggressive all the year round and should be culled.

## Grooming

Sheep rarely groom one another. While in full fleece, they find it difficult to scratch and usually seek a post or rail. In the absence of a suitable object to rub against, sheep will roll on to their backs. When in full fleece or heavily pregnant they may fail to get up and, if the stockperson does not notice them, they can die.

## Reproduction

When there is no ram present, ewes do not display the usual signs of oestrus. It is therefore imperative to use a male to monitor the incidence of oestrus. During oestrus, ewes stand quite firmly and may look backwards and display tail fanning. Ewes usually urinate when rejecting a ram. Ewes in oestrus may indulge in ram-seeking, but the male does most of the searching; sight and smell are the most important sensory cues. The initial approach may involve sniffing the perineal area, with or without lip curling and baring of teeth (flehmen response). Depending on the degree of stimulation, the ram may also simultaneously give a low-pitched gurgling sound, and nudge and paw with the front foot. Alternatively, the ram may extend his neck and rotate his head slightly in an exaggerated forward movement alongside the ewe. Some rams may show this behaviour at some distance. The ram may mount several times before thrusting, which usually signals ejaculation. After ejaculation, he dismounts and stands with his head lowered alongside the ewe before continuing to search.

The most vital aspect of behaviour to understand is that which occurs around lambing time, as it is vital to efficient stockmanship. Behavioural signs are crucial indicators of the imminence of lambing, the strength of the family bond, the well-being of the lamb, and the adequacy of the milk supply. Good stockmanship is greatly dependent on recognizing and acting on the behavioural signs.

In the field, ewes will isolate themselves from the rest of the flock to establish a birth site. They will stay around their site for a considerable period (up to six hours) and it is important not to disturb them if possible. The birth site becomes the focal point of the ewe. It is usually clearly visible because of her activity and she will recognize its smell. A period of conspicuous restlessness precedes the onset of full labour. When the ewe enters full labour, contractions become obvious and she will raise her head and point her nose to the sky. This is a signal that is identifiable from some distance. Labour usually lasts for less than an hour. When the ewe is disturbed during labour, contractions will normally resume after a short interval.

Irrespective of the system of management, the ewe must have every opportunity to lick her lamb(s) and to develop the social bond between herself and her offspring. Both sound and smell play an important part in establishing this bond.

Lambs attempt to rise within 20 minutes of being born and are attracted by the movement of their dam. This innate behaviour means a newborn lamb will accept any moving object and thus it is therefore relatively easy to imprint a substitute for its mother; this characteristic is essential in fostering.

After a period of exploration along the underline of the ewe, the lamb locates a teat and starts sucking within the first hour. During sucking the lamb usually wags its tail vigorous and occasional butts the udder. The ewe may help teat-seeking by manoeuvring herself into an advantageous position for the lamb. Also, as the lamb approaches the teat, she licks its hindquarters, particularly the perineal area, quite vigorously. The ewe nurses her lamb(s) 3-4 times an hour to start with and gradually less frequently as lactation advances.

Abnormal behaviour following lambing includes desertion, delayed grooming, mismothering and partial rejection of lambs. These behaviours arise from one or more of the following: the inexperience of lambing for the first time, fear of humans, fatigue after protracted lambing, overcrowding during housing, and a long interval between the birth of litter mates. Good husbandry can eliminate or reduce these effects.

A comprehensive account of aspects of sheep behaviour is to be found in Lynch *et al* 1992.

## REPRODUCTIVE BIOLOGY

The reproductive capacity of sheep may be described in terms of three main components: seasonality, fecundity (incorporating age at puberty, litter size, lambing interval and duration of breeding life), and fertility (the capacity to produce viable offspring). Basic data is presented in Table 4.2.

**Table 4.2**      **Reproductive data; adult ewes.**

|  | Data | Comments |
|---|---|---|
| **Seasonality** | | |
| Polyoestrous | 7–12 oestrous periods/season | Variation between and within breeds. Variation according to latitude. |
| Onset (mean dates) | Dorset Down (14 Aug)<br>Clun Forest (4 Sept)<br>Halfbreds (20 Sept)<br>Blackface (9 Oct) | Significant between and within breed variation, high year-to-year repeatability. Ewes responsive to introduction of ram near to onset of season. Dorset Horns & Poll Dorsets are aseasonal. |
| Oestrous cycle:<br>oestrus<br>ovulation | <br>24–36h<br>Towards end of oestrus | <br>Only displayed in the presence of male.<br>First ovulation of season 'silent' - not accompanied by oestrus. |
| Inter-oestrus interval | 16–17 days | |
| **Fecundity** | | |
| Live lambs born/ewe lambed | Welsh Mountain – 1.23<br>Blackface – 1.40<br>Clun 1.52<br>Mule – 1.78<br>Cambridge crosses – 2.40<br>Cambridge – 3.00 | Markedly affected by body condition and season. |
| Lambing interval | Annual<br>Dorset Horn and Poll Dorset have the capacity to produce 3 lambings in 2 years | Restricted by seasonality.<br>Aseasonal. With good management can achieve lambing interval of 240 days. Rarely practised. |
| Breeding life | Up to 15 years | Vast majority culled by 6 years of age. |
| **Fertility** | | |
| Number of ewes per ram:<br>hill<br>lowland<br>synchronised flocks | <br>40–50<br>50–60<br>10–15 | Vary according to size of flock, paddock size and terrain. |
| Percentage of ewes lambing | 92–95 | Lower in upland flocks. As low as 75 per cent in harsh hill flocks. |
| **Pregnancy** | | |
| Duration | 142–150 days | Variation due to breed and litter size. |

## Seasonality

All British breeds of sheep, except the Dorset Horn and Poll Dorset, have a clearly defined breeding season. It commences in the late summer/early autumn and continues until mid to late winter.

The main factor governing seasonality of breeding is the changing ratio of light to darkness. The pineal gland controls this photoperiodic response by producing a hormone, melatonin, only during the hours of darkness. Changing duration and/or timing of the melatonin secretion during a 24-hour period acts as a biochemical signal to the higher centres of the brain. They in turn control reproductive activity. The ability of sheep to respond and entrain to light rhythms was an important factor when establishing flocks in higher latitudes where the food supply is seasonal.

Rams of all breeds are less seasonal than ewes in their sexual behaviour. Breeds differ in mating activity during the summer, and seasonal fluctuations occur in semen quantity and quality. In practice, stockkeepers may counteract this by: reducing the number of ewes they allocate to each ram, closer supervision of matings and confining the sheep in a small field or paddock.

It is a common practice to use or sell ram lambs that have been born early, and thus have had every opportunity of growing well, as flock sires in their first autumn.

## Puberty

Data in Table 4.3 provides an indication of the reproductive performance of ewe lambs in different environments and at differing stages of development. The recent increase in the practice of breeding from ewe lambs is probably the result of economic pressures and an improvement in husbandry standards. However, this is only possible if they are sufficiently mature (ie 60–65 per cent of their adult bodyweight).

Ram lambs attain puberty at an earlier stage of their growth than ewe lambs, usually 40–50 per cent of their potential adult weight; this compares with 50–60 per cent for ewe lambs. The ram lamb should be capable of serving by 6½–7 months, but with fewer ewes than an adult animal.

## Litter size

The number of ova shed, fertilized, and developing to viable lambs at birth affects litter size. There is considerable genetic variation in ovulation rate, so fertility is, to some extent, a matter of breed selection. Nutritional and management factors also influence fertility. In addition it is necessary to consider age and stage of the breeding season in the assessment of performance. These factors seem to exert their influence up to the early stages of pregnancy. Subsequent reduction in viable foetuses usually occur with gross feeding mismanagement and from disease. The identification of a gene with a very large effect on ovulation rate in the British

**Table 4.3**      **Reproductive data: ewe lambs.**

|  | Data | Comments |
|---|---|---|
| Age at puberty | 7–9 months | Markedly photoperiodic. Must have reached 50 per cent of adult weight. |
| No. of ewe lambs/ram | 20–30 | Use mature rams and small fields. |
| No. of ewes born/100 to tup: |  |  |
|   lowland flocks | 56 | Markedly affected by management, |
|   upland flocks | 69 | particularly nutrition and stage of |
|   >60% of mature weight | 88 | growth and development at tupping. |
|   <60% of mature weight | 57 |  |
| Litter size/ewe/lamb lambed | 1.0–1.25 |  |
| Ewe lambs lambing/100 to tup | 49–67 |  |

'Cambridge' breed, has markedly improved the possibility of increasing performance through breeding. Prolific breeds require a high standard of care and nutrition throughout the production cycle, and particularly around lambing, when housing and intensive care is vital.

### Control of reproduction

There is considerable scope for adopting methods such as hormone (ie progestagen and pregnant mare's serum gonadotrophin) and/or light or melatonin treatments to control reproduction. Reasons for doing so may include the exploitation of superior genotypes, the induction of oestrus for out-of-season breeding, synchronization of activity to eliminate unacceptable variation within the flock, augmentation of fecundity in breeds/types with a low ovulation rate, or facilitating the use of AI (eg to control disease).

Further information on all aspects of sheep reproduction may be found in Williams (1995).

## MANAGEMENT

### Choice of management system

Geographical location, type of soil, type of crops and their seasonal availability, labour supply and marketing objectives largely dictate the most appropriate system of management. Generally, sheep units in the west and in the uplands and hills are almost entirely grass-based. In the east, where rainfall is lower, arable farming prevails, and crops other than grass are a feature of the feeding system at particular times of the year. Even in arable areas, leys (temporary grassland) maintain the flock for over half the year.

The timing of lambing will have a major influence on the seasonal pattern of management. This is because the sequence of actions and assessments that the flockmaster has to undertake during the sheep year revolves around the following key events: mating, lambing, shearing and weaning. Spring and autumn lambing are the main types of management systems. The example in Table 4.4 shows the timing of the main events that occur at contrasting times of the year, particularly in relation to the grazing season.

### Management of rams

Most flock owners acknowledge the benefits of a comprehensive clinical examination of rams well before the tupping period. It is vital to detect any physical abnormality or disease that affects the ram's ability to search for ewes on heat, to serve, and to yield good quality semen. Failure to do so may prevent flocks attaining high levels of fecundity and fertility, particularly if using only one or two rams, which is quite often the case. Ram lambs need to achieve 65–70 per cent of their mature weight by the date of the ram sale or breeding time (see Table 4.4).

**Table 4.4**   **The main events that occur at different times of the year.**

| Spring lambing flock | Key events | Autumn lambing flock |
|---|---|---|
| Mid October | Mating | Mid-July |
| Mid March | Lambing | Early December |
| Winter or early summer | Shearing | Early summer |
| Early August | Weaning | Late January |

Flockmasters normally maintain mature rams on grass, often in paddocks close to the homestead, and feed hay during the worst of the winter months. It is important to give rams access to reasonable grazing, and to routinely drench and vaccinate them along with the ewe flock. This will ensure that they are in good health and condition during the approach to the breeding season. Stockpersons should segregate those in poor condition and give them a higher plane of nutrition. By the start of tupping all rams should ideally have a body condition score of 3½ (see body condition scoring page 108).

All rams need some concentrates, partly to ensure an increased supply of energy, but also to provide minerals, vitamins and good quality protein. The level of feeding, however, will depend to a certain extent on condition. Rams with a score of 2½–3 should start at about 200g/day, gradually increasing to 500–700g/day. Training rams to respond to the sound of the feed bucket, and to accept bucket feeding, can be a great help during the six weeks or so of tupping. If the stockperson trains the rams in such a way it is possible to feed them individually. They are also easier to catch for such things as harness adjustment, change of crayon (see p 91) and general examination for soundness. Rams often lose weight during the tupping period, largely because they spend less time grazing.

It is important to assess all rams, irrespective of age, on breed type and soundness of limbs, particularly the conformation of the legs, pasterns and feet. The claws should be free of distortion, healthy and not have interdigital growths. Flockmasters must investigate promptly the slightest hint of lameness or any other disease condition any time before or during tupping. Anything that elevates body temperature can be disastrous for semen quality.

During examination of the mouth, the handler can establish the age of young rams and check the alignment of the incisors. They should also examine the eyes for signs of infection or scars from previous infections. There may be evidence of correction of inturned eyelids (entropion) and flockmasters should not buy, or retain, such animals in the flock.

Stockkeepers should inspect the body for vaccination abscesses, lesions and scars. They should give particular attention to the sternum and the skin leading up to the shoulder; this is because it is the part of the body that carries the keeling harness and is prone to superficial injury. This is especially a problem if the harness does not fit properly or the straps have become hard.

The stockperson also needs to examine the reproductive tract carefully, giving particular attention to the penis, appendage, prepuce, scrotum and testes. In the case of ram lambs, it is important to establish whether anatomical development is complete.

The physical examination of rams does not reveal their libido or fertility. Consequently, it is necessary to watch rams carefully to establish whether they go round teasing the ewes and serving those that appear on heat. Flockmasters cannot assess fertility until they know the proportion of ewes returning to oestrus. A return rate of over 15 per cent to first service and over 5 per cent to both first and second services needs investigating. Circulating rams between groups at least weekly mitigates the effects of an infertile ram.

Under certain circumstances there may be sound reasons for assessing the quality of the ram's semen. Under farm conditions there is no alternative to the use of electroejaculation. However, this technique is stressful and only a veterinary surgeon can legally perform it (MAFF 1995).

## MANAGEMENT OF BREEDING FEMALES
### Preparatory phase

Lowland flockmasters aim for a high lambing percentage (target of 170–190 per cent) and a compact lambing period. Correct feeding and management during the pre-service preparatory phase is the initial very important step to achieving this. Flockmasters will need to assess older breeding stock for soundness, and select or purchase replacement animals.

Ewes that are improving in body condition usually have a higher ovulation rate, and consequently produce more lambs. It has therefore been traditional to flush them by providing a generous level of feeding a few weeks before service. Ideally, however, the preparatory phase should begin early in the season, at weaning.

Mature ewes should be in good body condition (score 2½–3) a month from service. Those in poor condition need segregating and giving generous grazing and/or supplementary quality food. An improvement in the score by a half during the flushing period would require approximately 2kg liveweight gain per week in Scottish half-bred ewes. Target liveweight, expressed as the percentage of mature bodyweight (see Table 4.1) for the main age groups of lowland breeding ewes are given in Table 4.5.

Ewe lambs and two-tooth ewes require preferential treatment so that they continue to grow and develop satisfactorily. Ewe lambs should remain apart from the main flock. They need to achieve 60 per cent of their mature weight before they are ready for breeding in their first autumn. The flockmaster must therefore monitor their progress during the post-weaning phase and make forward projections of their probable mating weight. Two-tooth maiden ewes should easily achieve 80 per cent of their mature weight. If flockmasters are buying in such ewes they should do so early so that they have time to check and, if necessary, improve the animals' condition.

The preparatory phase spans different periods of the late summer/autumn depending on the geographical location and the breed or type of ewe. In very early lambing flocks it would occur in July/August, for traditional grassland flocks in September/October, and in upland and hill areas in October/November. Normally, the preparatory phase coincides with the very end of the grazing season, when either the results of good grassland management and/or the feeding of arable by-products and forage crops should fulfil the nutrient requirements of all age groups. It is unusual for shepherds to resort to feeding concentrates at this stage. In cases of a crop failure or prolonged drought, however, ewes will need 0.5kg/day for four weeks before service.

During the preparatory phase, all ewes should be dagged so that soiled wool does not impair oestrous detection and service. It is common to use vasectomized rams to initiate oestrous activity throughout the flock. Flockmasters normally introduce them four weeks before the start of tupping. This ensures that the entire ram, when put in with the ewes, will successfully mate most of them during their second or later oestrus when ovulation rate is high.

## Tupping

To check the progress of tupping during the first 17 days after the introduction of the ram and the incidence of returns to service, it is useful to fit rams with a keeling harness. Alternatively it is possible to smear their sternum with raddle. The coloured mark on the ewes' rumps will show which have been mounted. Stockpersons should fit harnesses a few days before the start of tupping. It is important to carefully adjust the harness when fixing the coloured crayon on the day the rams join the flock. More frequent change of colour (eg weekly) is useful for grouping ewes in late pregnancy according to lambing date.

Ewe lambs should be tupped after the main flock, using adult rams. Young ram lambs should be used on adult ewes and in a separate paddock ie no competition from maturer males. Avoiding the use of very large fields and keeping to the approved ratio of ewes to rams (see Table 4.2) helps to improve the likelihood that all groups will mate successfully. Approximately four weeks after removing the rams, it is useful to introduce a raddled vasectomized ram to identify non-pregnant cyclic ewes. The flockmaster can then segregate them and decide their future.

## Pregnancy

Implantation of the embryos occurs during the third to fourth weeks of pregnancy. This is the period when most embryonic mortality occurs. It is important to maintain a constant level of feeding, and avoid over-dependence on crops known to lower fertility, such as kale and rape.

Feeding management of different age groups changes during the second and third months. The feeding of ewe lambs and two-tooth ewes continues more generously than for mature females; during this period they should show an increase of 5 per cent in bodyweight. In practice, feeding the two younger age groups hay compensates for the fall in pasture quality. Mature ewes, on the other hand, can tolerate more severe treatment and, if in good condition at mating, can clear up pastures (finish off grass that remains after the end of the growing season). A loss of 5 per cent in bodyweight at this stage is not detrimental to foetal growth and development. Over fat ewes may benefit from this mild degree of under

**Table 4.5**    **Target liveweights of ewes at various ages and reproductive states, expressed as percentage of mature weight at conception.**

| | State of pregnancy | | | Lambing | | |
|---|---|---|---|---|---|---|
| | Conception | Day 90 | Day 120 | Pre- | Post- | Lactation (8th week) |
| Ewe lamb (bearing single) | 60 | 65 | 70 | 75 | 64 | 62 |
| Two-tooth ewe (bearing twins) | 80 | 85 | 90 | 97 | 80 | 77 |
| Mature ewe (bearing twins) | 100 | 96 | 100 | 112 | 90 | 83 |

Calculated from data given in MLC (1988)

nourishment, because it reduces susceptibility to metabolic disorders, in particular pregnancy toxaemia.

From the beginning of the fourth month, there is an increase in the rate of growth and development of the foetus and, consequently, in the nutrient requirements of the ewe. This phase coincides with winter and a severe reduction in grass. To safeguard the general health and welfare of ewes remaining at pasture, a regular supply of good quality basic feed should be available. This should comprise of hay and silage, or roots if obtainable.

The main problem during the last phase of pregnancy is to cater for the ewe's increasing demands that occurs at a time when her appetite is falling. This drop is due partly to the physical effects of the foetal burden and partly to physiological changes. In practice, it is possible to overcome the problem by feeding a basal diet of good quality roughage and supplementing it with concentrates (10–12MJ/kg). If the condition of mature ewes is satisfactory, further depletion of body reserves, resulting in a loss of five per cent basal weight, is acceptable. However, such weight loss needs careful control. The liveweight of twin-bearing ewes should increase by approximately 16 per cent during the last two months. Failure to achieve this increase can result in lower birth weights, less vigorous lambs, a delay in the onset of lactation, and poor development of mothering instincts. Table 4.6 presents a widely practised feeding plan for this stage of pregnancy.

Flockmasters should introduce new regimes, such as housing, before the last six weeks of pregnancy, and pen ewes according to their colour marks (which indicates those due to lamb together). They may separate outwintered ewes into lambing groups. It is crucial to accustomise ewes to alternate foodstuffs before adverse changes in the weather (such as snowfall) affects food supplies. It is also important to watch behaviour during feeding and identify any shy feeders. Outwintered ewes should be range-grazed near to the proposed lambing site. Lambing pens and aids required at lambing, and other equipment should be ready for use. This includes equipment such as that required for identification, docking and castration, and for fostering and artificial rearing.

**Table 4.6**    **Feeding regime (kg/day) for ewes in late pregnancy.**

| | Ewe weight and number of foetuses | | | | | |
| | 50kg | | | 70kg | | |
| | **Single** | **Twin** | **Triplets** | **Single** | **Twin** | **Triplets** |
|---|---|---|---|---|---|---|
| Six weeks before lambing | | | | | | |
| Hay[a] | 0.83 | 0.83 | 0.83 | 1.00 | 1.00 | 1.00 |
| or silage | 2.60 | 2.60 | 2.60 | 3.50 | 3.50 | 3.50 |
| plus concentrates | 0.18 | 0.30 | 0.34 | 0.24 | 0.37 | 0.44 |
| Four weeks before lambing | | | | | | |
| Hay[a] | 0.83 | 0.83 | 0.83 | 1.00 | 1.00 | 1.00 |
| or silage | 2.60 | 2.60 | 2.60 | 3.50 | 3.50 | 3.50 |
| plus concentrates | 0.28 | 0.45 | 0.51 | 0.36 | 0.56 | 0.66 |
| Two weeks before lambing | | | | | | |
| Hay[a] | 0.83 | 0.83 | 0.83 | 1.00 | 1.00 | 1.00 |
| or silage | 2.60 | 2.60 | 2.60 | 3.50 | 3.50 | 3.50 |
| plus concentrates | 0.37 | 0.59 | 0.68 | 0.48 | 0.75 | 0.86 |

[a] Assuming ME concentration in the dry matter of 10.0MJ for hay and 11.0 for silage and dry matter content for silage of 25 per cent. Roots could be used to replace 75 and 50 per cent of the concentrates on a dry matter basis at 6 and 2 weeks before lambing, respectively.    *Source: MLC (1988)*

## Lambing

Flockmasters should inspect flocks frequently, giving particular attention during the approach to lambing. Features to look out for include a long interval (more than 30 minutes) since first observing the ewe's nose pointing to the sky, and the appearance of the water bag. It is important always to maintain a high standard of hygiene and to correct abnormal presentations with great care. Appropriate medication is essential after all assisted lambings (see Eales and Small 1995).

Behaviour of all lambed ewes, and particularly those lambing for the first time, should be kept under close observation. There is no doubt that an important contributory cause of lamb mortality is abnormal maternal behaviour, which may include desertion, delayed grooming or rejection of one or more offspring. The consequence for the newborn lambs can be debilitating or even disastrous. Timid ewes increase the teat-seeking time and thus the chances of the lamb consuming pathogens lurking on the belly wool and other parts of the underline. Thorough grooming of the lamb, with a short interval to suckling, are important traits in the ewe.

## Lactation

It is important to check the udder and teats soon after birth and to dispose of the cleansings. After recording information regarding the ewe and her lambs, the stockperson should return them to a pen containing newly lambed ewes. If appropriate, they may move them to a paddock that has shelters in strategic positions (see p 98-99). A high level of surveillance must be maintained as early detection of ill-health or abnormal behaviour is a vital part of shepherding during the month following lambing.

Where late winter/early spring lambing occurs indoors, ewes and their lambs should be turned out of the sheep house within a week into nearby sheltered fields where surveillance can continue. Delaying this move can lead to cross-suckling and mismothering. Unshorn ewes may need dagging to reduce soiling and the risk of fly strike.

During early lactation, it is essential to encourage a rapid increase in appetite and to reach peak lactation quickly. This is easier to achieve in ewes that have been properly managed during pregnancy. Lactating ewes require a diet similar to that for late pregnancy, namely, concentrates supplementing a low daily allocation of good quality hay. The ewe needs to avoid undue loss of weight during the first 2-3 weeks. It is therefore important not to feed diets with low digestibility as they depress appetite. However, if appetite peaks early in lactation, a loss of approximately 5 per cent is acceptable. To meet the requirements of high yielding (2.5 litres) ewes suckling twins, they will need feeding 1–1.25kg of concentrates (12MJ/kg) containing good quality protein. Ewes suckling twins produce 40 per cent more milk than those with singles, and they also tend to reach peak lactation earlier. It is necessary to segregate such ewes from the rest of the flock, and give them preferential treatment to encourage high yield.

In most lowland flocks, the early phase of lactation coincides with the onset of spring grazing. The change in diet during the transition to total dependence on grass must be gradual. This is necessary to prevent metabolic disorders and to maintain a high level of nutrient intake (see MLC 1988). Effective grassland management at this time of the year calls for sound judgement of the quality and quantity of grass available. Stocking rate under good lowland conditions can be as high as 14–16 ewes/ha; regular dressings of nitrogenous fertilizers are, however, necessary throughout the grazing season to support these densities. Poor grassland management, with a low level of fertilizer use, will only support 6–10 ewes/ha. If flock owners overstock, the ewes will become thin. If they understock the pasture, grass will go to seed and it will become unsuitable for grazing. The most critical phase of rearing is during early lactation and when the lamb is totally dependent on its dam's milk. Once this period is over, it is possible to solve any unforeseen crisis by preferential feeding of the lamb or by separating mother and offspring.

## MANAGEMENT OF LAMBS

In most lowland flocks, preparations for lambing will include organizing some kind of protection for the newly lambed ewe and her offspring. During inclement weather, particularly combinations of high wind and rain, the body temperature of wet or semi-dry newborns can fall quickly; this will result in the animal becoming hypothermic. Particularly at risk are offspring from ewes in poor condition, from very old mothers, from those showing abnormal behaviour, from large litters and from difficult lambings. The first 24–48 hours are the most critical. It is vital that all staff are aware of the need for: resuscitation, disinfection of the navel cord, protection from the elements, early commencement of sucking and adequate intake of colostrum. In some situations, plastic coats for the newborn may be useful.

The newborn lamb requires adequate colostrum within about five hours of birth. If the amount is inadequate then there is a high probability of it becoming hypoglycaemic and hypothermic. Besides supplying energy for survival, colostrum is, at this time, the sole source of antibodies that protect against the diseases prevalent in the environment. It is also a laxative and aids the discharge of the foetal gut contents (meconium) soon after birth. This reduces the risk of disease such as watery mouth. Clean bedding can also minimize this risk.

With a large litter, it is advisable to hand-draw the first colostrum, which usually contains the highest concentration of immunoglobulins and to share it between members of the litter. This is a useful tactic in any situation where it is difficult to supervise progress during the newborn's first day. Lambs require approximately 50ml of colostrum per kg liveweight 4–5 times per day.

Flockmasters can take first-day colostrum from ewes with a surplus or that have lost their lambs and store it, deep frozen, for a year or more. Crossbred ewes receiving plenty of food can produce 2.5 litres of colostrum during the first day, if milked three times. Injecting cows with sheep clostridial vaccine before calving can provide an artificial alternative source of colostrum that is suitable for storing. However, some lambs receiving cows' colostrum develop severe anaemia. This is an unusual occurrence but, if it happens, it is necessary to discard the remainder of that cow's colostrum. Similarly, flockmasters may also use goats to produce a large volume of colostrum for storage.

## Hypothermia

Early detection is essential to treat hypothermia successfully. It is important to check immediately the body temperature of any lambs that are reluctant to follow their dams, or that seem even slightly abnormal in appearance or behaviour. The following ranges are a useful guide: 39–40°C normal; 37–39°C at risk; below 37°C critical – requires urgent action. Treatment consists of drying the lamb, keeping it warm and feeding it with colostrum, usually by means of a stomach tube.

A stomach tube consists of a rubber catheter (usually about 30cm long and 4.5mm in diameter), with rounded end and offset hole. The stockman uses this with either a graduated container or a 50ml syringe. The operator should sit, with the lamb across his/her lap (Figure 4.1). It is important to first lubricate the tube by drawing it across the lamb's tongue. Introduction of the stomach tube is relatively easy and, unless the animal is excessively weak, it is unusual for the tube to go down the wrong way. Coughing and rolling of the head are clear signs of incorrect insertion and, if these occur, it is necessary to withdraw the tube and repeat the procedure. In large lambs, only about 5cm of a 30cm tube should protrude. The handler holds the tube, to prevent the animal swallowing it, and connects the syringe or container to it. After giving the required amount of liquid (usually colostrum), the operator withdraws the tube while still attached to the syringe, and cleans it as quickly as possible. Having more than one tube will ensure that there is adequate time to soak them thoroughly in disinfectant (hypochlorite). It is advisable for those who have never stomach tubed a lamb before to perform the procedure under the guidance of an experienced operator.

The stomach tube is particularly useful for weakish lambs because it eliminates the risk of milk inhalation, which sometimes happens with bottle feeding. However, stockpersons should not tube feed very weak lambs – those unable to raise the head off the ground – and, in their case, an intraperitoneal glucose injection is more appropriate. Such treatment may also be useful for older lambs that are hypothermic due to starvation (see Eales and Small 1995 for full description of treatment).

## Fostering

Fostering lambs may be necessary for many reasons, including death or abnormal behaviour of the dam, lack of milk or mastitis. For it to be successful, an understanding of maternal behaviour in the ewe is essential. There are several methods, based on either

**Figure 4.1**        **Tube feeding colostrum.**

modifying the smell of the lamb, or physical restraint of the mother. Success rate can be quite high, if the recipient ewe has not had the opportunity of initiating a bond with its own offspring. The stockperson may immediately transfer lambs born simultaneously, if they are still in the wet state and the ewe is still recumbent after birth. If not, some means of imparting the smells of her own offspring on to the other lamb will be necessary. It is possible to improve the acceptance rate by intensifying maternal drive following gentle hand stimulation of the cervix.

The longer the interval to the first introduction of the foster lamb, the greater the likelihood of having to restrain the recipient ewe. Flockmasters can restrain it using one of the following: yolked stall (available as single and multiple units) (Figure 4.2), halter, or neck strap fitted with swivelling clip. When using multiple units, each stall must be self-contained, and have barriers to prevent lambs from straying between pens. It is important to have facilities for

feeding and watering, and to inspect regularly for skin damage. In the interest of the ewe's welfare, it is necessary to avoid prolonging the restraint. The stock keeper should test the reaction of the ewe towards the lamb(s) after the first day; there is little point in prolonging the attempt to foster beyond the third day. The ewe may be temperamentally unsuitable to be a foster mother, and it is important to be able to recognize this.

The presence of a sheep dog, when ewes first leave the fostering stall or when the stockperson initially presents the orphan, partially diverts attention and intensifies mothering instincts. Obviously the misuse of this tactic could impose unacceptable levels of stress on the ewe.

### Artificial rearing

Good quality ewe-milk substitutes, for use as liquid feeds during the first 6–8 weeks, are now available. Also available is a wide range of feeding devices for either *ad libitum* or restricted feeding of groups of lambs. Flockmasters can use these to rear orphan and mismothered animals, which they would otherwise have to foster. They also use them as an aid to management, allowing them to remove offspring early.

By changing the frequency and level of feeding (ie *ad libitum* or restricted) it is easy to manipulate the growth of the lambs.

Liquid feeding is expensive both in cost and labour requirements, so early introduction of palatable pellets is therefore advantageous. During the first three weeks, intake of liquids affects very little the intake of dry food. The level of performance during this stage should meet three essential requirements; it should maintain good health, prepare the lamb adequately for weaning, and use food economically. Growth rate should be not less than 200g/day. Since conversion efficiency at this time is approximately 1:1, a lamb requires some 200g a day of dry milk substitute.

It is important to offer good quality and highly palatable roughage and concentrates during the third and fourth weeks. This will accelerate rumen development and help to accustomize lambs to solid food. Artificial rearing provides every opportunity, through control of liquid intake, for adequate preparation for early weaning at 6–8 weeks.

**Figure 4.2        Yolked, single unit, fostering stall.**

**Autumn lambing**

The management of autumn born lambs is very different from those born in spring. The main features of autumn lambing are as follows:

1. It is restricted to Dorset Horn or Poll Dorset ewes, since all other types of sheep would require some kind of hormone treatment to lamb on a flock basis in the autumn.

2. Flockmasters can only outwinter ewes on very free draining land and if forage and root crops are available.

3. In many situations, flockmasters prefer to keep sheep indoors. Lambing occurs during low labour demand for field work, and the system is therefore most suitable for arable farms.

4. Ewes suckle their offspring for only 6–8 weeks. Weaning takes place at an early age and flockmasters feed the lambs high energy diets *ad*

*libitum* with small amounts of good hay.

5. Lambs born in late autumn can grow to slaughter weight (ie approximately 40kg) by 16 weeks of age. Marketing coincides with the season of high market prices (ie at Easter).

6. Many parasites are dormant during the winter. This, together with early weaning and dependence on grain-based diets, requires a different strategy for health control compared with spring-born lambs.

7. Flockmasters may stock ewes going out to grass in the spring without lambs at foot at very high rates as there is no danger of cross-infection between adults and offspring.

**MANAGEMENT OF FATSTOCK**

Spring lambing is traditional in the UK, and most of the variation in timing relates to the onset of grass growth. There is also a growing interest in late April/early May lambing on some lowland farms. Farms will differ in stocking rate, marketing policy, type of crops available for autumn/winter use, and whether they use winter housing.

Producers sell approximately half their spring-born lambs by the end of the grazing season. They sell the remainder in the autumn or as hoggets in the early months of the new year. It is usual to sell autumn-born lambs during the late winter/spring.

Generally, producers should market lambs at half their potential adult weight. At this stage the carcass has a composition that is most acceptable to the British market. Delay in marketing, and the consequent increase in weight, leads to an over fat carcass, and down-grading or rejection by the grading authorities. The current MLC carcass classification scheme (MLC 1998) gives particular emphasis to degree of fatness, muscle development, and carcass conformation and weight.

### Lambs sold off the ewe

Single lambs reared by their dams can grow to market weight by 10 weeks of age (ie approximately 20kg). Such lambs grow at 400–500g/day from birth, largely on milk and with a small amount of grass. However, lambs can only achieve this kind of performance when ewes sustain a high level of daily milk yield. The flockmaster selects lambs for market based on their liveweight and degree of finish (see MLC 1986), and then dries off the ewes.

Milk makes a smaller contribution to the growth and development of older lambs (ie those that remain with their dams until late summer before going to market) during their last few weeks. Their growth rate at this stage will largely depend on the quality of the grazing and freedom from internal parasites.

### Conventionally weaned lambs

Stockpersons usually wean lambs when they need to prepare the ewes for the next breeding season. In flocks with high stocking rates, many lambs can often be below marketing weight and finish (ie at a store stage). Weaned animals are usually 16–20 weeks of age and are almost totally dependent on grass. The target store market will dictate their post-weaning management, but the majority will follow a system that allows a moderate weight gain (200–250g/day).

Flockmasters usually fatten store lambs on aftermath and late summer grazing; they feed those that they keep throughout the autumn on grass/root crops or grass/arable by-products. Diminishing food supplies or adverse weather may force flockmasters to boost growth rate, at any stage, by feeding concentrates.

### Early-weaned lambs

It is advantageous, for a variety of reasons, to early-wean lambs that are: artificially reared; born in autumn/early winter and in housing; members of a large litter; or at risk of severe parasitism. The timing will depend on the opportunity to prepare the lambs for weaning. Where flockmasters provide palatable pelleted diets from three weeks of age, weaning can take place at 6–8 weeks of age and without check to growth and development. Subsequent performance will depend on the plane of nutrition.

## MANAGEMENT OF HILL SHEEP

Hill sheep can use uncultivated grasslands that would otherwise be unproductive. Lack of cultivation, no fertilizer or lime treatment, herbage of low nutrient value, a short growing season, harsh climate and minimal control of the grazing animal, all contribute to low productivity. Flockmasters are reluctant to supplement the ewe's meagre diet during crisis periods. Coupled with the seasonal nature of the nutrient supply, it means that even the small hardy hill breeds are subject to prolonged stress from early pregnancy to early lactation. The gradual loss of basal weight reaches its lowest point during early lactation, although this is a time of increasing grass supply. Most of the deaths of hill ewes occur around this time. High survival rates in hill flocks receiving little or no supplementary rations are dependent on good recovery of body conditions during summer/early autumn. This forms the basis of the reserves that the animal mobilizes in late winter and early spring. It is very important that hill farmers condition score all their ewes before the tupping period. Those with a score of less than two are not suitable for breeding unless the farmer confines them to the in-bye areas. At a later stage of pregnancy, flockmasters with several sheep at a condition score of less than 1½ are providing inadequate care and welfare (FAWC 1994).

Producers can significantly improve welfare and performance by using in-bye land and fenced-off improved hill areas, during the key periods around mating and lambing. Confining ewes during these periods also provides the opportunity to feed supplementary diets, including concentrates and minerals.

In many hill areas it is common to away-winter (on more favourable sites) ewe lambs during their first autumn/winter; some hill farmers use cheap housing as an alternative.

## Organic Production

Organic systems of production endeavour to exclude the use of inorganic fertilizers, pesticides, fungicides, growth regulators, livestock drugs and feed additives to enhance productivity or to control animal diseases. Organic systems are akin to those which prevailed prior to the late 1930s before most of the products listed above became available. Up to that time farmers relied on a combination of crop rotation, soil cultivation and the use of animal manure to maintain and improve soil fertility. This approach sustained a large number of relatively small non-intensive low stocking rate animal production systems.

The current standards for organic production are set and monitored by organisations such as the United Kingdom Register of Organic Food Standards (UKROFS). Great reliance is placed on the adoption of sound animal husbandry and land management to control, for example external and internal parasites and as alternatives to the routine use of vaccines and anthelmintics. Where veterinary advice deems if necessary drugs can be used to safeguard the health and welfare of the animals. Animals so treated can usually be marketed as organic provided extended withdrawal periods are strictly adhered to.

## ENVIRONMENT AND HOUSING
### Hill flocks

Hill farmers do not attempt to provide the same degree of protection against the weather as their lowland colleagues. Instead they place greater emphasis on the genetic fitness of the breeds they use. In hill and harsher upland areas flock owners rely on a system of pure breeding and the selection of home-bred stock for replacement purposes. When farms change hands, it can be advantageous to transfer part or all the flock to the new owner. These policies ensure that the native stock provide the lead, particularly regarding seeking shelter, the source of water and the flock or subflock grazing territories. Flockmasters are making increasing use of weather forecasts. It gives them time to act appropriately, such as moving the flock off exposed ground, delaying the return of the flock to the hill, or delaying shearing. In recent years, better communication and cooperation between the Forestry Commission and the farming community and their advisers has helped to generate interest in the provision of shelter belts in hill and upland areas.

Flockmasters who continue to use traditional practices are also able to ensure protection from the elements. For example, in many areas stockpersons gather all breeding stock into the in-bye land (ffridd) for a few weeks around lambing. Stone walls and banked hedges can provide very effective shelter. A few flock owners use degradable plastic jackets for newborn lambs, which provide good protection during the first three days or so. The transfer of ewe lambs to lowland farms over their first winter ensures that their growth and development continues unimpeded. This well-established practice is becoming more difficult to pursue in some areas, due to the cost and to the difficulty in finding lowland farms with surplus grazing. Consequently, flock owners are providing cheap housing in the hill and upland areas for such animals, and having to transport some feedstuffs from lower areas.

### Lowland flocks

During inclement weather the availability of buildings and the proximity of the flock on lowland pastures is advantageous. These factors enable lowland flockmasters to act quickly and provide shelter so much more easily than in hill regions. It is extremely important for the sheep and for the system's success to have shelters available during either part or the whole of the winter/early spring period. The longer the period sheep require shelter, the more permanent the facility needs to be.

At lambing time, temporary or short-term shelter usually caters for the flock; it provides the ewes with protection from the weather for 24–48 hours after parturition. Such accommodation consists of individual pens using hurdles, straw bales and netting wire, or cheap boarding with some type of temporary weatherproofing covering most the pen. These pens should also have a quick and easy front access (which should face away from the prevailing wind). These shelters may form part of a corral where large straw bales form the perimeter wall and straw litter covers the whole area. Such shelters may also form part of a small section of a paddock or field in which stockpersons may keep those they expect to lamb overnight. By placing straw bales strategically around the main field, flockmasters can provide useful shelter for the ewes and lambs they will release from protective accommodation.

In recent years, long-term winter housing has become commonplace. This trend may be a result of: higher stocking rates, the need to rest land overwinter and the adoption of autumn and early winter lambing. To some extent, it also acknowledges that well-trained staff deserve good facilities to show their competence. Some flock owners provide purpose-built housing, but many use spare capacity in multi-purpose buildings and fodder barns; housing is therefore very variable. In the interest of the flock's general welfare, and irrespective of the type of building, producers must give due regard to the following: the maintenance of good health, adequate feeding and watering arrangements, ample space for resting, good working conditions for stockpersons, limitation of group size, easy flock inspection, and means of segregating individuals or small groups of sheep. The newly-lambed ewe and those undergoing fostering and artificial rearing require special arrangements. Sick animals and artificially reared lambs are the only sheep requiring some local heat. Flockmasters usually use infrared lamps, which they place well clear of the animal, for this purpose.

Fully covered buildings for shorn and unshorn sheep should not attempt to provide an artificial level of warmth. The general house temperature follows that of the local environment fairly closely. The design should allow movement of air, but be free from draughts and always provide dry conditions. In the interest of health and welfare, it is important to avoid hot and humid conditions. Webster (1994) has reviewed heat and cold tolerance of sheep.

## Winter housing

### Partly covered yards

Flockmasters may adapt cattle courts or use open yards linked to limited covered space for wintering sheep. Such systems expose the sheep to the weather some of the time, and this may lead to a very damp humid atmosphere in the covered area. It is essential that the open yard has good drainage, to keep feet sound. There should also be free movement of air in the covered area to reduce the risk of respiratory disease. With an apex roof, the ridge should have an opening 1.5–2.3cm deep along most of its length to ensure good air flow. Similarly, monopitch buildings should have an opening 2.3–3cm in depth along the back wall.

### Fully covered buildings

There is a wide range of types in this category. These include multi-purpose climate houses, single row and face-to-face monopitch buildings, and shelters clad with various types of plastics (Figure 4.3).

Purpose-built, ridge-ventilated, wide-span buildings provide ample scope for an efficient layout of pens. Flockmasters can use relatively cheap materials to clad the lower and upper halves of the sides and end. In some locations a completely open space may be left between a low side wall and the eaves; however, some kind of baffle against wind, rain and snow is usually desirable. Traditional Yorkshire boarding in combination with fine wire mesh, or other materials would also give a comparable effect.

### Internal features

It is very important that stockpersons can feed, water, litter and supervise their animals easily. At key times, such as the peak of lambing, they need to be able to segregate lambed ewes and regroup them with the minimum of disturbance. The choice of floor is largely between slats and litter. Slats have the advantage of keeping feet in good condition and providing a dry lying area enabling the fleece to remain clean. Such a system requires less floor space per sheep (Table 4.7) and eliminates both the time a stockperson requires for littering, and the problems that accompanies rising levels of litter.

**Figure 4.3     Fattening lambs in cheap plastic-clad (Polypen) housing.**

**Table 4.7         Recommended floor space (m²) for sheep on slats and straw.**

| Type of sheep; unshorn* | Slats | Straw |
|---|---|---|
| Large ewe (68–90kg) in lamb | 0.95–1.1 | 1.2–1.4 |
| Large ewe (68–90kg) with lambs | 1.2–1.7 | 1.4–1.85 |
| Small ewe (45–68kg) in lamb | 0.75–0.95 | 1.0–1.3 |
| Small ewe (45–68kg) with lambs | 1.0–1.4 | 1.3–1.75 |
| Hoggs (32kg) | 0.55–0.75 | 0.75–0.95 |
| Hoggs (23kg) | 0.45–0.55 | 0.65–0.95 |
| Lamb creeps (2 weeks) | – | 0.15 |
| Lamb creeps (6 weeks) | – | 0.4 |

\* Shorn sheep require 20–25 per cent less floor space.

The main disadvantages are the high capital cost of installation and the unsuitability of slats for ewes with young lambs at foot. Unsuitable materials and poor construction can lead to warping and splitting of slats, with consequent risk of injuries to legs and feet. Straw, on the other hand, may be readily available at low cost, and arable crops can use it afterwards as manure. It is more versatile than slats for all age groups, but regular littering is laborious, and flockmasters must thoroughly clean out sheds annually. As litter builds up there is the danger that foot ailments, such as footrot, may spread through the group. Care and attention to feet before housing is an important aspect of management.

For ease of handling and selection, and to reduce the risk of injury and crushing under stress, the number of sheep in a group should be low, preferably less than 30. The feeding arrangements and trough space (Table 4.8) requirements dictates the general layout of the pen. Stockpersons should feed from the outside; long narrow pens with a depth of not less than 3m are the most suitable as they allow free movement of sheep during feeding time.

Sheds may have "walk-through" troughs (see Figure 4.4); these allow feeding from one or both sides, and may also act as divisions between pens and along gangways. Autumn-lambing ewes may be housed in cheap plastic clad housing (Figure 4.5).

**Table 4.8    Recommended trough and hayrack length (mm) for housed sheep.**

| Type of sheep | Trough | Hayrack |
|---|---|---|
| Large ewes | 475–500 | 200–225 |
| Small ewes | 375–425 | 172–200 |
| Lambs (36–45kg) | 350–400 | 200 |
| Lambs (up to 35kg) | 300–350 | 150 |

Stockkeepers may segregate ewes giving birth by building portable hurdles within pens. Alternatively they can move ewes and lambs soon after parturition to more permanent individual mothering pens, each measuring 1.8 by 1.2m.

## NUTRITION AND FEEDING
### Diet
Most sheep rely purely on grass during the grazing season, and hay or hay and silage during winter. If flockmasters do feed concentrates and/or supplements at all, it is during the last month or so of pregnancy and the very early stages of lactation. Their use, however, depends on availability of spring grass. Certain classes of sheep, such as store lambs, need feeding for lengthy periods on arable crops and by-products. Producers may rear and fatten early-weaned, autumn-born lambs exclusively on concentrates, which they usually give *ad libitum*, with only minimal roughage (100–200g/day).

It is essential that flockmasters adopt some method of monitoring feeding, particularly during critical phases of the production year. This will enable them to take corrective action if necessary. The most practicable are weighing and body condition scoring (see p 108). Details of nutritional values of a very wide range of foodstuffs are in MLC (1988).

### Grass
Grass supplies 90–95 per cent of the annual energy requirement of most of the national flock. The level of dependence on grass varies according to local resources and husbandry systems. Producers keeping flocks on extensive low input/low output regimes (these include many in hill and upland areas) rarely provide anything other than the available uncultivated rough grazing. They do however supply some hay during very severe weather in late winter and early spring. In contrast, grass is only available for sheep on intensive farms, where stocking densities are high, from the advent of good quality grazing in the spring to late summer or early autumn.

Being ruminants, sheep are anatomically and physiologically able to graze and digest a range of foodstuffs. Successful feeding management requires a sound understanding of the nutrient requirements and a sensible assessment of the quality of the available foodstuffs. The latter involves subjective judgement; it is rare for a flock owner to resort to laboratory analysis to a certain nutritional quality of grasses and other crops. Subjective judgement usually involves noting stages of growth, density and type of sward, and the ratio of grass to clover. Flockmasters assess hay, the most common winter feed for sheep, on the basis of colour, and absence of dust, which often includes deleterious moulds. This provides a reasonable indication of the efficiency of curing and storage. They may establish the potential feeding quality of the crop at the time of mowing according to the botanical make-up of the hay and stage of growth. This also applies to silage and, in addition, colour, texture and smell will indicate how efficient the manufacture and storage was.

### Forage and root crops
As stocking densities increase, the need for alternate forms of forage arises to bridge the gap between the end of the grazing season and the start of full winter feeding reliant on hay and/or silage. During the autumn months, before the winter frosts, producers feed: green fodder crops (kale, forage rape, fodder radish), root crops (swedes, turnips, mangolds) and arable by-products (sugar beet tops, leafy brassica wastes). These crops contain sufficient levels of energy and protein to meet the requirements of breeding animals and store lambs.

Soil type plays quite an important part in decisions concerning the provision of forage and root crops during autumn/winter. In high rainfall areas, and on heavy land, the use of tillage and root crops

**Figure 4.4    Sheep house with 'walk-through' troughs.**

**Figure 4.5    Winter-shorn ewes and young lambs with access to a creep area.**

leads to: considerable food wastage, heavy soiling of fleeces and a high incidence of foot ailments. The higher the clay content of the soil, the more difficult it becomes to prevent poaching and damage to the crops. It is customary to control the grazing of these crops using electric fencing or folds. The use of fall-back pasture, or feeding hay during the introductory period in autumn/winter, helps to control digestive upsets. Over-dependence on fodder crops can lead to loss of appetite and general unthriftiness due to the presence of brassica anaemia factors.

It is possible to lift and store root crops for use later in the winter and around lambing time. Using the crop *in situ* may be difficult for young breeding animals at the four-tooth stage. Roots are low in dry matter content, 10–12 per cent compared with 25 per cent in silage and 85 per cent in hay. It is therefore advisable to use fall-back pasture or hay at this time.

During the autumn, vast quantities of crop residues are available. Lowland producers buy in store lambs, to use such by-products, which they then sell as finished (fattened for slaughter) lambs later in the season. Sugar beet tops need wilting for at least a fortnight, otherwise the oxalic acid content of the leaves will lead to digestive upsets. If flockmasters feed breeding ewes tops for lengthy periods they can easily become overfat by mid-pregnancy and therefore their intake needs controlling. If they do provide tops through to late pregnancy, they should supply balanced concentrates during the last 6-8 weeks. Fattening lambs may need a supplement of mineralized cereals (0.25kg/lamb/day) during the last stages of fattening.

## Compound feeds and cereal supplements

During late pregnancy and early lactation the nutrient requirement of breeding ewes increases. Flockmasters usually provide these animals with compound feeds or cereal grain with added protein, mineral and vitamin supplements. Store lambs during the late stages of fattening also require this type of diet. Farms that produce their own compounds use a blend of cereal grains that they balance with vegetable and approved animal proteins, such as fish meal, in the ratio 80:15:5. The final mix also contains 2.5 per cent mineral/vitamin supplement. To avoid wastage, it is advisable to cube or pellet concentrates. Proprietary compounds cover a range of requirements.

Flockmasters may feed complete diets to housed sheep. These can range from cubed fortified dried grass, to complete compounds that also include ground roughage. Where winter housed ewes receive only compound diets, it is important to feed clean barley straw *ad libitum* to satisfy their requirement for roughage. Failure to supply clean straw will encourage ewes to eat soiled bedding and will increase the risk of digestive upsets. During winter housing ewes will consume approximately 1.5kg of straw a day.

Compound feed blocks are also available to supplement the basic fodder diet during winter. The blocks weigh 25–30kg, resist weathering and supply a good source of energy (usually from cereals), they also contain urea, minerals and, sometimes, anthelmintics. Blocks are useful in hill and upland areas where transport is difficult.

## Minerals and vitamins

The mineral/vitamin ingredients of sheep diets will, to some extent, compensate for the known deficiencies of the region. In most situations sheep require extra magnesium in the spring, and therefore concentrates for breeding ewes should generally include it. Where it is impractical to include magnesium in a feed, then mineral blocks (with a high magnesium content), or the spraying of it in solution on to herbage, are alternatives.

Flockmasters may give trace elements (such as copper and cobalt) as a drench or by injection. This form of supplementation has the advantage of providing a direct and quantifiable dosage to every animal in the flock. In the hill and upland areas, where concentrate feeding is nil or minimal, it is the only reliable method. Using slow-release boluses containing the element required may reduce the frequency of drenching.

Vitamins are usually an ingredient of compound feeds or supplementary concentrates. Injecting an appropriate proprietary product will normally correct cases of acute vitamin/mineral imbalances.

## Water

Water should be available *ad libitum* in all situations. Flockmasters should provide housed sheep with raised water tanks, in such a place where there is little danger of contamination from foodstuffs. Buckets are usually adequate for small groups and for animals in individual pens if the stockperson frequently tops them up. It is important to protect the water supply from frosts, but this is not easy under paddock and field conditions in severe winters. Flockmasters with fields without a natural source of water need to provide tanks that they refill periodically. Under certain conditions grass and forage and root crops may satisfy the water requirement. However, it is inadvisable to rely totally on these foodstuffs to supply the needs of the sheep.

### Practical feeding

For details on practical feeding see under sections on management of: rams, breeding females, lambs and fatstock.

## HEALTH AND DISEASE

Irrespective of the system of production in the UK, all flocks rely on hygiene, preventive treatments and early detection of ill-health, for the maintenance of healthy stock. A good stockperson will recognize the early symptoms of abnormality and will act without delay. Vigilance and powers of observation are essential to pick up subtle changes in behaviour, the early stages of lameness and other physical symptoms during regular inspections. The signs that indicate health include: general alertness, free movement and absence of lameness, good uniform fleece and absence of rubbing, active feeding and rumination, and no visible wounds, abscesses or injuries. It is therefore important to know and understand the behaviour of the animals. The signs of ill-health in sheep include listlessness, abnormal posture and behaviour, persistent coughing or panting, absence of cudding, poor condition and, in some circumstances, separation from the flock.

Good hygiene and preventive treatments are vital aspects of sound flock management. In extensive systems there is little scope to practise hygiene except at times of stock treatments, which may be minimal anyway. In contrast, lowland systems involving high stocking rates and winter housing will need good hygiene measures and preventive treatments to safeguard the health and welfare of the sheep.

### General hygiene

Stockpersons can ensure a high standard of cleanliness in handling areas and houses by simply brushing and washing surfaces and by using clean dust-free litter. After destocking a building it is important to thoroughly clean and disinfect everything, including all fittings and food and water containers. Similarly, collecting pens and handling yards require cleaning and disinfection periodically. Flockmasters need to keep the facilities they use for housing sheep during the lambing season extremely clean and thoroughly disinfected. Hygiene in the pens for mothering, fostering, and artificial rearing of orphan lambs is particularly important. They also need to carefully clean, and disinfect or sterilize the equipment they use for routine procedures and treatment after use.

### Preventive and other treatments

Preventive treatments are available for conditions caused by a variety of organisms, including viruses, bacteria and other microorganisms, and external and internal parasites.

The flockmaster usually draws up a programme of treatment following consultation with the veterinary surgeon. The plan will place particular emphasis on the previous disease history of the flock and the prevalent diseases of the locality. Treatments involve vaccination, dosing (drenching), spraying or dipping, foot bathing and foot treatment. Many are seasonal and some relate to the key events of the sheep year – mating, lambing, shearing and weaning. The timing and type of treatments will therefore vary according to the production system (see Henderson 1990).

An example of a plan of action for a spring lambing flock is in Table 4.9. A flockmaster may draw up similar regimes for rams, ewe lambs, fatstock, and replacements purchased from markets or other sources. A flock health plan is likely to become a requirement for all flocks when the new sheep welfare code is introduced.

### Vaccination

Most vaccines are injected subcutaneously, preferably high on the side of the neck or over the lower rib cage. The needle should be inserted parallel to the body surface and into a cavity formed by raising a fold of skin. If the needle is inserted at any other angle, it can easily penetrate the underlying muscle.

Hygiene is important in vaccination. It is important to use the syringe according to the maker's instructions and to change the needle periodically. Incorrect or careless injection techniques produce abscesses.

Some vaccines provide protection against many diseases. Most flocks in this country use a multi-vaccine, which provides protection against as many as eight diseases.

**Table 4.9**     **Relationship between key events and preventive treatment in a closed spring-lambing flock.**

| Month | Key events | Type of preventive treatment |
|---|---|---|
| August | | Spraying/dipping (all stock). Intensive foot care (all stock). |
| September | | Vaccination[a] (all stock). |
| October | Mating | Vaccination[a] (all stock). |
| November | | |
| December | | |
| January | | |
| February | | Vaccination[a] (all stock). |
| March | Lambing | Drenching (ewes before released to clean pasture). |
| April | | |
| May | Shearing | Spraying/dipping (depending on season, ewes only). |
| June | | Spraying/dipping (all stock). |
| July | Weaning | Drenching (depending on availability of clean grazing). |

[a] 8 in 1 vaccine.

Giving a multi-vaccine in late pregnancy provides protection to the ewe, and ensures the presence of antibodies in the colostrum, which initiates passive immunity in young lambs. A vaccination programme for breeding stock involves two injections in the first year and an annual booster injection. Flockmasters usually give this during late pregnancy.

**Internal parasites**
Flockmasters may dose their animals to control internal parasites such as stomach worms, nematodirus, tapeworms, liver fluke and coccidia. There is a wide range of drugs available; some are effective against more than one of the above parasites. The choice of anthelmintic and the timing of dosing is important to reduce the risk of resistance. It is advisable to consult a veterinary surgeon before dosing. Grazing cattle and sheep alternately, and including conservation (hay/silage) in a programme of grassland management, helps in reducing the use of anthelmintics and the frequency of dosing.

The development of various kinds of plastic drenching devices has made routine dosing a relatively easy task and not too time consuming. It also helps to treat the flock in a race. For dosing, the handler should hold the head at back level, or slightly higher, using an open hand under the jaw and then insert the mouth-piece at the corner of the mouth, with the tip resting on the back of the tongue.

Anthelmintics and other medicines can sometimes be given to the sheep in pill form or in boluses. Flockmasters may use special devices (eg balling guns) for administering these. Ideally this should be a two-person operation, with the handler concentrating on keeping the sheep still. The doser then opens the mouth and positions the device so that the pill or bolus drops on to the back of the throat.

**External parasites**
Flockmasters should control external parasites, such as ticks, lice, keds, mange mites and blowfly larvae, which can cause prolonged stress and discomfort, by dipping and spraying. If they do not detect infestations it can lead to serious losses. Effective control relies on dipping and spraying at particular times of the year. Stockpersons require a certificate of competence and approved protective clothing before using organo-phosphorus dips. These requirements recognize the dangers to human health and the health and welfare of sheep. Both dipping and spraying involve some stress to the animal; the handler should therefore perform the procedure with calm efficiency and should not treat sheep which are hot, tired, wet, thirsty or fully-fed. Avoiding dipping during very hot weather or during the hottest part of the day reduces the risk of poisoning through

absorption. Dipping and spraying may coincide with the breeding ewes having lambs at foot, and stockpersons must pay particular care when treating young lambs. The level of noise when ewes and their lambs become separated during handling reflects the degree of general upset. It is advisable to check for mismothering after the flock returns to the field. Spraying is less stressful than dipping, for it allows the ewe and her lambs to walk through together, and there is less risk of injury. Spraying is not as thorough, and consequently the period of protection is shorter than that following dipping.

## Lameness

Lameness is probably the most widespread cause of pain and suffering in sheep and attention is rightly drawn to it in the *Codes of Recommendations* (MAFF 1989). It requires that stockkeepers have experience and are competent in the prevention and treatment of foot-rot.

Although foot-rot is the main concern when it comes to foot inspection, care and treatment, lameness may be due to one of several other causes. Such causes include injuries to the sole of the foot from embedded stones, glass or thorns, and to the interdigital skin from walking through corn and brassica stubble. Interdigital growths can become infected leading to painful lameness. Only a veterinary surgeon can treat these growths by surgical removal under local anaesthesia. Congenital deformity of the claws can occur; it is best to pare them to the best possible shape and then cull the animals as fatstock at the first opportunity. Sometimes acute laminitis occurs in all four feet after introduction to a very high plane of nutrition, such as lush pasture, and this may immobilize the animal. Transfer to a poorer diet is the only effective course of action. Stockpersons normally carry out inspections and treatment in the handling pens leading up to the footbath, but they may deal with individual lame sheep in the field. It is important to prepare essential items beforehand. The equipment includes: secateurs and a sharp knife for paring away infected tissue, a brush (if the sheep have muddy feet), a bucket with disinfectant and fresh solution in a footbath.

Flockmasters should treat sheep with extensive damage to their feet with an antibiotic aerosol or cream after paring and before bandaging. They will then need to keep the animal on dry straw in an isolation pen for 4-5 days and then re-inspect. Healing may take 2-3 weeks. For animals that are less seriously affected it may be possible to simply use a footbath or spot-treat individuals with an aerosol antibiotic. Stockhandlers may separate the lame animals requiring treatment from the rest of the flock and hold them until all the sound sheep have gone through the footbath.

Sheep should be put through the footbath in batches, and must stand in it for the recommended time (usually 1–2 minutes). Irrespective of the chemical, it is important to follow the manufacturer's instructions regarding safety and preparation. Currently the choice is between formalin and zinc sulphate. Workers should handle formalin with great care, preferably wearing eye protection to avoid it splashing in their face and eyes. If this occurs, they should wash immediately, using plenty of clean water. It is important to note the recommended concentration for routine preventive treatment and for the treatment of clinical cases. Excessively high concentrations will be particularly painful to animals that have had their feet pared. Many flockmasters now prefer zinc sulphate to formalin because it penetrates better, is more effective and is less distressing to the sheep.

## Mastitis

Mastitis is an inflammation of the mammary gland caused by infectious bacteria. Most cases occur soon after lambing and during the first month of lactation. Very few cases occur after weaning. The incidence amongst ewes varies in the range of 1–15 per cent. Lowland flocks have a higher incidence than hill flocks.

In severe cases of mastitis the ewe may separate from the flock, movements may become stiff and the lambs will show signs of starvation. The affected gland will be very swollen and painful and the skin over the udder may become red or purple. Mastitis in its worst form is an acute disease leading to the death of the ewe. If the ewe survives, the gland may slough and the healing process may take several weeks. In less severe cases a proportion of the gland may become hard and lumpy and this may not be detected until the pre-mating examination.

Most of the causal organisms are sensitive to several antibiotics. Parenteral administration will usually lead to the recovery of the ewe but the udder is often permanently damaged and the ewe may have to be culled.

Some forms of chronic and subclinical mastitis result in reduced milk production and poor growth rate of lambs.

There are no effective vaccines and the most effective means of prevention relies on maintaining a high standard of hygiene. This is especially necessary in systems involving housing and/or penning for long or short periods. There should be the liberal use of straw in the main and lambing pens; segregation of ewes with mastitis; thorough disinfection of pens used by affected ewes and, provided weather permits, the turning-out of the ewe/lambs to grass as soon as possible after lambing.

## Notifiable diseases

Notifiable diseases that may occur in sheep include anthrax, foot-and-mouth disease, maedi-visna, scrapie and sheep pox. Flockmasters must report these diseases, even if they only suspect them, to the local MAFF animal health office, local authority or police. They must isolate animals they suspect and immediately stop all movement of livestock. Comprehensive information regarding these diseases and general health care is in Henderson (1990); Martin and Aitken (1991).

## Zoonoses

These are diseases that effect man following contact with infected sheep, or their tissues, fluids and faeces, or ingestion of infected meat, offal or meat products. Widespread zoonoses are enzootic abortion of ewes, toxoplasmosis, orf and hydatidosis.

### Enzootic abortion

*Chlamydia psittaci* causes enzootic abortion. The infected animal usually aborts, and the evidence is clearly visible in the cleansing. Infection in the cotyledons is severe, and the areas between them are thickened and have a leathery appearance. Fluids and membranes are highly infectious for at least two weeks. Personnel can develop flu-like symptoms. The most serious risk is to pregnant women, and several cases of human abortion have been a result of too close contact with infected sheep.

### Toxoplasmosis

This is a widespread cause of abortion in sheep. It usually occurs during the last month of pregnancy but the lambs may go to term and may be born dead or weak. Infection is very common in humans, but there is only a low incidence of clinical disease, although pregnant women should be particularly careful to avoid infection. Cats are generally the main hosts for this protozoan parasite *Toxoplasma gondii*. Humans become infected if they consume raw or insufficiently cooked meat, or handle infected meat, foetal membranes and fluids, or material contaminated by cat faeces.

### Orf

This viral disease is sometimes known as contagious pustular dermatitis. It primarily affects the lips of lambs and the teats and udders of ewes. It may also effect the genitalia and the area adjoining the hooves in older animals. Handlers should wear protective gloves during treatment and vaccination.

### Hydatidosis

This disease has a cyclic transfer between dogs and sheep. The mature worm (*Echinococcus granulosus*) lives in the gut of the dog and the sheep acquires the parasite by ingesting the eggs which have been passed out in the dog's faeces. These eggs develop into cysts within the sheep and dogs become infected when they eat raw offal containing these cysts. Humans acquire the disease from accidentally ingesting eggs from dog faeces. This leads to the development of cysts in the lungs, liver and other organs and advanced cases may require surgical treatment.

## ROUTINE PROCEDURES
### Identification/marking

Sheep can be marked in several ways. They can, for example:-
1) have their faces/fleeces temporarily marked with scourable-out paint dyes
2) have their ears
    a) notched or hole punched (permanent) – individuals can be identified via simple recording codes

b) ink tattooed with unique number (permanent) – easier done on pale coloured skins

c) tagged with uniquely numbered, coloured plastic or metal tags – can be permanent (tamper proof).

There is a move, in the UK, towards a statutory marking system by which all sheep will have a permanent, unique number put on them such that they can then be traced back to their farm of origin.

## Soundness, culling and selection

Flockmasters inspect their sheep for soundness annually, usually after weaning the main flock and before the preparatory phase for tupping. Inspection involves the close examination of all classes of stock of breeding age and may include an assessment of their productive performance.

The first stage is a general appraisal, which is a normal part of flock inspection at any time; this includes looking for signs of ill-health, abnormal behaviour, posture or gait, poor body condition and ragged fleece. The second stage involves closely examining the udder, teeth and feet.

When examining the udder, it is important to look carefully at both teats for any evidence of injury during shearing or the previous lactation. The stockperson should palpate both glands for hard or abnormal tissue, which is usually a sign of mastitis.

When examining teeth, important features to look for are alignment and number, type and soundness. Correct alignment between the incisors and the upper pad is necessary for efficient grazing. When selecting rams and breeding ewes, producers should look for this feature during the transition from temporary to permanent incisors. The eruption of the first permanent incisors (the central pair) occurs at approximately one year and three months, and the adjacent pair at one year and nine months. The next two pairs erupt at two years and three months and two years and nine months respectively. Counting the number of permanent incisors (or broad teeth) and fitting it within the above range, is a part of sheep nomenclature (see Williams 1995); it is the normal method of ageing sheep. Two-toothed ewes are approaching their first year of full production potential and usually attract the best prices at auctions. The industry refers to ewes showing four

pairs of permanents as full-mouthed; this description remains until one or more teeth are lost, when they become broken-mouthed. The age at which this occurs is variable. Broken-mouthed ewes often lose condition, particularly on hard grazings, and producers usually cull them.

The stockperson should closely examine feet for clinical symptoms of disease, distortions of the hooves and excessive growth of horn.

## Body condition scoring

The method of body scoring involves five grades - 1 (very thin) to 5 (very fat) - and is easy and quick to learn. It requires no equipment and a stockperson may score the sheep wherever it is convenient to pen them. However, it is preferable to condition score in a race, so that it is easy to regroup the ewes according to their score. This regrouping allows stockpersons to give those in poor condition extra feed to restore them to the level they require.

The assessment involves palpation immediately behind the last rib, using the tips of the fingers to determine:

(a) the prominence of the two parts of the lumbar vertebrae, the spinous and transverse processes,

(b) the amount of muscular and fatty tissue underneath the transverse process, and

(c) the fullness of the eye muscle and its fat cover, by pressing the fingers into the angle between the spinous and transverse processes.

A full description of the five grades is in MAFF (1994).

## Weighing and selection of lambs

There are several types of walk-on weigh scales for sheep. Siting the scales at the outlet of the race enables stockpersons to complete the task without the need for handling. Lambs quickly become familiar with the weighing procedure.

Flockmasters select lambs for the fatstock market at approximately 50 per cent of their potential adult weight. When all lambs are within 10 per cent of the desired weight, they should physically assess the level of fatness and determine the fatness score. The MLC (1986) describes the criteria for this in detail.

## Docking and castrating

Docking or tailing is the amputation of part of the

tail. It is important not to dock during the first 24 hours, or the first 2-3 days for weak lambs, so that it does not interrupt suckling. It is legal to tail dock a lamb before 12 weeks of age without an anaesthetic. In hill areas it may involve only a few centimetres at the tip to prevent wool balls forming around the tip of the tail. Under lowland conditions, stock keepers remove the greater part of the tail to reduce faecal contamination, which attracts blowflies and leads to fly strike. It also eases management at mating and lambing.

While it is possible to induce effective anaesthesia by epidural injection, it is a technically difficult procedure and therefore requires a veterinary surgeon to perform it. With this procedure there is a high risk of it not working properly because the injection site is very precise. There is also considerable danger of introducing infection into the spinal canal. Injection of anaesthetic is not fully effective and is probably more painful than the tailing procedure itself (FAWC 1994).

Docking and castrating of lambs less than a week old normally involves two operators. The first holds the right legs in one hand and the left legs in the other, with the lamb against their chest. The second operator is then able to carry out the actual procedure. When the operation has to be done single-handed, the handler may restrain the lamb by placing it between the knees and with the hindquarters facing them. Handlers usually perform the operation on older lambs by holding the front legs and resting the rump on a rail or bale at a convenient height. There are several methods of docking: the use of a rubber ring that restricts blood flow to the tail; cutting with a knife; crushing with a Burdizzo bloodless castrator before cutting or the use of a sharp-edged hot iron. The first method involves putting on a rubber ring with an Elastrator applicator. It takes 7–14 days for the tail to drop off and normally there is no bleeding. It is important to position the ring at the end of the wool-free skin on the underside of the tail; this will ensure that the remaining stump covers the anus and vulva. This is a legal requirement under *The Welfare of Livestock (Prohibited Operations) Regulations 1982* (these regulations also prohibit tooth grinding) (MAFF 1995).

It is also possible to use the rubber ring method for castrating. The stockperson places the ring at the neck of the scrotum and in front of the rudimentary teats and with both testicles within the scrotum. It is an offence to use the rubber ring method to either dock or castrate lambs more than seven days old.

The other methods of castration involve the use of a bloodless castrator to sever the spermatic cord (vas deferens), and the use of a knife to remove the testicles. When using a knife to dock or castrate it is important to maintain a high standard of hygiene and to treat the wound to prevent infection, particularly fly strike. Stockpersons should not attempt these techniques without prior training. It is illegal to castrate male sheep by any means without the use of an anaesthetic if they are over three months old. Only a veterinary surgeon may castrate a ram of this age.

A recent report on the welfare of sheep (FAWC 1994) draws particular attention to the pain and distress of castration and docking irrespective of the method. It implores flockmasters to consider carefully the need to adopt these procedures particularly in systems where they market the lamb at an early age. The current legislation appears to take the probably erroneous impression that lambs less than a week old feel less pain than older ones. When further scientific data on the effects of method and age on pain and behaviour becomes available it is highly likely that new legislation will change to include greater controls.

**Dagging**

Dagging consists of the removal of excess wool around the tail and down the inside of both thighs, to prevent soiling during early spring and before shearing. It reduces the risk of fly strike early in the season, and the need to trim fleeces before rolling. Flockmasters may dag unshorn lambs in the late summer and adults before tupping and lambing. The choice of whether to use mechanized or hand shears will depend on the amount of wool the shearer needs to remove. *The Welfare of Livestock (Prohibited Operations) Regulations 1982* prohibits freeze dagging (MAFF 1995).

**Shearing**

Producers usually shear their sheep from late May in the more favourable areas of the lowlands, to mid July in the harsher environment of the north west of Scotland. Only a small proportion shear lambs; most animals therefore are shorn for the first time at 14–15 months old, and then annually.

When shearing is done early in the season, wet, windy, and cold conditions can result in severe chilling, a high incidence of mastitis and, sometimes, death. Flockmasters can provide some protection by using sheltered paddocks, or by using combs that prevent shearing too close to the skin. Usually it is only some hill farms that adopt this practice. It is particularly important to pay attention to the weather forecast before shearing. During unsettled weather it is always advisable to house batches of ewes overnight so that work can start the next day. In wet and windy conditions, it is necessary to keep newly shorn sheep in sheltered paddocks or housing overnight.

Many flockmasters, particularly in the south west of England, now shear their sheep in winter because of overwinter housing. They should only shear those ewes housed before early January and which are not turning out until mid March or later. To limit the stress of housing and shearing, both should coincide; this may help to reduce the incidence of wool slip (partial baldness, caused by hormonal changes associated with stress) during the weeks following shearing. Shearing before early January allows sufficient regrowth by turning-out time. Operators should always use deep combs for winter shearing. Initially, the main benefit of winter shearing was the reduction in pen and trough space requirement, but others have since become apparent. The respiratory rate of shorn ewes is always lower than unshorn ones. Shearing leads to easier surveillance at lambing, easier teat-seeking by lambs and a lower incidence of crushing, particularly in the post-lambing mothering pen. Lambs from shorn ewes have a higher birthweight and a higher survival rate. The incidence of lambings that need assistance however may be higher. The gestation period is also one day longer in shorn ewes.

Contract shearing is now well-established throughout the UK. Contractors are usually well equipped and bring everything they need, including a person to roll the fleeces in the approved manner. The flockmaster needs to provide catching pens close to the shearing area and to keep a flow of sheep into these pens. Shearing is a highly skilled operation and requires a period of training. Lantra (formerly Agricultural Training Board Landbase) courses in combination with considerable practice under the guidance of a fully trained instructor provide suitable training.

## GENERAL CARE AND HANDLING
### Stockmanship
The Codes of Recommendations (MAFF 1990) for sheep draw particular attention to the importance of stockmanship in sheep husbandry. It is a term that is difficult to define, but it encompasses stock sense and skill in stock tasks. Further emphasis to this and other aspects of husbandry will be given in the revised sheep code which is to be published in 1999. Stockpersons must undertake a wide range of tasks during a sheep year and complete them with minimal stress to the sheep, which involves a great deal of skill. Fortunately, there are several sources of tuition available to the new entrant, such as Lantra and local agricultural colleges. The Lantra course provides hands-on training in small groups, which is most effective for practical stock workers and farmers, and welfare matters form an integral part. Acquiring stock sense is somewhat more difficult. It involves recognizing, at an early stage, abnormal behaviour and changes in the appearance of animals. Stock sense can only come by working with animals and alongside good stockpersons. A sound understanding of the sheep's behavioural characteristics should improve the standard of stockmanship during gathering, driving and other routines around the sheep pens. This will help the stockperson to minimize stress during these occasions.

### Gathering and driving
There are many occasions during the year when the stockperson needs to handle every sheep in the flock. These occasions include: disease prevention, weighing and condition scoring, drafting and regrouping, and shearing.

The initial task of moving ewes from a field or gathering a flock off the hill will probably involve the use of a sheep dog. Sheep management in upland and hill areas, where the flock ranges freely, would be near impossible without a dog. The outstanding harmony between the shepherd and his/her dog is unparalleled in other forms of livestock husbandry, as it involves communication from a considerable distance. A sheep dog needs to be able to perform all the movements necessary to move a flock and segregate it into subgroups. Although the collie has been selectively bred as a working dog, it is proper training and ability to undertake all the manoeuvres that is more important. Nothing can be more stressful and damaging to a flock than a dog that is not under proper control.

Sheep usually follow one another and form a group when they are alarmed. During the early phase of the suckling period a lamb will follow its dam; gradually this tendency develops into a group pattern of behaviour, particularly during gathering and driving. Sheep also become conditioned to the various signals, such as the sight of a bag or sound of a bucket, indicating feeding time. The stockperson can use all these aspects of behaviour to reduce stress during the movement of sheep between fields, and from the field to handling and loading areas. A decoy (Judas sheep) can very effectively help to move groups around the pens and buildings and to ease the task of loading (see Williams 1993).

### Design and use of pens

The arrangement and design of pens and gates should always allow easy movement and good working conditions, and generate the least stress. Interaction between sheep while going through the pens is quite important. Unflustered older sheep, familiar with the various procedures, will have a calming influence on younger animals and those using the facility for the first time. Pens that funnel towards narrow passageways and races, and the absence of sharp corners and protrusions, which slows movement and may cause injury, greatly helps the flow.

A sheep handling unit (eg Figure 4.6) usually provides a large reception pen that leads through to smaller pens and races or narrow walkways, depending on the task. It also has pens for subgrouping or regrouping in the area leading back to the pastures or loading bay. Where the unit has links to some kind of housing, the stockperson can keep sheep under cover in preparation for activities such as shearing, dipping and weighing. Where the unit is away from the farm buildings, it is important to ensure that procedures do not pollute the water course. This may particularly be a problem when

using large quantities of chemicals, such as in pesticide sprays and dips.

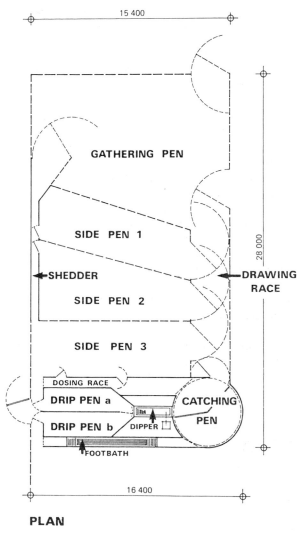

**PLAN**

**Figure 4.6**     **A plan view of a sheep handling unit.**

Table 4.10 lists husbandry tasks and the types of facilities a stockperson requires to carry them out. Several of these tasks do not involve restraining or manipulating the individual animal. Others involve contact and the use of restraint; for these it is best to confine the animals in a small pen or race.

## Handling and restraint

In a pen, the handler can restrain the animal by placing his/her open hand under the jaw and holding the head above the back line. It is then possible to manoeuvre the sheep backwards or sideways towards a corner of the pen. To restrain the animal further the stockperson can place their other hand high on the neck behind the ears. For close examination of the head and mouth of adult horned breeds, it is possible to restrain them by holding both horns while standing astride the back.

When segregating sheep in the corner of a field or at a gateway, away from the handling unit, the stockperson may use a crook. They can either use it to hook the animal immediately above the hock, or move forward quickly to grasp a hind leg above the hock. Only skilled personnel should use a crook. It is important not to restrain sheep by any part of the fleece, or a young animal by its horns because of their fragility.

## Casting

Full restraint often involves casting and this allows the examination and/or treatment of the feet and any part of the underline. Sheep handlers also use this position for the first stages of shearing, and for dagging and the removal of excessive wool around the udder.

There are several methods of casting; the size of the sheep usually decides this choice. Casting light sheep is usually by the lift method, and heavy sheep by unbalancing the hindquarters and thus getting them into a sitting position. In both methods it is important to get the sheep standing quietly. Standing close to the body, with the knees close together and behind the shoulder, the handler places their left hand under the animal's jaw. In the lift method, the handler uses their right hand to grasp a fold of skin and wool low down on the right flank region. This hand then lifts the hind leg off the ground, while simultaneously a nudge of the knees prevents the sheep bracing on the left hind. Finally, the handler gently turns the animal so that it sits on its rump with the withers resting between the knees. For heavier sheep, the handler

**Table 4.10    Restraint of sheep.**

| Husbandry Task | Restraint | Type of pen |
|---|---|---|
| **Preventive treatments** | | |
| Dosing or vaccinating | Minimal | Small pen or race. |
| Spraying | None | Spray race or shower pen. |
| Dipping | Minimal | Dip with slide-entry. |
| Dipping | Full | Entry to dip controlled manually. |
| Foot bathing | None | Race with permanent or portable footbath. |
| Foot trimming | Full | Small pen. |
| Docking and dagging | Full | Small pen. |
| Castration | Full | Small pen. |
| | | |
| **Weighing and condition scoring** | | |
| Weighing | None | Near outlet from race. |
| Condition scoring | Minimal | Race or small pen. |
| | | |
| **Drafting and regrouping** | | |
| Tattooing, tagging, ear notching | Minimal | Small pen or race, depending on age group. |
| Examination for soundness | Full | Race (teeth), or small pen (feet and udder). |
| Drafting and regrouping | None | Race with 2- or 3-way outlet for drafting. |
| Marking for marketing | None | Race between weighbridge and outlet gates. |
| Loading | None | Small pen leading to ramp of loading gate. |
| Shearing | Full | Small catching pens adjacent to shearing floor, with sheep released to 'counting' pens. |

places their right hand at the base of the tail; by bringing the head and tail towards each other, and, with their right forearm exerting some pressure on the rump, the hindquarters drop to a sitting position. The handler can then bring up the fore-end so that the withers rest between the knees (Figure 4.7). Placing the hand just above the hock and pulling on the off-side hind leg, the stockperson can then throw the hindquarters off balance. There are several devices that intend to aid this type of handling, such as cradles to hold sheep in the cast position. Mechanical casting and restraining devices usually hold sheep in the upside-down position (Figure 4.8). During late pregnancy it is important not to hold sheep in the casting position for long periods.

## Loading

The MLC have highlighted the need to reduce the injuries and bruising to fatstock that occurs during their transfer to markets and slaughterhouses. Such injuries consequently causes carcass damage. Attention to detail during the design and construction of handling pens can contribute a great deal to the reduction of stress and injury

Loading at the farm usually takes place directly from the collecting pen (Figure 4.9). To get the sheep to a particular level within the vehicle, the handlers use the lorry's own ramp and side hurdles. A decoy animal greatly helps to start the flow of animals during loading. When loading sheep the following recommendations are important:

1. Drive the sheep quietly.
2. Do not use dogs in confined spaces.
3. Never prod sheep with sticks.
4. Avoid dragging or lifting by the fleece.
5. Do not crush lambs as they go through shedding gates.

Where it is necessary to lift animals, the method will depend on size. Stockpersons can lift and carry lightweight lambs and fatstock by placing one forearm underneath the animal, in the region of the flanks, and the other arm firmly in front of the chest. They will need to lift older and heavier animals by firmly grasping the flank area, while holding the head and neck firmly against the body. The handler may use their knee as a lever during the lift, they then turn the body slightly so that they take the weight through the flank and neck.

## CASUALTY SLAUGHTER

Irrespective of the method of killing, it should be quick and humane. The handler should not make the animal frightened, excited or apprehensive. Shooting with a rifle or shotgun is usually the only satisfactory and humane on-farm method for killing sheep older than 2–3 weeks. For further details of the casualty sheep see the Sheep Veterinary Society (1994).

### Stunning

It is possible to kill lambs younger than 2–3 weeks of age (ie up to 5kg) by stunning and bleeding. The stockperson should hold the lamb by the back legs, and swing it through an arc to hit the back of the head with considerable force against a solid object. It is essential that the handler delivers the blow swiftly, firmly and with absolute determination. After stunning, and while the animal is unconscious, it is important to bleed the lamb by cutting a large blood vessel in the neck. The animal dies through blood loss.

### Shooting

A flockmaster may kill a sheep with a rifle or shotgun. For adult sheep it is possible to use a shotgun (12–bore, 16–bore or 20–bore), but with shot not smaller than no.6. Giving the animal some food will help to keep the animal still. The operator should hold the gun 15-30cm from the front of the head, shooting the sheep just above the eyes and pointing down the length of the neck. It is possible to use a rifle but this should be as a last resort. The operator should hold the gun in the same way with the muzzle held 3–5cm from the animal's head.

### Disposal of carcases

It is important not to dispose of foetal material and small carcases by putting them on the manure heap. It is essential that the method of disposal should be vermin-proof and that it should remove the possibility of passing infection to humans or other animals.

The most suitable method is to use a deep pit fitted with a fly-proof and smell-proof manhole cover, which is also child-proof. It should be approximately 3m deep and 1.5m in diameter. The sides of the pit will need lining, but the bottom can be left as earth to allow drainage (see MAFF 1980). It is important to choose a well-drained site, far away from a water

(a)

(c)

**Figure 4.7**   **A method of casting. (Three stages (a), (b) and (c) - described in text)**

(b)

**Figure 4.8**   **A mechanical restraining device being used during footcare.**

course and above the water table. It is also necessary to clearly mark the site and there should be some means of keeping heavy machinery away from it.

Stockpersons should deeply bury large carcasses or, if appropriate, sell them to a kennels or a similar outlet.

## THE WAY AHEAD

Sheep enterprises have not followed the pattern and scale of intensification seen in other food animals and therefore, in terms of public perception, sheep production remains traditional, 'green' and animal welfare friendly.

It is imperative that this image is preserved, for it is abundantly evident that the British public will increasingly demand the highest standard of care for food animals. It is to be hoped that this attitude will also apply throughout the European Union.

There is evidence to show that flock size and stocking rate have increased significantly during the last decade in lowland units. Housing for prolonged periods during the winter is an established practice in many areas and this is often due to the need to prevent damage to grassland as a result of high stocking densities. To safeguard the welfare of sheep in such units, highly efficient management and skilled stockmanship are called for. Consideration should be given to the suitability of the genotypes for such systems. Breeds show clear differences in behaviour and the ease with which they may be conditioned to the procedures and practices adopted in such units.

There is no doubt that increasing the number of stock per person can easily lead to deterioration in the

**Figure 4.9    Loading sheep.**

level of surveillance: hence the early recognition of changes of behaviour may be missed. This aspect cannot be over emphasized particularly in relation to early detection of diseases such as foot-rot and mastitis that cause severe pain and discomfort.

Increasing the number of stock per person can also lead to overdependence on contract workers. In such a situation the manager should give a great deal of attention to the assessment of competence in carrying out the procedure and also to the general organisation of the livestock prior to and after the procedure.

In the hills and uplands, sheep farming will remain as the most extensive system of livestock production practised in the UK. It is well established that subsidies make up an increasing proportion of gross output in these sheep units. Since it is acknowledged that sheep play a significant role in the economy of these areas the current level of support will probably continue but with closer monitoring of the breeds used and the stocking rate adopted. In some areas consideration may be given to the keeping of wether (castrated male) flocks as an alternative to breeding ewes, thus removing the class of stock which undergo considerable stress during late pregnancy and early lactation.

As a result of the structure and organisation of the UK sheep industry, vast numbers of sheep (in excess of 20 million) are transported annually. Journeys include farm to farm, farm to sales and markets, and markets to slaughterhouses. Due to the reduction in the numbers of slaughterhouses in the UK, fatstock now have to travel longer distances than in the past. However these are relatively short compared with the distances travelled by exported fatstock to slaughterhouses as far afield as Southern France, Italy and Spain.

There has been a great deal of public concern and violent demonstrations against the export trade and this has been directed mainly at the export of live fatstock. Reference to a significant reduction in the number of livestock exported, mainly fat lambs, has been made earlier. This is partly the result of public concern but also the publicity given to the unanimous view of the veterinary, welfare and agricultural organisations, that slaughter in the areas of production is preferable.

There is a great need for effective communication between farmers and the general public. In so far as the welfare of sheep is concerned, emphasis should be given to the consequences of disturbing and frightening groups of animals, the seriousness of the dog problem in some areas, and the need to respect fences and gates. These problems are not confined to the fringes of urban areas; there is no doubt that hills and moorlands will continue to attract more people and it is imperative that due regard is given to the countryside code and to the welfare of the animals which are such an important part of that environment.

## REFERENCES AND FURTHER READING

**Croston D and Pollott G** 1993 *Planned Sheep Production, 2nd edition.* Blackwell Science: Oxford, UK

**Eales A and Small J** 1995 *Practical Lambing and Lamb Care.* Blackwell Science: Oxford, UK

**Farm Animal Welfare Council** 1994 *Report on the Welfare of Sheep.* FAWC: Surbiton, UK

**Henderson D C** 1990 *The Veterinary Book for Sheep Farmers.* Farming Press Books: Ipswich, UK

**Jones J E T** 1991 *Mastitis.* In: Martin W B and Aitken I D (eds) *Diseases of Sheep,* pp 75-77. Blackwell Science: Oxford, UK

**Kilgour R and Dalton C** 1984 *Livestock Behaviour.* Granada: London, UK

**Lynch J J, Hinch G N and Adams D B** 1992 *The Behaviour of Sheep.* CAB International: Wallingford, UK

**Martin W B and Aitken I D** (eds) 1991 *Diseases of Sheep.* Blackwell Science: Oxford, UK

**Meat and Livestock Commission** 1988 *Feeding the Ewe.* MLC: Milton Keynes, UK

**Meat and Livestock Commission** 1986-98 *Sheep Yearbooks.* MLC: Milton Keynes, UK

**Ministry of Agriculture Fisheries and Food** 1980 *A Simple Disposal Pit for Foetal Material and Small Carcases.* L648. HMSO: London, UK

**Ministry of Agriculture Fisheries and Food** 1989 *Codes of Recommendations for the Welfare of Livestock: Sheep.* Leaflet 705. (Amended 1990) (Being revised). MAFF Publications: Alnwick, UK

**Ministry of Agriculture Fisheries and Food** 1994 *Condition Scoring of Sheep.* MAFF Publications: London, UK

**Ministry of Agriculture Fisheries and Food** 1995 *Agriculture in the UK: 1994*. MAFF Publications: London, UK

**Ministry of Agriculture Fisheries and Food** 1995 *Summary of the Law Relating to Farm Animal Welfare,* (reprinted 1998) *PB 2531*. MAFF Publications: London, UK

**Ministry of Agriculture Fisheries and Food** 1997 *Survey of Agriculture: 1 December 1996*. Government Statistics Service: York, UK

**National Sheep Association** 1994 *British Sheep*. NSA: Malvern, UK

**Sheep Veterinary Society** 1994 *The Casualty Sheep*. British Veterinary Association Animal Welfare Foundation: London, UK

**Webster A J F** 1994 *Heat and Cold Tolerance*. In: Proceedings of 3rd International Sheep Veterinary Society Conference 1993, pp 123-128. Sheep Veterinary Society: Edinburgh, UK

**Williams H Ll** 1993 *Facilities for Handling Intensively Managed Sheep*. In: Grandin T (ed) *Livestock Handling and Transport,* pp 159-178. CAB International: Wallingford, UK

**Williams H Ll** 1995 *Sheep Breeding and Infertility*. In: Meredith M J (ed) *Animal Breeding and Infertility,* pp 354-434. Blackwell Science: Oxford, UK

# 5   Goats

## A J F Russel and A Mowlem

### INTRODUCTION

Evidence of the domestication of goats has been found in Neolithic sites at Jericho dating from 7,000 BC. The domesticated goat *Capra hircus* is now found throughout the world and is only absent from the extreme polar regions. It almost certainly derives from the so-called Persian Wild Goat (bezoar) *Capra aegagrus*) found in Turkey, Iran and Western Afghanistan. The total world population of goats is estimated at around 490 million.

The question of the difference between goats and sheep is often raised and indeed some breeds of goat and sheep look very similar. The most telling difference, although not visible, is that goats have 60 chromosomes and sheep have 54. Anatomical differences are not so obvious. Goats generally hold their tails up whereas sheep let them hang down. Male goats have a very characteristic smell which is quite different from that of rams, and a ram, but not a goat, has a secretory gland on each of its hind feet.

The most obvious difference between goats and any other species is in their behaviour. Some earlier accounts still ring true today, such as that written by Professor Low in his *Domesticated Animals of the British Islands* published in 1845. He wrote "the goat although obeying the law to which all domesticated animals are subject, and presenting itself under a great variety of aspects, retains many of the characters and habits that distinguish it in the state of liberty. It is lively, ardent, robust, capable of enduring the most intense cold and seemingly little incommoded by the extremes of heat. It is wild, irregular and erratic in its movements. It is bold in its own defence, putting itself in an attitude of defiance when provoked by animals, however larger than itself.....When the Goat is kept apart from the flock

he becomes attached to his protectors, familiar and inquisitive, finding his way into every place and examining whatever is new to him. He is eminently sociable, attaching himself to other animals however different to himself."

It is almost a paradox that animals of such an independent spirit take so well to relatively intensive conditions in a farming situation. The problems of internal parasite control has resulted in most of the large dairy goat farms in the UK and in other European countries, such as France, keeping their goats housed rather than going out to graze (Figure 5.1). As far as one can tell, as long as goats are comfortable and adequately fed they are perfectly content in such conditions. Goats kept for fibre production do not have the same problem of milk withdrawal after treatment with anthelmintics and therefore more extensive systems, where there are fewer problems associated with parasite control, are used.

**Figure 5.1   Many large dairy goat herds are housed throughout the year.**

# THE GOAT INDUSTRY

Goats are reputed to have come to Britain with Neolithic man some 5000 years ago and there have been many importations from the time of the Vikings to the present day. At one time goats were more numerous in parts of Scotland than any other form of livestock but, with changing agricultural systems, numbers declined markedly in the nineteenth century.

There has been an interest in pedigree goat breeding in the UK for over 100 years. This has recently been strengthened by a renewed awareness of goats' milk as a health food, by an increased demand for speciality foods such as goat cheese, and by an increase in goat fibre production due to increasing demands for both the raw fibre and the end products manufactured from it.

## Structure of the industry

Until recently goats were not recognised as an agricultural species and were not included in the annual agricultural returns. The exact size of the UK goat population is not known, but has been variously estimated from about 75,000 (Locke 1985) to 100,000 (Mowlem 1990).

Until very recently goats were the province of the smallholder. However, in the dairy sector, increasing legislation, initiated by the European Commission, has made it more difficult for the small or hobby farmer to produce goat dairy products for sale. This has resulted in an increase in the percentage of the goat population in a relatively small number of large goat farms. The economics of fibre production are such that anyone wishing to derive an income from this type of goat is likely to keep comparatively large numbers.

Very few herds are maintained specifically for meat production although the demand for it is increasing. At present meat kids are a by-product of goats kept for milk or fibre production.

## Productivity

Goat breeding has not been subjected to the degree of commercial pressure which has influenced the development of other livestock enterprises. However, selection for milk production has probably been associated with some loss of hardiness, and modern European dairy breeds require a high level of expert management to maintain productivity. The highest recorded milk yield is from a European breed, the Saanen, in Australia, where one animal produced over 3500kg of milk in its second lactation. Most commercial dairy goat farmers aim to produce 1000kg of milk per goat per annum.

Goat hair was used in the UK in the seventeenth and eighteenth centuries to make a wide variety of products ranging from ropes to wigs for the judiciary. In recent years there has been a renewal of interest in the commercial production of goat fibre, from two quite different types of animal, but both aimed at producing high value products – cashmere and mohair.

During the last ten years there have been significant developments in the breeding of cashmere goats in the UK and substantial improvements in both the weight and quality of cashmere produced per goat have been achieved. Average levels of production are of the order of 100–120 grams per goat. The top 10 per cent produce more than 300g per head and the highest level of production recorded to date has been over 420g of first quality cashmere.

Angora goats typically produce around 4.5-6kg mohair per adult per year, harvested in two shearings at six-monthly intervals.

## Breeds

The principal dairy breeds in the UK stem from importations from other countries. There are two pure Swiss breeds, the Saanen and Toggenberg. The British Saanen, British Toggenberg and the British Alpine are the result of cross-breeding using these Swiss breeds. The British Saanen is the highest yielding breed, closely followed by the British Toggenberg. The Anglo-Nubian breed has been derived from crossing breeds from the Middle and Far East, which arrived in the UK on ships returning from the Empire during the latter half of the last century, with the indigenous English goat.

Of the fibre goats, the mohair-producing Angora (Figure 5.2) originates from Turkey although the largest numbers are now farmed in South Africa and the southern states of the USA. The Angora goats in the UK are mostly based on stock imported from Australasia in the early nineteen-eighties, but the production and fibre characteristics of the present population have been improved by the introduction of bloodlines originating in South Africa and Texas.

**Figure 5.2    The number of Angora goats in the UK has increased substantially in recent years.**

Recent breeding work in the UK has led to the development of the Scottish Cashmere goat. It was evolved from crossbreeding native Scottish feral goats with genetic material from Iceland, Tasmania, New Zealand and Siberia. Superior progeny from a wide range of crosses were selected and interbred to produce this new breed.

There are more than 200 identifiable breeds of goat in the world and although the majority are kept for meat there is only one improved meat breed , the Boer goat from South Africa. A few of these are to be found in the UK, originating from animals bred in Germany.

*Improvement schemes*

The small and only comparatively recent interest in commercial goat production in the UK means there has not been the same stimulus for improved productivity that has been the case with other sectors of livestock farming. However, small-scale pedigree breeders interested in the competitive aspects of showing goats, milk recording schemes and milking competitions, organised by the British Goat Society, have produced some goats that are among the best in the world. There are some British Saanens, for example, that yield over 3000 litres of milk per lactation.

Although artificial insemination is available, this has not yet been much used as a tool for genetic improvement. Other techniques, such as embryo transfer, have also been tried, particularly to multiply goats of superior characteristics, such as the best of the Angora and the cashmere types.

Significant improvements in the quantity and quality of cashmere production have been achieved in a group breeding scheme and genetic selection programme operated by Cashmere Breeders Ltd and managed by the Macaulay Land Use Research Institute. The main selection objectives in this programme are cashmere value, cashmere quality and resistance to helminths (gastro-intestinal parasites).

The small interest in goat meat hardly justifies any major efforts to improve meat production but nevertheless a little work has been carried out crossing Angoras with dairy breeds (Butler-Hogg and Mowlem 1985) and more recently there has been interest in using the introduced Boer breed to improve carcase conformation.

## ENVIRONMENTAL REQUIREMENTS

Goats kept outdoors must have access to shelter at all times. Where herds are kept for meat and fibre production under extensive systems of management on rough hill country or with access to wooded areas, they may find adequate natural shelter. In most cases, however, it will be necessary to provide roofed shelters designed to afford protection from rain. All goats dislike wet weather and their coats provide less protection against wetting than does a sheep's fleece. Goats also seek shade on very hot days and will make use of shelters in all weathers. Unless the shelters can be moved frequently, their floors and surrounding areas become fouled and muddy very quickly. Where possible, shelters should be equipped with slatted floors to ensure that the animals have a clean and dry lying area (Figure 5.3).

The main environmental requirements for goats kept indoors, whether in purpose-built accommodation or adapted buildings, are that the housing should be clean, dry, well lit and well ventilated. Heating is not necessary, but the building must be free from draughts. Good draught-free ventilation can be provided by use of spaced or Yorkshire boarding between the top of a solid wall 1.5-2.0m high and the eaves. In some locations, the space between such a wall and the eaves can be left completely open on the most sheltered side of the building, or covered with a light mesh or netting screen to keep out driving rain and snow. Any smell of ammonia in the atmosphere when the building is first entered in the morning is an indication of inadequate ventilation.

**Figure 5.3    Shelters should ideally be moveable and have slatted wooden floors.**

Goats must always have a dry floor to lie on. This can be bedded with clean straw, sawdust, wood-shavings, peat or some combination of these or other suitable materials. Alternatively, a raised solid or slatted wooden lying area is suitable and provides an object on which kids will play and obtain exercise.

Goats are sociable animals and, where individually penned, should be able to readily see other animals in adjoining pens. If no other goats are kept in the house they will accept animals of other species as companions. Where a goat has to be kept on its own it should be housed in a building from which it can see other animals or people.

## NUTRITION AND FEEDING SYSTEMS

Goats are ruminants (for details of ruminant digestion see Chapters 2 and 3). Under natural conditions the principal difference in the feeding behaviour of goats and sheep is that goats will readily browse on leaves and small branches of trees and shrubs, whereas sheep prefer to graze plant material close to the ground. Given a wide choice of plant species, goats will tend to concentrate on those plants which are commonly regarded as weeds, such as willowherb,

rushes, thistles and gorse, in preference to sown grass or clover. There is considerable scope for the complementary grazing of goats and sheep on certain types of pasture, and particularly on the indigenous plant communities found on unimproved hill land.

On sown lowground pastures, which are frequently perennial ryegrass-clover mixtures, goats require a higher sward surface height than sheep if they are to achieve the levels of herbage intake required for good performance. While sheep will perform well on pastures with a sward surface height of 4-6cm, goats require a sward height of around 8-9cm. Sward management for goats is thus more akin to that needed for cattle than sheep. Although optimum performance may not be achieved by grazing sheep and goats together in the same field at the same time, there is advantage to be gained, particularly in terms of sheep performance, by grazing the two species sequentially. Recent research has shown, for example, that weaned lambs gain substantially more weight on swards previously grazed by goats than on pastures grazed by sheep; this benefit arises from the goat's discrimination against clover which results in an increase in the clover content of swards grazed by goats.

Despite their liking for roughage and plant material not readily grazed by other animals, goats are fastidious eaters and will refuse food which is not clean and palatable. They will not eat mouldy hay or food which has fallen from a rack or trough onto the ground. Neither can they be expected to produce well from diets comprising only low-quality roughage.

As with other productive ruminants, the goat's diet must be balanced in terms of energy, protein, vitamins and minerals. It should contain at least 40 per cent roughage and must be fed in relation to the goat's nutritional requirements, which in turn are related to its size and weight and to its level of production, whether that be of milk, live-weight gain, foetal growth or fibre. In general, the requirements of the dairy goat are virtually the same for those of dairy cattle when compared weight for weight in terms of body size and milk yield. The requirements for fibre and meat goats are more comparable with those for sheep.

Most of the larger dairy units house their goats all the year round. The goats are fed *ad lib.* forage, of which an increasing proportion nowadays is maize silage. In some cases this is the sole forage source. Grass silage is also used, although most farmers are aware of the risk of the disease, listeriosis, in goats or sheep fed baled or bagged grass silage.

Goats are better able to select different plant materials than any other domesticated ruminant and this allows quite high levels of production from fodder of low overall nutritional value. This is only possible, however, if these materials, which are often by-products such as various types of straw, can be offered in large quantities. In this situation the goats are effectively grazing or browsing in a stationary position, but are more or less doing what they would do in nature, namely selecting the most palatable and nutritious parts of plants.

It is particularly important that clean fresh water is available to all goats at all times.

For detailed information about the composition of diets and the levels of feeding for different types of production the reader is referred to the sources given in the **General Reading** and **Further Information** sections at the end of this chapter.

## REPRODUCTION AND BREEDING

Goats are seasonally polyoestrous, with females coming into heat at regular intervals during the breeding season which, in the northern hemisphere, extends from about August to February. However, it is not safe to assume that females will not exhibit oestrus or conceive at other times of the year. Males are generally capable of breeding throughout the year, but libido and sperm production are likely to be poorer during the summer months.

Female kids become sexually mature and show oestrus at about 6 months of age, but are often not mated until approximately 18 months old, as yearlings (goatlings). However, if they are well grown and have attained about 75 per cent of their expected mature weight, they can be mated earlier. Male kids are generally believed to attain sexual maturity at about 6 months, but there is ample evidence of 3-month-old kids being fertile. It is therefore important, where uncastrated kids are being kept for subsequent breeding, that a close watch is kept for signs of sexual activity, and that they are weaned and segregated from female stock before reaching sexual maturity.

The length of the oestrous cycle is about 21 days and oestrus generally lasts 1-2 days. The onset of cyclical activity in the autumn can be stimulated by the smell of the male. Oestrus is readily detected by behavioural changes including bleating, tail wagging and by the swelling and reddening of the vulva. Where males are run with a herd of females, one male can be expected to serve 30-40 females. Where females are brought to the male for service, one male can serve 100-150 females or more in one season.

Artificial insemination can be used with goats (Mowlem 1983). Goat semen freezes as well as cattle semen although the method of processing it is more difficult. Insemination is carried out visually with the inseminator looking into the vagina, through a lit speculum, in order to locate the cervix. Experienced inseminators would expect to achieve a 60-70 per cent conception rate for the first insemination.

Gestation length varies from about 146-156 days, and averages some 150 days. Most females can be bred annually and there are reports of some animals kidding twice a year. This, however, is unusual. Dairy goats have an extended lactation, milking continuously for more than 18 months, and in some cases are mated only once every 2 years.

The number of kids produced varies according to breed, with feral stock in the UK seldom giving birth to more than one kid each year, while most dairy breeds commonly produce twins and some, such as the Anglo-Nubian, have litters up to four or more.

Pseudo-pregnancy, false pregnancy or "cloudburst", in which the female shows all the external and behavioural signs of being in kid but is, in fact, not pregnant, can occur. The type of ultrasonic scanning in which the uterus and its contents can be viewed on a TV-type screen, and which is now routinely used to determine foetal numbers in sheep, can be used to distinguish between true and false pregnancies as well as to count the number of kids carried.

### Management of breeding males

Breeding males are too often left to look after themselves for much of the year, without any particular thought being given to their nutritional requirements. Special attention must be given to their feeding, particularly before and during the breeding season. They must not, however, be allowed to become too fat.

Male goats are not difficult to manage and, if handled with care and respect, should not be aggressive. They are, however, extremely strong and thus require substantial housing. Block-built pens with metal gates are preferable. During the breeding season (August to February), they tend to go off their food and consequently lose condition. It is therefore important that they start the season in good condition. The have a repugnant smell, which is particularly noticeable in the breeding season.

## Management of breeding females

Prior to the breeding season, female goats should be in good condition but must not be too fat. Fat goats tend to have lower fertility and lower milk yields. Young non-lactating goats have a tendency to become fat and care should therefore be taken when feeding concentrates. The increasing possibilities for commercial goat farming have resulted in an interest in the ways of overcoming some of the natural limitations influencing reproduction. One major problem is the seasonality of breeding and the consequent seasonal lactation. Major retail outlets for milk require a year-round supply.

Out-of-season breeding is already practised by many commercial goat milk producers, and is attracting increasing interest. The two methods most commonly used are the use of intra-vaginal progesterone-impregnated pessaries or sponges (Table 5.1), and the use of artificially extended daylight during the winter. Ashbrook (1982) has recommended an extended light regime in which animals are exposed to 20h of light per day for 60 consecutive days during the period January to March. Oestrus will occur approximately 10 weeks after return to ambient lighting.

## Mating

Although the majority of goats are mated naturally, artificial insemination is becoming more common. In Britain most matings will be carefully selected, because many goats are pedigree animals. If a male is run with the females, a sheep brisket-raddle (crayon) harness can be used to record mating dates. Artificial insemination using either fresh or frozen semen presents opportunities for extending the use of

superior males. The female is restrained for insemination in a similar way to sheep, ie head down with the hind-quarters lifted up towards the inseminator (Figure 5.4).

## Embryo transfer

The technique of transferring fertile embryos from donor to recipient females is now being practised in goat farming. It is of particular value where a rapid increase in the numbers of a desirable breed or type is required. Up to 10 or more embryos may be removed from a donor female to be implanted, usually in pairs, into non-pedigree recipient females which act as surrogate mothers. This considerably increases the reproductive potential of one donor female. The technique is complex and must be carried out under close veterinary supervision. It requires the synchronization of the recipient and donor females' oestrous cycles using intra-vaginal progesterone impregnated pessaries and the injection of follicle stimulants, prostaglandins and pregnant mares serum gonadotrophin (see Table 5.1).

## Feeding of pregnant females

The requirements of non-lactating females can be met from good quality forage. Milking animals will require concentrate supplements and, unless grossly overfed, are unlikely to become too fat.

During late pregnancy it is important to feed sufficient nutrients to meet the needs of the growing foetuses and to ensure a good subsequent milk yield. It is generally recommended that the energy and protein intakes should be increased during the last month of gestation (Oldham and Mowlem 1981). Also, animals encouraged to consume a high level of forage during pregnancy will be conditioned to a high forage intake during lactation and this will benefit both milk yield and quality. An average dairy goat of around 65kg live-weight should be offered a good quality forage *ad lib.* and, 2 months before parturition, 250g of dairy concentrate/day rising to 450- g/day 1 month before parturition.

All goats require clean water which can usually be provided in a self-filling trough.

## Housing of pregnant females

Related goats or long-established groups can be housed together. In fact, being companionable

**Table 5.1** **Programme for synchronization of oestrus and ovulation for artificial insemination superovulation and embryo transfer technologies.**

| Day | Time | Donors | Recipients |
|---|---|---|---|
| 0 | | insert pessaries | insert pessaries |
| 10 | am | inject FS and prostaglandins | |
| | pm | inject FS and PMSG | |
| 11 | am | inject FS | |
| | pm | inject FS | |
| 12 | am | inject FS | |
| | noon | | remove pessaries |
| | pm | remove pessaries | |
| | | inject FS | |
| 13 | am | inject FS | |
| | noon | withdraw food until after AI | |
| | pm | inject FS | |
| | all day | record oestrus | record oestrus |
| 14 | pm | AI | |
| | all day | | record oestrus |
| 19 | noon | withdraw food until after ET | withdraw food until after ET |
| 20 | | embryo transfer | embryo transfer |

FS     = follicle stimulant
PMSG   = pregnant mare's serum gonatrophin
ET     = embryo transfer
Dose rates of proprietary drugs will vary with size of goat and season and must in all cases be prescribed by a veterinary surgeon.

animals, they usually benefit from this. Much time will be spent lying together, particularly during late gestation. Care should be taken if some animals have horns, as injuries can occur, particularly during feeding. It is important that hooves are well trimmed at this time. Projections which could injure heavily pregnant animals should be avoided.

**Parturition and suckling**
Ideally, when parturition is due, goats should be housed separately in a clean straw-bedded pen 2-2.5m². The first sign that parturition is imminent is the increased engorgement of the udder. Twenty four house before parturition the relaxation of the pelvis can be felt as hollows on either side of the tail. Sometimes a discharge will be seen and, providing this is clean and odourless, it is quite normal. The goat may become restless and may spend a lot of time pawing the ground, alternately lying down and standing up.

Most goats kid lying down and, once they begin to strain, regular checks should be made about every 15 min. If a goat has been observed to strain for more than 1 hour without anything apparently happening it can be assumed that there is a problem. Unless the stockman is experienced at coping with difficult births, expert help from a shepherd or veterinarian should be sought. A great deal of damage can be done by inexperienced people attempting to assist an animal at this time.

Once the kids are born, it is important to ensure that the mother has the opportunity to lick them dry immediately. This provides essential stimulation for the kids and helps establish the mother-offspring bond. It is also important that the kids feed within 2-3h of birth. Kids that are too weak to suck from their mothers should be fed colostrum/milk via a stomach tube, after which they will usually gain enough strength to suck on their own. Some goats have such pendulous udders that it is impossible for the kids to feed. In these cases the kids should be artificially reared, but it is most important that they are fed colostrum for the first 24h. It is useful to save some colostrum in a deep freeze for emergencies.

**Figure 5.4   Restraining a goat for artificial insemination through the cervix.**

### Rearing of kids

In commercial dairy units, most kids, whether intended for herd replacements or for meat, are taken from their mothers after receiving colostrum for 24h. They can then be reared on a proprietary milk replacer. Many milk replacers formulated for calves and lambs, and a few for goats, are available - they are all based on skimmed cow milk, whole cow milk or whey powder and, for large-scale rearing, the less expensive calf milk replacers are often used. There is evidence, however, that the use of calf milk replacers can cause toxicity in Angora kids (Humphries *et al* 1987) and it is therefore recommended that a lamb milk-replacer be used for kids of this breed.

Kids can be weaned from liquid food at 6-8 weeks (Mowlem 1984). A suitable feeding regime is

outlined in Table 5.2. For replacement female kids, it may be considered worthwhile to wean a little later, at 8-10 weeks, to ensure a good early growth rate. Some herd replacements may not go out until the following spring as kids grow less well when put out to grass early in life.

**Table 5.2   Feeding regime for artificial rearing of goat kids (figures in parenthesis show ages for a later weaning system).**

| Age | Milk feeds |
|---|---|
| Weeks 1-4 (1-6) | Supplied *ad lib*. |
| Weeks 5-7 (7) | ½ amount consumed end of week 4 (6) |
| Week 6 (8) | ½ amount consumed end of week 5 (7) |

For *ad lib.* feeding it is best to offer the milk in a "lamb bar" type feeder where several kids can suck from several teats in one milk container (see Figure 5.5). Feeders should be fitted with lids to prevent the milk becoming fouled by bedding, faeces, etc. Automatic calf-rearing machines can also be used. Milk replacers can be fed either warm or cold, but the latter usually gives better results.

**Figure 5.5   Kids feeding from a 'lamb-bar' type feeder.**

126

Alternatively, kids can be reared using shallow troughs in which a ration of milk is offered. One litre of milk replacer per kid per day in two feeds will ensure adequate growth and if the kids are weaned early, ie 6-8 weeks, any loss in early growth performance will be made up when eating solid food. Kids reared in this way should be equal in weight to kids reared by other methods by the time they reach 6 months of age. Feeding limited amounts of milk replacer like this considerably reduces the rearing costs, a particularly important aspect when rearing kids for meat.

A rearing concentrate containing 16-18 per cent crude protein, with good quality hay and clean water, should be offered from 2 to 3 weeks of age. Any dirty or fouled food should be regularly replaced. At weaning, kids should be eating 400-500 g/day of the concentrate feed if a growth check is to be avoided.

Castration of males not required for breeding, and disbudding should be carried out during the first week of life (see p 132 for further details).

Goats kept for fibre production may be run as a suckler herd, with the kids remaining with their mothers until weaned at around 12-14 weeks of age. Any male kids left entire should be weaned before attaining puberty and segregated from all female stock before they are 12 weeks old.

Kids enjoy the company of others, and should never be reared in single pens. Housing need not be elaborate – pens that are free from draughts and are provided with plenty of clean straw are all that is required. If the building is not particularly cold, kids penned in groups of 12 should not require supplementary heating.

Kids are very adept at finding their way out of pens, and therefore solid partitions should be used and pen sides should be about 1.2m high. Approximately 0.5m$^2$ of floor space should be allowed per kid up to 2 months of age, increasing to approximately 1.5m$^2$ at 6 months of age. Care should be taken to ensure that there are no projections in the pens, or gaps where legs or feet might become trapped. Hay nets are not recommended as kids may jump up on these, catching and breaking their legs.

Infection is likely to spread through groups of intensively reared kids, and it is important to observe a high standard of hygiene and husbandry. Kids should be checked constantly for signs of ill health,

and appropriate action taken as soon as disease or infection is suspected.

## MILK PRODUCTION

The choice of milking system depends largely on the scale of the enterprise. Those who keep just a few goats will probably milk by hand, while those with large herds will use a milking machine. The choice of milking machine system is as varied as for dairy cows, ranging from simple small-scale bucket units to large sophisticated parlours (Figure 5.6). Milking machines not only make the milking of large numbers easier, but, if well-managed, may be instrumental in producing better quality milk containing few spoiling microorganisms.

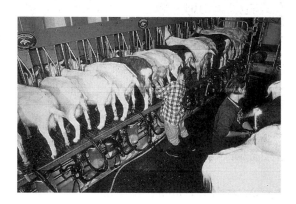

**Figure 5.6    Goats being milked in double-sided Fullwood Lo-jar parlour.**
*(Photo courtesy of Holt Studios International)*

### Feeding for milk production

While it is not necessary as part of the milking routine, it is convenient to feed concentrates to goats when they are being milked, and it also allows more control over individual feed intakes. Automatic, in-parlour feeding systems are available for dispensing concentrates, but on economic grounds are probably not justified in other than the largest herds. Forage is fed in the yards, pens or paddocks.

Whether a small bucket system or a full parlour is operated, it is likely that a yoking system will be used to restrain the goats during milking, and feed troughs or buckets can be attached to this.

Water is also an important requirement of milking goats and they should be provided with clean water at all times.

## Housing for milk production

It is most economical to house goats in groups. This saves space, labour and bedding.

For loose housing, approximately 2m² should be allowed per goat. If the animals are familiar with each other, there appears to be no upper limit to the number that can be housed together. New herds made up of unrelated or unfamiliar animals should be allowed more space until they have settled down, otherwise the resulting stress may affect their health and almost certainly their milk yields.

Deep-litter straw bedding is suitable, but other materials, such as wood chips, can be used, except at parturition. Such materials stick to newborn kids and may dissuade the mother from cleaning them.

When designing goat buildings or modifying existing ones it should be borne in mind that goats can be destructive. They will chew soft material such as wood, rubber or plastic, and may reach up on their hind legs to a height of more than 2m. Timber parts of buildings such as doors and door frames are best clad in sheet metal to avoid such damage. Feed troughs should be constructed of metal, and hay racks of heavy weld mesh or wooden slats. Routes in and out of buildings, and particularly the milking area, should be simple and straightforward for efficient operation.

## Milking systems

Anyone contemplating setting up a dairy goat enterprise is advised to assume that, even if they begin with only a few milking animals, the herd will increase in size. At its simplest, a milking area can comprise a clean space away from bedded pens, where the goat can be tethered and milked. Some goat-keepers construct a raised milking platform with a feed bucket at one end and hole at the other to take the milking bucket. For machine milking, a raised platform on which a number of goats will stand is required. As goats readily climb ramps or steps it is not necessary to have an operator's pit, although, if adapting a cow parlour, one may already be available. Steps leading up to the standings should have treads 54cm deep with 14cm risers and those leading down should have 80cm deep treads with 41cm risers. Care should be taken to ensure the steps and standings do not become slippery when wet.

Milking machines for goats are in principle similar to those for cows. However, the mechanical settings are different (see Table 5.3), and it is important that this is noted if adapting cow-milking equipment. Goats milk quickly, and with one operator in the parlour there is bound to be some over-milking. This does not seem to cause problems but too much would suggest that the parlour is not operating efficiently if the machine is running for periods when the flow of milk has ceased. In commercial units one operator may handle as many as 18 milking points at one time. Little work has been done with automatic cluster removers with goats, and it is possible that the use of these could increase the number of units that one person could manage. Without automatic cluster removal, one person can milk approximately 100 goats/h in an 18:36 side by side milking parlour (Figure 5.6).

**Table 5.3**  **Mechanical settings for milking machines for goats.**

| Vacuum | Pulsation rate | Pulsation ratio |
|---|---|---|
| 37 kPa (11"Hg) | 70-90 p.p.m. | 50:50 |

Like cows, goats become used to particular routines. Changes in these, or continual changes in milking personnel, cause stress which may in turn cause a fall in yield. Production often suffers when new goats are introduced into larger herds and, because of this, some farmers have adopted a policy of herd replacement only from home-bred animals.

To prevent contamination of the milk, or infection of the udder, it is advisable, immediately before milking, to wash the udder with an approved disinfectant solution and to dry it with a disposable paper wipe. It has been argued that this procedure is unnecessary with clean goats coming straight from pasture. Certainly washing with plain water, not drying after washing, or drying all udders with the same cloth are likely to cause more problems with contamination of the milk than if the udder is not washed at all. Another sensible precaution is for the operators to wear protective overalls and hats reserved solely for working in the milking area.

In some countries, the milking equipment is often in the same areas as the bedded pens. This is not recommended as the milk may be contaminated with dust, faeces and bedding. Indeed, it is not allowed in the UK under The Dairy Products (Hygiene) Regulations 1995. The milking parlour and milk-handling room should be sited some distance away from the goat pens and in a separate air space.

Immediately after a goat has finished being milked, its teats should be dipped into a suitable disinfectant solution as a precaution against mastitis. A 4 per cent solution of sodium hypochlorite or a proprietary iodophor-based teat dip will be suitable for this purpose. A useful check for early signs of mastitis is to milk a stream of milk from each teat at the onset of milking in to a strip cup which contains a black rubber disc which will show any clots in the milk. Clots are usually, but not always, the first signs of udder infection.

## Milk handling

It is important to transfer milk from each animal into a cooling system as quickly as possible. Ideally this should be a refrigerated tank, but in-churn coolers operated by running mains water are also quite effective. If the milk is not quickly cooled, the lipolytic enzymes present will break down the fats to produce off-flavoured fatty acids, giving the milk what is, to some consumers, an undesirable strong goaty taint. This lipolysis can also be brought about by excessive mechanical agitation when raw milk is pumped through elaborate systems of pipes.

All surfaces which come into contact with the milk, including milking machines, churns, buckets, tanks and pipe lines, must be cleaned thoroughly to prevent build up of bacteria and subsequent spoilage of the milk (Cousins and McKinnon 1979). Proprietary dairy solutions are available for this purpose and the manufacturers should be consulted for the most appropriate cleaning routine.

## FIBRE PRODUCTION

Most goats have two coats, a coarse, hairy outer or guard coat and a fine, soft undercoat which is cashmere (see Table 5.4). The quantities of cashmere carried by most modern breeds of dairy goats are negligible, but the feral goats found in many of the mountainous parts of the UK have significant quantities of very fine cashmere. Levels of production, averaging about 50 g/animal, are too low for commercial production but are being improved by selective breeding and by crossing with imported stock. The Angora, on the other hand, has only a single coat, which is mohair. This fibre is coarser than cashmere but is produced in greater quantities, averaging about 4.5 to 6 kg/goat/year. In contrast to wool, which is generally much coarser than either cashmere or mohair and often has a marked crimp to tight curl, goat fibres are straight or only lightly crimped. That, and their different structure, enables them to be used in the manufacture of fine, light-weight garments which are much in demand and command high prices.

**Table 5.4    Goats and their hair types.**

| Goats | Coats | | End product |
|---|---|---|---|
| | coarse outer guard hairs | fine undercoat (cashmere) | |
| *Dairy goat* | much | little | |
| *UK feral goat* | much | significant quantities | |
| *Cashmere goat* | much | much | Cashmere |
| *Angora goat* | – only single coat– (coarser than cashmere but greater quantities) | | Mohair |

## Mohair

Although the importation of Angora goats to the UK in 1981 was not the first attempt to establish the breed in this country, it appears at this stage to be the most successful, and numbers have expanded rapidly.

Mohair is marketed in two ways. Many producers add value to the raw mohair by having it processed into knitting yarn, or indeed in some cases into finished garments. Others sell their mohair through a marketing co-operative, British Mohair

Marketing, to the processing mills which are situated mainly in Bradford. As there is not a well developed goat meat market it is necessary to keep a substantial flock of Angora goats to be viable from raw mohair sales alone.

The principal difference between Angora goats and sheep is in their requirement for shelter, if not housing, because the Angora's fleece does not afford the same protection against wet conditions as the sheep's more greasy wool.

Mohair is a continuously growing fibre which is harvested by shearing every 6 months. In some herds the age structure will probably be similar to that of a sheep flock with the fibre being harvested from the adult stock and yearlings, and with perhaps one shearing being taken from those kids not required as herd replacements, before they are slaughtered for meat. In other situations the herd may contain a high proportion of adult castrates kept primarily for their fibre and sold for meat at an older age. (For rearing of kids, see p 127.)

### Cashmere

The feral goats in the UK, from which the Scottish Cashmere goat has recently been developed, are thought to originate from stock which were among the first animals to be domesticated in the country. Scottish Cashmere goats retain many of the attributes of their feral ancestors and are particularly well suited to hill farms, although herds are now established across the spectrum of land types in the UK and in a wide variety of climatic conditions in southern Europe.

The guard hairs in the fleece of cashmere goats afford greater protection against rain than does the coat of dairy breeds or Angoras, but where there is no natural shelter it is advisable to provide some means of protection from the worst of the weather. Artificial shelters, such as those illustrated in Figure 5.3, need not be elaborate, but, unless on a very freely-draining site, they should have slatted floors to prevent fouling and poaching.

Cashmere can be harvested either by shearing, which should be carried out in February before any of the valuable fibre is lost by natural shedding, or by combing, which can be done only after the fibres have been released from the hair follicles, generally sometime between March and May. The time of

shedding varies between individual animals and is influenced by factors such as level of feeding, body condition and date of kidding. It is common practice to comb each goat twice at an interval of two to three weeks. An indication that a goat is almost ready to comb is the appearance of tufts of loose cashmere around the eyes and ears. Combing is most readily carried out with the animal held by a head restraint, as illustrated in Figure 5.7.

Where cashmere is harvested by shearing it is imperative that the goats are housed in a draught-proof building and given a substantial increase in level of feeding (at least 1.75 times maintenance requirements) until there is a sufficient growth of new coat to give adequate protection against the prevailing weather conditions. Newly or recently shorn goats cannot be kept outdoors in the UK during February.

Combing takes longer than shearing and is therefore more expensive in terms of labour. This additional cost can, however, be more than offset by the fact that combed goats can safely be outwintered and thus require only supplementary feeding rather than the provision of their total nutrient requirements, as is the case with housed goats. Combing is not the daunting task it might appear to those who have not had experience. In practice a team of two people can comb 40 to 50 goats per day.

Cashmere will be harvested from adult breeding stock, and most probably also from the previous season's kids at about 10-12 months of age. Those young stock not required for breeding may be sold for slaughter at that point or kept for a further 6 months, being used, for example, to control or eliminate weeds from sown pasture (Russel *et al* 1983, Grant *et al* 1984) and to make further live-weight gains from relatively inexpensive grazing. It is also probable that, in some situations, herds of castrates may be maintained for purposes of grazing management and the production of cashmere and meat.

### MEAT PRODUCTION

Goat meat is always lean and, at a time when there is discrimination against animal fat, is being consumed in increasing amounts. Most of the goat meat presently produced comes from kids, usually males, which are surplus to dairy herd requirements. Although there is a limited specialist market for entire

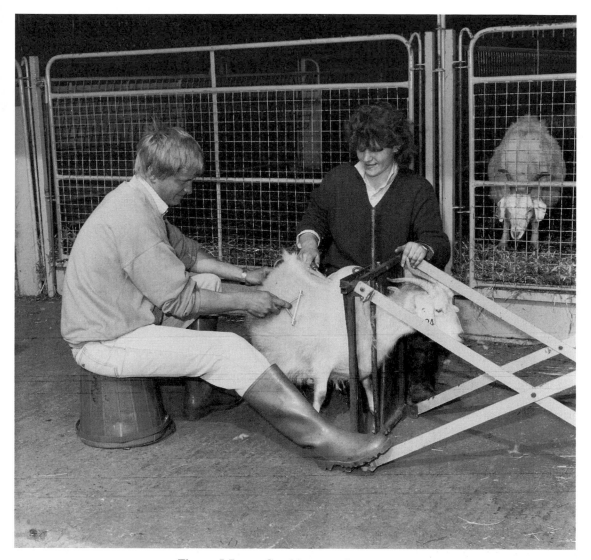

**Figure 5.7    Combing a cashmere goat.**

males, their meat is liable to carry the strong characteristic taint of such animals, which most people find unpleasant. Castrated males are more generally acceptable for meat production and are also more easily managed.

## GENERAL CARE AND HANDLING

Goats require similar environments, feeding and husbandry to other types of domesticated ruminant. Their independent nature and behaviour, however, is different from that of other livestock and it is only after much experience that the understanding necessary for successful management is acquired. Anyone contemplating starting a goat unit, however small, is advised to seek out another goat farmer who would allow them to gain some experience by working with their goats, even for a few days.

### Handling and restraint

Dairy goats are usually naturally inquisitive and will crowd round their handler. It is thus relatively easy to catch individual animals. Once caught they can be restrained by holding firmly but carefully around the neck just behind the head. It is advisable to attach a collar and lead if an animal is to be restrained for some time or is to be moved.

Unlike sheep, groups of dairy goats cannot be driven, and are most easily moved by being led. It is, however, advisable to have a second person walk behind the group to move along those stopping to investigate points of interest. Feral goats which have been taken from their natural habitat are more difficult to handle, but they too learn very quickly and can be worked like sheep by dogs. Feral kids reared as pets or in small groups are as tractable as dairy goats. Cashmere goats herds are managed like sheep flocks, but the sides of pens and handling yards require to be higher and more secure than those designed solely for sheep.

## Routine procedures

During the first week of life, males not required for breeding can be castrated using the rubber ring method. Care must be taken to ensure that both testes have descended into the scrotum before the rubber ring is applied.

If goats are to be dehorned it is desirable to disbud the kids within one week of birth. Goat kids' horns grow more vigorously than those of calves and more care is needed. The kids are anaesthetized, ideally with a general anaesthetic, and the horn bud is cauterized using an iron with a flat head approximately 2.2cm in diameter that has been heated to become red hot (Buttle *et al* 1986). In Britain, goats of any age may be disbudded or dehorned only by a veterinary surgeon.

Goats' feet, and particularly those of dairy breeds, require to be trimmed regularly to prevent lameness due to overgrowth, especially with heavily pregnant animals. The technique for hoof trimming is the same as for sheep, using foot-trimming shears and/or a sharp knife. Unlike sheep, dairy goats should not cast, but should be tethered and have the feet trimmed in a "horse shoeing" position.

Identification marking of goats, particularly white ones, can be a problem because ear-tags and other identification aids tend to be chewed. On commercial dairy farms numbered collars and/or leg bands are used for identification in the milking parlour when individual milk yields are recorded. Ear tattooing is

the only permanent method of identification and is a requirement of the British Goat Society for registered pedigree stock. Coloured animals can be freeze-branded.

Oral drenching is carried out in the same manner as for sheep and using the same type of drenching gun. Vaccinations are relatively straightforward because of the thin covering of hair and lack of subcutaneous fat, and are generally given in the loose skin around the neck.

The jugular vein is easy to locate in the goat, making the collection of blood samples also relatively simple.

Goats are able to reach almost all parts of their body with their mouths and it is therefore difficult to apply dressings in such a way that they will not be pulled off. The only completely satisfactory method is to restrain the goat so that it cannot turn round. This can be achieved by short tying the animal using a halter or by temporarily confining it in a narrow pen which restricts movement.

## Transport

If goats are to be moved long distances, a suitable vehicle must be used. Goats generally travel well, being able to maintain their balance easily, provided they do not slip on the floor. The ideal floor covering for a truck or van is a rubber mat covered with a thin layer of straw. Space should be adequate but not excessive, otherwise the animal may fall if the vehicles stops or turns suddenly. In the interests of safety, a partition must be fitted between the driving cab and the area of the truck occupied by the animals.

## HEALTH AND DISEASE

The maintenance of a healthy herd demands a high standard of stockmanship and the ability to recognize diseases and disorders in their early stages so that proper attention and treatment can be given promptly. All animals must be observed at least daily by an experienced person who can recognize signs of ill-health. It is important not only to recognize the obvious signs such as diarrhoea, lameness and laboured breathing, but also to know when animals are showing atypical behaviour or signs of discomfort. The stockman must be able to judge what conditions are amenable to treatment by non-veterinarians, and when it is essential to call on the services of the veterinary surgeon.

Adherence to the Ministry of Agriculture, Fisheries and Food's (MAFF 1989) *Codes of Recommendations for the Welfare of Livestock: Goats,* proper feeding, the provision of clean, dry and well ventilated accommodation, and strict adherence to a preventive medicine programme of vaccination and dosing, designed in consultation with the local veterinary practice to suit the particular herd requirements, will minimize disease problems and make a positive contribution to herd health.

## Bacterial and viral diseases

All adult and young stock should be vaccinated regularly to provide protection again clostridial diseases, such as enterotoxaemia, pulpy-kidney and tetanus. Where pneumonia is a persistent problem, tests should be carried out to determine whether the causal organism is one for which there is a vaccine.

Foot rot should be controlled through good management including regular foot examination, hoof trimming, timely treatment of affected animals and the frequent use of footbaths. Where these measures fail to control the problem, vaccination should be considered.

Louping ill is a disease caused by a virus transmitted by ticks. In dairy goats there is a potential danger to human health, because the virus can be secreted in the milk of apparently healthy animals. The disease can be prevented by vaccination, and advice on the occurrence of louping ill in a particular area should be sought from the local veterinary practice.

Mastitis can be caused by any one of a number of organisms which gain access to the udder through poor hygiene, bad milking techniques or injury. In severe mastitis the udder is destroyed and the animal may die. If diagnosed in the early stages, most cases respond to antibiotic treatment applied directly into the teat, although sometimes this may require to be supplemented by intramuscular injection.

Johne's disease generally occurs in adult goats and is characterized by a general ill-thrift and weight loss progressing to emaciation and death. It is possible to vaccinate kids and although this will not eliminate the causal organism it will considerably reduce the incidence of the clinical disease. Testing is unreliable as false negatives occur. However, if goats react positively to skin and serum tests and the organism is found in faeces this is conclusive evidence that the animal is a carrier, even if not obviously diseased. Whole herd testing is probably impractical for the larger dairy herds with 500-600 goats.

## Parasites

Internal and external parasites are a common cause of ill-health and poor performance in goats and, in some cases, can lead to death. There is some evidence to indicate that goats do not develop as good resistance to internal parasites as do cattle and sheep. Recent research with cashmere goats suggests that resistance to gastro-intestinal parasites is genetically determined and amenable to selection. Virtually all stock harbour various species of parasitic worms, ingested from the pasture as larvae, in the stomach and intestines, and all adults and kids should be dosed regularly with a suitable anthelmintic. The more intensive the stocking, the more frequent will be the requirement for dosing, and in some cases it may be necessary to dose kids every 4-6 weeks during the summer months.

It should be noted that many drugs (and vaccines) prescribed for use on sheep are not licensed for goats. An important difference in the health care of the two species is that the effective dose rates of certain anthelmintics are higher for goats than for sheep, and that the incorrect use of such anthelmintics will not only be ineffective in controlling worm burdens in goats, but can lead to the development of anthelmintic resistance.

The problem of endoparasites is the main reason for the system of management commonly practised by most dairy farms where the milking animals are kept housed or yarded. Without access to grazing the worm burden can be considerably reduced, thus removing the need for regular drenching with anthelmintics and the consequent withdrawal of milk for three days afterwards. Veterinary advice should be sought when treating any lactating goats for health problems as few products are licensed for goats, and advice about withholding milk after treatment will be required.

Other internal parasites which should be considered when designing a routine dosing programme include lungworm, coccidia and, particularly in wetter areas, liver fluke.

Of the external parasites, lice are particularly common on goats but are readily controlled by dipping, dusting with an appropriate powder or treatment with "pour-on" preparations. Mange is rarer, but much more difficult to control, and requires frequent treatment at short intervals over a period of months. Veterinary advice should be sought.

## Metabolic disorders

Metabolic disorders such as pregnancy toxaemia, hypocalcaemia (milk fever) and hypomagnesaemia (grass tetany) are the result of dietary inadequacies, although this does not necessarily imply that the level of feed *per se* has been unduly low. Pregnancy toxaemia is most likely to occur in the final weeks before kidding, and is most common in overfat females carrying multiple foetuses. To prevent excessive mobilization of body fat reserves, energy intakes should be increased as kidding approaches, using, for example, good quality concentrates. To prevent hypocalcaemia, a low calcium diet should be fed prior to kidding to stimulate mobilization of the calcium reserves in the skeleton; normal feeding with a calcium-supplemented diet should be resumed after kidding. Hypomagnesaemia is most like to occur in goats grazing lush pastures in the spring, and can be prevented by supplying additional magnesium, either in a magnesium-fortified concentrate or by addition to the drinking water. Hypocalcaemia and hypomagnaesaemia are much less common in grazing goats than in either sheep or cattle. Feeding some hay or straw can also be effective in the prevention of these conditions.

## On-farm slaughter

For those who do not wish to rear kids, particularly males, a suitable method of euthanasia will be required. Kids can be killed satisfactorily using an overdose of barbiturates or a humane killer. These may not be available to many goatkeepers, in which case it will be necessary to seek the help of a veterinary surgeon. It should be remembered that if barbiturates are used, the carcase must not be fed to dogs or other animals.

If an animal is to be slaughtered to provide meat for home consumption, or has to be destroyed because it is seriously ill or injured, this should be carried out by a veterinary surgeon or by a person from an abattoir which provides an emergency service.

## THE WAY AHEAD

There is a small but growing market in the UK for goat dairy products produced under as near as possible "natural" conditions, and the future for specialist, small-scale dairy-goat enterprises appears to be reasonably secure (Mowlem, 1990). As farmers seek alternatives to conventional agricultural commodities, there is likely to be an increase in larger-scale, more intensive enterprises concentrating solely on production and relying on others to undertake the processing and marketing of their product. There appears to be a sufficient demand for goat diary products for both types of enterprise to co-exist, so that in the future the range of sizes of dairy-goat enterprises is likely to increase.

The rapidly expanding interest in mohair and cashmere production is likely to result in a further substantial increase in goat numbers.

At present there is a potential annual supply of some 2000 tonnes of goat meat in the UK, but many of the animals which could contribute to this are destroyed at birth because goat farmers either cannot identify a market, supply a market on a regular basis, or obtain an economic price for this product. At the same time, there is a considerable unsatisfied demand for goat meat. There would thus appear to be a need for special goat-rearing enterprises which would bridge the gap between supply and demand. These would pay producers a fair price for surplus kids and would be of a sufficient size to be able to supply markets on a regular basis and thus command realistic and reasonable end-product prices.

All these factors combine to point to a significant increase in the goat population of the UK over at least the next decade. Such an expansion will bring in its train both problems and opportunities, and it is important that goat husbandry and welfare be incorporated into the curricula of college and universities, and included in the teaching not only of those who will be farming goats, but also of those who will be expected to provide special research, advisory and veterinary support.

## REFERENCES

**Ashbrook P F** 1982 Year-round breeding for uniform milk production. In: *Proceedings of 33rd International Conference on Goat Production* pp 153-154. Dairy Goat Journal: Scottsdale, Arizona

**Butler-Hogg B W and Mowlem A** 1985 Carcase quality in British Saanen goats. *Animal Production 40*: 752 (Abstract)

**Buttle H, Mowlem A and Mews A** 1986 Disbudding and dehorning goats. *In Practice 8:* 63-65

**Cousins C M and McKinnon C H** 1979 Cleaning and disinfection in milk production. In: Thield C C and Dodd F H (eds) *Machine Milking* Technical Bulletin No 1. National Institute for Research in Dairying: Reading, UK

**Grant S A, Bolton G R and Russel A J F** 1984 The utilization of sown and indigenous plant species by sheep and goats grazing hill pastures. *Grass and Forage Science 39:* 739-751

**Humphries W R, Morrice P C and Mitchell A N** 1987 Copper poisoning in Angora goats. *Veterinary Record 121:* 231

**Locke D G F** 1985 *Goat Husbandry Survey Report.* Islay and Jura Goat Society: Argyll, UK

**Mowlem A** 1983 Development of goat artificial insemination in the UK. *British Goat Society Yearbook* pp 4-6

**Mowlem A** 1984 Artificial rearing of kids. *Goat Veterinary Society Journal 5(2):* 25-30

**Mowlem A** 1990 *The UK Dairy Goat Industry - a Feasibility Study.* Food From Britain: London

**Oldham J D and Mowlem A** 1981 Feeding goats for milk production. *Goat Veterinary Society Journal 2(1):* 13-19

**Russel A J F, Maxwell T J, Bolton G R, Currie D C and White I R** 1983 A note on the possible use of goats in hill sheep grazing systems. *Animal Production 36:* 313-316

## GENERAL READING

### Books

**Devandra C and Burns M** 1983 *Goat Production in the Tropics.* Commonwealth Agricultural Bureaux: Farnham, UK

**Dunn P** 1994 *The Goatkeeper's Veterinary Book, 2nd edition.* Farming Press: Ipswich, UK

**Gall C (ed)** 1981 *Goat Production.* Academic Press: London, UK

**Guss S** 1977 *Management and Disease of Dairy Goats.* Dairy Goat Publishing Corporation: Arizona, USA

**Hetherington L** 1992 *All About Goats.* Farming Press: Ipswich, UK

**MacKenzie D** 1993 *Goat Husbandry, 5th edition.* Faber and Faber: London, UK

**Matthews J** 1991 *Outline of Clinical Diagnosis in the Goat.* Blackwell Scientific Publications: Oxford, UK

**Mowlem A** 1992 *Goat Farming.* Farming Press Books: Ipswich, UK

**Ministry of Agriculture, Fisheries and Food** 1989 *Codes of Recommendations for the Welfare of Livestock: Goats* (reprinted 1991) PN 0081. MAFF Publications: London, UK

**Salmon J** 1981 *Goat Keeper's Guide.* David and Charles: Newton Abbot, UK

**Thear K** 1985 *Commercial Goat Keeping.* Broad Leys: Saffron Walden, UK

**Wilkinson J M and Stark B A** 1987 *Commercial Goat Production.* BSP Professional Books: Oxford, UK

## FURTHER INFORMATION

**European Association for Animal Production** 1991 *Goat Nutrition,* Pudoe, Wageningen.

**Boden E (Ed)** 1991 *Sheep and Goat Practice.* Baillière Tindall: London, UK

**The Nutrition of Goats** 1998 AFRC Technical Committee on Responses to Nutrients. CAB International: Wallingford, UK

**European Fine Fibre Network Publications**
(EFFN Macaulay Land Use Research Institute, Aberdeen)

> **Genetic improvement of fine fibre producing animals** (Eds J P Laker & S C Bishop) EFFN Occasional Publication No 1 (1994)

> **Hormonal control of fibre growth and shedding** (Eds J P Laker & D Allain) EFFN Occasional Publication No 2 (1994)

> **The nutrition and grazing ecology of speciality fibre producing animals** (Eds J P Laker & A J F Russel) EFFN Occasional Publication No 3 (1995)

> **An analysis of the welfare of animals kept for the production of speciality textile fibres in Europe.** EFFN Report

# 6 Pigs
## I J Lean

## THE UK INDUSTRY

Modern pig production is a blend of business ability, good husbandry and the rapid adoption of new technology. There is considerable variation between producers and between their products; efficient pig farmers can maintain a margin even when the market is poor and uncertain.

### Trends within the industry

Since the Second World War, the number of individual pig producers has fallen, but the national herd has risen almost by a factor of four. Units have become specialized, using relatively small areas of land and relatively sophisticated intensive housing. In the late 1940s Government price incentives, followed in the 1950s by a move to more economic output that took advantage of technological change, were responsible for this increase in herd size and specialization.

In the 1950s, the government sponsored National Pig Industry Progeny Testing Board (later the Pig Industry Development Authority - PIDA) developed and improved the traditional structure of the industry. It encouraged research into all aspects of pig production, amassed data on all aspects of the industry through a series of recording schemes and tested stock superiority. In 1968 the Meat and Livestock Commission (MLC) took over these recording schemes; they also maintained some of the research interests of PIDA.

The quick reaction of the industry to fluctuations in market price has led to a cycle in pig production. This involves erratic swings between over-supply with low prices, and under-supply with high prices. Producers can keep on pigs they intended for slaughter as breeding animals with relative ease. This,

together with the rapid and prolific breeding cycle of the pig, makes it easy to enter the industry or to increase production. The pig cycle, together with the increasingly stringent requirements of the pig meat market, has speeded the decline of the small non-specialist pig producer. Short-term fluctuations greatly affect small herds. The Government attempted to stabilize prices in the 1960s and early 1970s by use of a flexible guarantee scheme, but they abandoned this following entry into the European Economic Community (EEC) in 1973.

Public pressure on pig producers to alter intensive systems has led to legislative change. *The Welfare of Livestock Regulations 1994* prohibited the use of dry sow stalls and tethers from 1 January 1999. This has put increasing pressure on producers and experimental workers to re-evaluate a range of loose housing systems. The rapidly accumulating experimental data on housing, behaviour and production should help in the evolution of modified and improved systems. The challenge is to find replacement systems that allow sows to maintain performance. Currently there is no single alternative way of keeping pigs to which producers are switching. Producers are showing considerable interest in systems that still allow them to control feed intake.

Outdoor pig production has expanded greatly in recent years. Currently, around 20-25 per cent of the UK breeding herd is outdoors and this proportion is rising. The industry is keeping increasing numbers of the national breeding herd outdoors for all or part of the breeding cycle. The reasons for this are twofold. Firstly, outdoor production for breeding stock offers a cheap alternative to expensive indoor accommodation, if land is available. Secondly, the public view outdoor systems as more animal welfare friendly. Other costs such as water, power and maintenance are also lower. While productivity from indoor systems is generally higher, outdoor production gives a better overall return on capital. With good stockmanship and similar management to that for weaning under intensive conditions, sow performance is comparable with indoor production.

The public are also concerned about the possible effects on human health of eating pig meat that contains large quantities of fats. They also worry about drug residues from products that intensive fattening systems require to maintain pig health.

### Structure of the industry

On a world-wide basis people consume more pig meat than any other type. The British national breeding herd of some 794,000 sows and boars produced about 1,098,000 tonnes of pigmeat in 1997 – the bulk of which was bought through supermarkets (in 1997 pork 68%, bacon 75%). Competitive retail prices on pig meat and concern about animal fats, as well as the BSE crisis in the beef industry have resulted in a switch by consumers away from beef and lamb. Pig meat consumption continues to rise, and various promotional campaigns encourage sales. To expand export markets, farmers need to produce a heavier carcase of 70-80kg with low backfat; this requires some modifications in production.

The traditional breeding structure of an animal industry is that of a pyramid. At the apex are a few breeders (nucleus breeders); they supply multipliers with male and female animals, and the commercial producers in the lower stratum of the pyramid with some male animals (Figure 6.1). Thus, over a period, genetic improvements percolate through the industry.

In the immediate post-war period, pedigree pig breeding had a long and successful record of improvement. In recent years, commercial breeding companies have dominated the upper strata of the breeding pyramid for pigs. They maintain and improve top quality pure-bred animals by selection under rigorous conditions, and by combining various genetic lines to take full advantage of hybrid vigour. These companies now supply a very large proportion of the gilts to commercial producers.

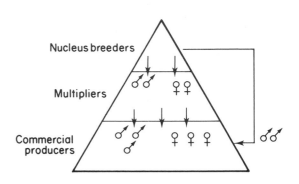

**Figure 6.1**    **Structure of the pig breeding industry.**

Producers sell pigs for slaughter between 55 and 125kg liveweight. Consumers demand lean meat; genetic, feeding and environmental aspects are therefore all important in producing, in the optimum time, a carcase with a low quantity of subcutaneous fat. Efficient producers have to manipulate these factors continuously to ensure that when they send animals to slaughter, the carcasses are destined for the most financially attractive markets. Table 6.1 shows the four main categories that slaughter pigs fall into in the UK.

Most of the fat in a pig is subcutaneous. This has led to the development of comprehensive payment schemes for carcases based on subcutaneous fat measurements, weight and, for bacon pigs, a minimum length (see MLC 1997 for details of a typical carcase classification system).

Outdoor herds are mostly in the southern part of England and East Anglia. However, units can thrive in certain areas of Nottinghamshire, Yorkshire, Northumberland and north east Scotland. This is because they can provide the necessary conditions in terms of soil type, topography and weather.

### BREEDING AND GENETICS

Breeders have genetically manipulated the pig, such that the present-day type reaches a more advanced weight for age than previously before becoming fat.

**Table 6.1**  **The four main types of slaughter pig in the UK.**

| Type of pig | Slaughter weight (kg) | Market |
|---|---|---|
| Pork pig | 55-62 | Small joints on the bone, sold as fresh meat |
| The cutter | 64-82 | Trimmed of fat and skin, prepared for multiple retail shops and supermarkets, with some of the carcase used for manufacturing purposes |
| Bacon pig | 90-100 | Traditionally cured |
| Heavy hog and culled adult stock (both sows and boars) | 100-125 | Often trimmed of skin and fat and most frequently used for manufacturing into sausages, pies and processed meats |

**Breeds**

There are three principal white breeds of pig in the UK: Large White, Landrace and Welsh. Breed differences are not obvious other than in relation to the position of the ears; Large White have pricked ears while in the other two breeds the ears point forward. The Large White is usually slightly more prolific and faster growing, with better quality meat. British Landrace stock are descended from Scandinavian importations, but breeders have also imported Landrace from other European countries. Many other British breeds, such as the British Saddleback, Tamworth, Gloucester Old Spot and Large Black, have a more limited following. These are sometimes kept by farmers with outdoor systems.

In recent years producers have imported exotic breeds. These are generally coloured animals which have a shorter length, greater eye muscles and larger hams. Such breeds include Pietrain, Hampshire, Duroc, Lacombe, Poland-China and Meishan. Breeders have crossed these with British stock and produced progeny with better reproductive potential, growth rate and carcass quality. They have subsequently marketed the useful genetic strains resulting from these programmes to commercial producers.

About 75 per cent of UK pigs are crossbred, and these animals usually perform better than the pure-bred lines. Traditionally, commercial producers crossed only two breeds of pig; for example, they crossed the Large White and Landrace to produce high-quality piglets showing the beneficial effects of hybrid vigour. Pig breeding companies have improved on this technique by selecting high performance lines from various pure breeds. They then combine these to produce first or second cross gilts for selling on to commercial producers. They also select meat-type male lines to mate with these females. The resources of a large breeding company allow considerable control of all aspects of selection, and guarantee the producer high performing animals always.

The characteristics that an outdoor pig requires are different from those of an animal living indoors under intensive conditions. To limit the risk of lameness, farmers should use animals with good foot and leg conformation. Selection of pigs with thicker hair and pigmented skin may help against sunburn. Sows must have strong mothering qualities; they must be agile so that they can control lying and avoid crushing their piglets. Sows should be quiet and easy to handle and train. There may also be some advantage in selecting for smaller litters of stronger more viable piglets. The outdoor sow also needs to have a thicker layer of backfat than the white hybrid sow to improve hardiness. No single breed can meet all these requirements; many producers use Saddleback Landrace crosses. The Hampshire and Duroc are other breeds that cross well with the Landrace to produce an outdoor sow. The choice of boar for use on the outdoor sow is nearly always the Large White, which gives better meat qualities to the slaughter generation.

*Genetic selection*

Commercial breeders and rearers purchase appropriate stock for their production methods and chosen market. Breeders select boars according to their growth rate, feed conversion and carcass quality of their siblings. Gilts are selected for breeding based on their growth rate, feed conversion ratio and reproductive history of their dams and sires, and on their own anatomical and production characteristics. Selecting gilts in this way is questionable since the

139

qualities a breeding animal requires are different from those required in a meat animal. Breeding sows will have a longer lifespan and will undergo the rigours of multiple pregnancies.

The usual method of selecting pigs of high genetic merit is by feeding them a known diet under controlled environmental conditions. After a fixed period, the breeder selects pigs on liveweight gain and low back fat thickness. This latter is measured with an ultrasonic scanner. The top performing band of these animals go forward into a continuing breeding programme.

In the breeding herd, the number of piglets a sow successfully rears per year is a major indicator of efficiency of production. In the feeding herd, efficiency of lean meat production is most important. Many biological and management factors limit both parameters.

Pig breeders have largely achieved genetic improvements by natural mating of superior stock. Deficiencies in artificial insemination (AI) technology, and problems of handling and storing semen on the farm, has meant that AI has played only a minor role in breed improvement in this country. Improved techniques are beginning to alter this situation and increasing numbers of pig producers are taking advantage of access to superior genetic material available through pig breeding companies.

## NATURAL HISTORY AND BEHAVIOUR

The wild pig is a forest animal, which will seek shelter from adverse weather conditions. Pigs are inquisitive animals. They have poor eyesight, but highly developed senses of smell, taste and touch. Domestication, genetic change and husbandry practice have produced marked alterations in the behaviour of the present-day pig.

### Social structure

The social unit of wild and feral pigs comprise a sow and her litter. The European wild pig farrows once annually and has a litter of between one and four piglets. At weaning, which usually occurs at about 12 weeks, two or three sows and litters will join together. Sows often form hierarchical matriarchic groups, which hold until the onset of the breeding season. Old boars generally live alone, but join the groups at this time, driving off the young boars and

serving the sows. Young boars will live together during this period. Breeding females and sub-adult animals usually regroup after breeding.

Domesticated, weaned, growing and finishing pigs live in a comparatively barren environment, which provides them with few choices. Farmers usually put these animals together in groups of the same age, sex and weight. In this situation they have to learn to recognize feed, water and dunging sites and to fit into a social group. Newly-mixed pigs will fight to establish a linear hierarchy. Fighting may be intense for up to 48 hours after mixing and will continue intermittently. The social rank that develops from fighting usually reflects the weight of individual pigs within it; the heaviest is most dominant, and the lightest is least dominant. Management techniques, such as providing adequate feed and living space, can ease this settling period after mixing. Aberrant behaviours, such as tail-biting and ear-biting, which may occur, suggest that management is at fault in being unable to satisfy the behavioural needs of the pigs.

### Reproduction

To achieve optimum reproductive performance, it is important to create conditions that stimulate sexual behaviour. Young boars should not grow up in isolation; they benefit from the stimulus of being in the company of others. Gilts of breeding age and weight react positively to the sudden introduction of a boar into their close environment by showing oestral behaviour within seven days.

Oestral sows and gilts are restless and may have a variable appetite and a swollen, red vulva. They sniff the genitalia of pen mates and allows pen mates to mount them, or will ride others. Females grunt characteristically, and will actively seek the boar. If he is receptive, she will adopt a rigid, immobile stance.

During mating, the boar approaches the sow, giving a characteristic series of grunts. He noses the vulva vigorously, champing his jaws and frothing at the mouth. Most boars mount once, and there is a rapid thrusting phase that ceases at ejaculation. The process lasts about 4-6 minutes.

As parturition approaches, the sow becomes restless and will attempt nest-making even in the absence of bedding. Under very intensive conditions, the sow has little opportunity to exhibit behaviour that occurs in more natural situations. During labour the sow may lose some fluid from her vulva and she may twitch her tail during abdominal contractions.

*Lactation*

Suckling piglets respond to the sow's milk let-down call by nuzzling the part of the udder they identify as their own. A change in note is associated with milk release; sucking piglets prefer the anterior teats of the sow, which offer more protection and possibly have a greater milk supply. Piglets establish teat order quickly after birth.

*Feeding*

The wild pig feeds on a wide range of plants, small animals and insects. It consumes some herbage and can despoil crops in its soil-turning search for feedstuffs. The tactile, highly sensitive snout is essential for rooting behaviour. Pigs in intensive housing maintain this pattern of behaviour, and will persist in nosing the surface of concrete pens. Outdoor pigs can spend many hours a day exploring their environment, food-seeking and eating. In contrast, indoor pigs spend a relatively short time eating and have little opportunity for exploratory behaviour. In intensive situations, pigs compete for food. Feeding methods and the total availability of food and water affects appetite and intake. Pigs that have only limited access to feed will often spend the time drinking. Adequate feeding space and watering points are therefore essential.

## Stockmanship

An awareness of the normal behaviour pattern of animals is essential for good stockmanship and management. The Preface to the Ministry of Agriculture, Fisheries and Food's (MAFF) *Codes of Recommendations for the Welfare of Livestock: Pigs* (MAFF 1983) supports the view that pigs should be able to show as many behavioural patterns as are acceptable to efficient management. Commercial conditions do not necessarily always allow pigs to show a full behavioural repertoire.

One important objective of good stockmanship is to maintain a good relationship with the animals; the presence of the stockperson should not stress the pig. It is vital that the handler does not behave adversely to the animals, such as by hitting, kicking or goading them.

Management can exploit the inquisitive and learning behaviour of pigs in several ways. Perhaps the most common is to take advantage of the pig's demarcation of dunging and urinating places. It is common practice to drive pigs into a pen via the dunging area and hold them there for a short time. This is because after moving, young animals often dung and/or urinate; pigs rapidly associate the area they first fouled with these functions and will keep the rest of the pen clean. Overcrowding or inadequate control of temperature leads to a breakdown in this system of demarcation, and soiling becomes general. Once this has occurred, pigs seldom return to their previous clean state.

## Reproductive biology

Boars generally become sexually active between 5-8 months of age. Both fertility and the total quantity of semen a boar produces builds up gradually over a long period. Ejaculate volumes peak at about 30 months of age, while densest sperm count might occur later. A three year old male is therefore more fertile than an animal of 8-12 months.

The age that gilts reach puberty can vary between 135 and over 250 days. Breed, genotype, nutrition and environment affect puberty in the gilt. Contact with the male by sight, sound or smell is important in triggering first oestrus. Ovulation rates in sows are very variable. The number of ova shed per cycle increases over the first three parities (pregnancy/parturition) peaks for a further two or three parities, and then falls away. Energy intake may affect ovulation rate, and short-term increases in the ration post-weaning can increase the number of ova shed.

Fertilization approaches 100 per cent of ova shed, but embryonic mortality is high. Sows may lose as many as 40 per cent. If the number of fertilized ova falls below five, then the pregnancy usually terminates naturally, and the sow will return to oestrus some 25-30 days later. After 35 days of pregnancy sows usually retain dead foetuses until parturition, when these foetuses are born mummified.

Farrowing normally lasts for 2-3 hours; gilts take longer than older sows. Piglets are born at about 15 minute intervals. The greater the litter size, the more

**Table 6.2**     **Summary of reproductive data.**

| Sow | Mean | Range |
|---|---|---|
| Age at puberty | 200 days | 135-150 days |
| Age when first bred | 221 days | 156-271 days |
| Oestrous cycle length | 21 days | 18-23 days |
| Duration of oestrus | 53 hours | 12-72 hours |
| Time of ovulation | 38-42 hours after the onset of oestrus | |
| Rate of ovulation | 16 ova | 10-25 ova |
| Gestation period | 114 days | 110-120 days |
| First normal ovulation after parturition | 5 days post weaning | 4-12 days |
| Cycle type | Polyoestrous all year | |
| Fecundity | Gilts average 9 live-born | |
| | Sows average 11 live-born | |
| | Approximately 2-2.5 litters/sow/year | |
| Breeding life | 3-5 years | |

| Boar | Mean | Range |
|---|---|---|
| Age at puberty | 200 days | 150-250 days |
| Age when first bred | c245 days | |
| Seasonality | Long days and high temperatures reduce sperm production and viability but are not considered of practical significance in temperate conditions | |
| Breeding life | Usually culled between 3-4 years of age | |
| Sow/boar ratio | 25 sows per boar is a practical ratio for natural mating | |

variable is individual piglet weight at birth. This is because of intra-uterine competition. The number of teats a sow has (usually 14), which secrete varying quantities of milk, affects piglet weight at weaning.

After weaning, the sow will need no special treatment for drying off, apart from a reduction in rations; she will, however, still need water. The removal of the suckling stimulus is adequate to begin gland reversion. The sow reabsorbs secreted milk and the process is complete within about three days. Reconditioning of the reproductive system is complete within about three weeks after parturition and the sow is then ready to return to oestrus.

**Management of boars**

Farmers usually introduce breeding males into a herd at 6-8 months of age. Commercial holdings usually buy-in boars from breeders or breeding companies who specialize in their production and evaluation. It is important to quarantine bought-in boars for 21-28 days. Animals should be physically sound and have a good libido. At this age, the boar is not yet fully sexually mature, so it is important not to over-use him. A young boar should serve only 4-6 sows per 21-day breeding period. Stockpersons will need to supervise inexperienced animals during mating because bullying activities by females may reduce their libido. One adult boar per 25 sows is usually adequate.

Farmers frequently group-house young boars in straw yards. This is of benefit to both behaviour and management. These animals are more tractable to handle and less likely to have a lower libido than individually housed males. Producers tend to pen older boars singly, but this is a reflection of the small numbers that most units retain. As these males will be of varying ages, and strangers, individual housing avoids the risk of antagonistic behaviour.

### Management of breeding females

Farmers bring gilts into the breeding herd for some time before service to expose them to the diseases present there. It is important to quarantine bought-in gilts for 21-28 days, while simultaneously exposing them to the manure of the indigenous stock.

*Oestrus*

Traditionally, producers did not breed at the first oestrus. However, commercial pressure is changing this, and many producers now attempt to mate gilts as early as possible. Farmers should manage gilts so as to suddenly expose them to the boar, as this is important in triggering first oestrus. They should cull those that fail to come into heat by the age of about eight months, produce poor litters, or show unsocial behaviour to their young, such as savaging or cannibalism. Farmers should also cull those that are unable to cope with housing conditions or suffer arthritic limb or foot problems.

A receptive female in oestrus is usually easy to detect, since she will stand rigid if a handler applies pressure on her back. Sows respond to touch, sound and smell. Lacking a male, synthetic boar odour aerosols, containing a steroid similar to that which the boar produces, are available to help trigger this behaviour.

Producers can use gonadotrophins to stimulate ovulation, or progestagens for synchronization. However, it is relatively unusual to synchronize oestrus in batches of sows and gilts with hormones or their derivatives.

*Mating*

The optimum time for insemination is 6-12 hours before ovulation. There are different ways of managing pigs at mating. In 'hand mating', the boar serves the sow twice, with the stockperson supervising the procedure. The initial service is on the first day of standing oestrus, and the second 12-24 hours later. This technique spans the ovulation period and increases the chances of successful insemination. With 'pen or batch mating', the boar remains with a group of females; this allows him to detect behavioural oestrus as it occurs. However, this has the disadvantage that the male may continue to mount a particular female as long as she is receptive. Consequently, other females in the group may miss his attention, or be inseminated with an inadequate quantity of semen. Alternatively, farmers may choose to artificially inseminate females.

*Artificial insemination (AI)*

Commercial producers in the UK have used AI for sows since about 1955. In 1965, PIDA introduced a semen delivery service, which the MLC has maintained and improved. AI is available to any pig producer through this system and through other schemes run by commercial organizations.

In the UK, survey evidence suggests that up to 20% of producers use AI for about 50% of all services and its use is gradually increasing, though it is still far below that in some other European countries. The reason for the comparatively low use of AI in the UK might lie in the wide distribution of pig holdings; this makes the delivery of semen difficult. The rise in the use of AI reflects improvements in the handling and storage of semen, and interest in 'boar plus AI' matings and "pooled" (multi-boar) AI matings.

The major reason for using AI is to improve performance and carcass quality of slaughter pigs. Only the top 5 per cent of tested boars enter stud. Their offspring have better lean meat production and superior feed conversion characteristics. The fact that producers do not need to keep breeding boars if they use AI, allows them to either save money or slightly increase their number of breeding sows. Also, by not having to buy in boars it reduces the risk of bringing in disease. Disadvantages of AI lie in the inconvenience of having to order semen and carry out inseminations at the optimum time. Also, the rather high rates for return to oestrus make AI less attractive. Conception rates of 60-70 per cent are good. This may reflect lack of skill in oestrus detection in the absence of a boar, and in the insemination technique itself, although this is straightforward and easy to learn. The greatest single factor responsible for poor conception rates is probably poor timing of insemination. Not all animals show the standing reflex; the presence of teaser boars, aerosols or synthetic pheromones and even the taped "love song" of a boar will all help in detecting service time. Stockpersons should look for oestrus signs at least twice a day. Two inseminations about 12 hours apart should improve the chances of conception (Figure 6.2).

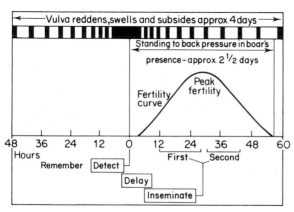

**Figure 6.2    Optimum time for insemination of the sow.**

*Source: MLC recommendations for the AI of sows.*

Farmers use a clean, spirally twisted catheter that they lock into the sow's cervix. They allow the semen to run into the sow under gravity. The process takes 5-15 minutes and it is important not to hurry. For a full account of artificial insemination in pigs see Hughes and Varley (1980).

*Pregnancy*

It is important to avoid all stresses in the first 35 days of gestation, since this is the period when loss is most likely to occur.

Producers can diagnose pregnancy using a variety of techniques. These include ultra-sound measurements through the uterus wall, oestrone sulphate estimations in blood or urine, enzyme-linked immunosorbent assay (ELISA) using blood, or rectal palpation. Close supervision of service, together with the relative precision of the sow's oestrous cycle (18-23 days) and accurate service records, should mean that the efficient stockperson only needs these complex techniques in a few cases of infertility. Of the diagnostic tests available, ultra-sound is currently the most common. A range of equipment exists, from the relatively inexpensive device that identifies the pulsations of the uterine artery, through to sophisticated scanners that produce an image of the uterus. Efficiency of interpretation is poor before 30 days of gestation.

*Parturition and lactation*

Farmers should treat sows for worms, lice and mange during gestation. They will then be clean to move into clean farrowing accommodation some four days before parturition. Boosting the sow's immune status, such as with *Escherichia coli (E. coli)* vaccines, should also be complete by this time. As parturition becomes imminent, milk is usually present in many or all the teats and the udder is firm to the touch.

Gilts and sows seldom require expert help; intervention is only necessary if the interval between piglets is more than an hour. Only an experienced person should assist with farrowing. The longer the duration of birth, the greater the risk that the restless sow will tread on early-born piglets and that those born later will die of anoxia. Provision of a warm, well-lit creep area goes some way to reducing this risk for early-born animals, but, on average, up to one piglet per litter may be born dead. Farrowing generally requires little supervision, apart from regular checking to ensure that the airways of the piglets are clear and that no animals are chilled.

If the unit keeps precise records of service, the pig person can synchronize farrowing to some degree. This is achieved by injecting prostaglandin on the 111th day of gestation. Farrowing occurs within 30 hours with no deleterious effects on the piglets. By synchronizing farrowing it is possible to largely avoid the need for supervision during the night or at the weekend; this has consequential savings in labour costs. As piglets are born at almost the same time, the stockperson is able, where necessary, to balance up litter numbers by fostering animals. Batch farrowing allows depopulation of indoor intensive houses, and the opportunity to clean and disinfect the farrowing area properly.

Under intensive housing conditions, it is usual to farrow sows in crates. These can reduce piglet mortality, particularly in the 48 hours after the birth. Crates also enable the pig person to keep a close watch, it makes routine handling of piglets simpler, and it allows higher stocking density within the space available.

## Management of outdoor production
The success of the outdoor pig enterprise depends upon the sensible use of modern equipment and techniques. Such equipment includes electric fencing for paddocks and large plastic ear-tags for identification. Also, the keeping of sows in relatively small groups (15-20), and skilled enthusiastic stockpersons are an advantage. The three most important factors to ensure success and maintain the welfare of pigs in outdoor systems are: suitable site location, appropriate genetic make up of the animals, and good stockmanship.

Integration of an outdoor sow herd into an otherwise arable rotation can have beneficial effects on soil and crop performance, saving on artificial fertilizers. Such an enterprise is also a useful outlet for surplus straw. Pig production provides a regular cashflow to the arable farmer. Normally, sows will occupy a pasture of, for example, a short-term grass ley for 12 months, although it is possible to move them out onto stubble. Producers move the animals on to reduce the risk of enteric disease and internal and external parasites, and before they have damaged the soil texture. After one year's grazing by sows, it is common to grow winter wheat. Stocking rates under these systems will vary with soil type and with land values, but densities of 10-15 sows/ha are common.

Farmers usually wean piglets at 3-4 weeks and transfer them to standard weaner accommodation, where they rear them under more controlled conditions. In contrast to most indoor units, which carry out all aspects of the production process, outdoor producers frequently sell on piglets for fattening at other units.

### Stockmanship
Outdoor pig keeping demands a high quality and quantity of stockmanship. Skills for outdoor production are different from those for indoor/intensive systems. Dedication and enthusiasm are important; such persons need to be committed, if only to cope with the weather, and be eager to record data and maintain stock and equipment under difficult conditions. Large herd sizes are common, with one thousand and more sows under one management scheme.

Training is important to ease stress of handling procedures, so the stockperson should start when gilts or young boars are first brought in. Training makes animals less timid and so it enables the handler to check them more closely.

### Predator attack
The presence of foxes in the area does not necessarily mean an increase in piglet mortality. If losses to foxes and/or birds are a problem, it may be useful to place some sort of barrier, such as plastic flaps, over the entrances to the arks. This will dissuade the predator from entering and stealing the young. Pest control programmes may also be necessary.

## Management of piglets
### Suckling period
Piglets are born naked, with little hair, no fat, little liver glycogen reserves and poor disease immunity. It is essential that they are born into a warm, dry, clean environment, so that they escape chilling. This will also help them to rapidly find their mother's supply of colostrum. Colostrum is a source of immunoglobulins that the new-born piglet absorbs directly prior to gut wall closure. In addition, it is a rich energy source and gut stimulant. Piglets will find their way to a teat directly after being born. It is imperative that colostrum is available; the stockperson should therefore check the sow's teats to ensure this. Most piglets that die in the first three or four days of life have never sucked. Stockpersons therefore should supervise piglets to ensure that they are obtaining colostrum/milk. The time they invest in transferring weakly piglets to other sows, or feeding colostrum through a stomach tube, is probably well spent. A weakly piglet will soon cope for itself and thrive if it survives through this initial period.

Piglets need further attention during the first 24 hours of life. The umbilicus may need trimming and require disinfecting. It is also common practice to ear-notch the piglets with a simple coding system for purposes of identification.

Sow's milk is low in iron, so in situations where the piglets have limited access to soil, they need supplementary iron to prevent anaemia. Farmers usually give this as an injection, but alternatively they may increase the iron content of the sow's feed. This will give a consequential rise in the iron content of her milk, so reducing the need for injections. Piglets may run the risk of becoming anaemic at any time before they are eating solid food.

*Weaning*

In terms of intensive production allowing a sow to wean her piglets naturally is lengthy and inefficient. The lactating sow seldom comes on heat regularly.

The earlier a producer weans, the greater the need for sophistication in housing, management, feed supply and disease control. Although it is possible to wean piglets at one day of age, such animals obviously need a very clean, warm environment. Also, they need a manufactured food that is similar to sow's milk. Later-weaned, naturally-reared pigs are cheaper to produce, because the sow satisfies their nutritional requirements, and they need a less controlled environment. The costs of very early weaning systems and the skills - both mechanical and husbandry - necessary to run them successfully, have reduced their popularity with producers.

*The Welfare of Livestock Regulations 1994* ban routine weaning earlier than three weeks. Presently, most farmers wean their piglets between 24 and 32 days of age. This is a compromise between economic reality and welfare recommendations. Piglets have acquired a fair degree of immunity by this age, have begun to adjust to solid food, are easy to look after and will be outgrowing the crate and creep area in the farrowing house.

While sows can rear more pigs per year the earlier that weaning occurs, to sustain production requires high management costs. Farmers also lose production time in waiting for sows that they wean earlier than 21 days to show oestrus. The *Codes of Recommendations* (MAFF, 1983) state that producers should not wean piglets earlier than three weeks other than because of lactation failure or disease in the dam, or their need for special treatment.

**Production**

Table 6.3 gives some examples of production data for pigs.

**Environment and housing**

Outdoor production units have moveable field shelters where the animals can take refuge when they choose; most husbandry systems, however, confine pigs inside buildings. Indoor pigs are removed from the effects of adverse weather conditions but they may have a social stress imposed on them because there is insufficient space for them to avoid their companions. The producer needs to be as generous as possible in allocating space, but cost is a constraining factor.

Pig producers use a wide range of buildings, and they vary in the control the stockperson has over environmental parameters. The design should meet both the pigs' and operator's needs. *The Welfare of Livestock Regulations 1994* stipulate requirements for buildings and lighting. Table 6.4 lists the environmental requirements of the various classes of stock in indoor housing.

Until the late 1990s most producers will have housed dry sows in stalls, or tethered them. In intensive situations, the use of individual feeding arrangements or housing in stalls, can reduce social and competitive problems. Such systems may eliminate problems such as bullying, but this severe restriction greatly limits the animal's freedom of movement. Pigs often develop abnormal behaviours, which may in turn adversely affect production.

*Temperature*

Small piglets with a high ratio of surface area to body mass run the risk of chilling. Large pigs can tolerate cold but have difficulty in keeping cool in high temperatures.

Pigs may lose body heat by convection and evaporation, the rate of loss relates to air temperature and air movement. The degree of floor insulation affects the rate of heat loss to the floor. The stocking density in a pen affects air temperature and the movement of heat between pigs. Heat-stressed growing pigs eat less and, in breeding stock, it may reduce foetal growth or sperm quality. Cold stress is obvious, because animals huddle more, erect their hair and, over a period, hairiness will increase. Rapid loss of body condition occurs if housing, and possibly feed supply, remain poor, and pigs use feed for heat production instead of growth.

**Table 6.3**    **Examples of production figures derived from the MLC recorded herds in 1996.**

*(Information adapted from MLC Pig Year Book 1997)*

| | Bottom Third | Top Third |
|---|---|---|
| **Breeding stock** | | |
| Number of herds recorded | 85 | 85 |
| Average number of sows and gilts | 217 | 338 |
| Sow replacement (%) | 43.2 | 45.9 |
| Sow mortality (%) | 6.2 | 3.7 |
| Sow performance | | |
|   Litters per sow per year | 2.11 | 2.34 |
|   Non productive days | 55 | 26 |
|   Pigs born alive per litter | 10.45 | 11.12 |
|   Mortality of pigs born alive (%) | 12.9 | 10.1 |
|   Pigs reared per sow per year | 19.2 | 23.4 |
|   average weaning age (days) | 26 | 24 |
|   Sow feed per sow per year (t) | 1.306 | 1.343 |
| **Rearing stock** (weaning to 20-40 kg range) | | |
| Number of herds recorded | 51 | 51 |
| Average number of pigs | 664 | 1284 |
| Pig performance | | |
|   Wt of pigs at start (kg) | 6.8 | 6.6 |
|   Wt of pigs produced (kg) | 30.3 | 36.6 |
|   Mortality (%) | 3.2 | 2.0 |
|   Feed conversion ratio | 1.92 | 1.75 |
|   Daily gain (g) | 405 | 488 |
| **Feeding herd** | | |
| Number of herds recorded | 60 | 60 |
| Average number of pigs | 1113 | 1108 |
| Pig performance | | |
|   Weight of pigs at start (kg) | 21.8 | 16.4 |
|   Weight of pigs produced (kg) | 86.3 | 85.9 |
|   Mortality (%) | 4.5 | 2.39 |
|   Feed conversion ratio | 2.86 | 2.39 |
|   Daily gain (g) | 558 | 631 |

*Ventilation*

Ventilation affects temperature, relative humidity and air composition. It is important to avoid over ventilation as this can lead to a loss in production. Minimum recommended rates probably show the thresholds of heat and odour acceptable to the stockperson, rather than being of any particular advantage for the pig. The problems of noxious gases, dust and air-borne microorganisms are often the result of inadequate ventilation. Ammonia and hydrogen sulphide are obvious, even before they affect the pig. As carbon dioxide is odourless it is not obvious; under normal conditions it does not occur in dangerous concentrations. As feeds (meal or pelleted) can be dusty and pigs (and humans) are susceptible to pulmonary infection, it is important to take all practical steps to limit dust.

There are several ways to ventilate a building. Air movement and temperature may naturally displace air, or fan systems may force it in and out. Air movement, however, can take heat with it and move dust and microorganisms around. Most ventilation systems use temperature-control mechanisms, which switch fans on and off.

*Relative humidity*

Pigs can tolerate a wide range of humidity. Exceptionally high humidity levels may reduce the incidence of pneumonia. In very hot, humid environments condensation is so extreme that it precipitates out microorganisms. Low humidity drys out the pig's skin and increases the dustiness of the atmosphere. This may aggravate respiratory conditions for both pigs and stockpersons.

*Light*

The pig can adapt to a wide range of lighting regimes with little effect on growth rate. Long periods of darkness sometimes seem to adversely affect breeding females. Both increasing day length and total darkness reduce semen production and quality in boars. Ideally, breeding animals should have 12-14 hours of light a day.

The *Codes of Recommendations* (MAFF 1983) advise that, 'Pigs should not be kept permanently in darkness. Throughout the hours of daylight the level of indoor lighting, natural or artificial, should be such that it is possible to see all pigs clearly. Adequate lighting for satisfactory inspection should be available at any time.'

*Site location of outdoor units*

A successful outdoor system for breeding stock requires light to medium, free-draining soils, such as chalk, gravel or sand. Heavy clay or silt soils are totally unsuitable because excessive poaching causes major problems for pigs, stockpersons and machines. Equally, potential producers should avoid sites with very sharp flints underlying thin soil, as the top soil will blow away and the stones can cause lameness. It is important not to choose a site at high altitude or exposed to high winds, as it will be too cold. Also, the ground should be level or gently sloping; steep slopes present problems of access and soil erosion. In farrowing paddocks a high gradient causes straw bedding to gravitate to the lower end of the hut. Areas of low annual rainfall (ie less than 76cm) and/or those where the winters are mild are preferable. Good infrastructure of roads and tracks and an adequate supply of water and electricity, are essential.

*Floor type for indoor accommodation*
Floors should provide maximum comfort for the pig to lie, stand and walk. They should not be so rough as to cause skin abrasions, but must provide sufficient grip for the pigs to stand easily. Slatted floors allow heat loss from the pig; the house insulation and ventilation are important in controlling this. Slotted metal, slatted or mesh floors must cater for the foot shape and size of the animal using it. Metalwork should have no rough edges or protrusions to injure the animal. Comfort and cleanliness of the pig needs balancing against installation costs. Pigs between four and eight weeks of age can use slats with gaps of up to 15mm. Over this age the gap can be up to 20mm.

Tethers confine the front two-thirds of the body on concrete and the hind end on metal slats. In some tether systems the whole of the area is slatted. Farrowing crates may also have slatted floors, in which case there may be a risk of piglets trapping their feet. Slats for sows should be at least 100mm wide, with 25mm gaps. They should have rounded edges and there should be no sharp projections.

The *Codes of Recommendations* (MAFF 1983) stress the importance of good floor design and maintenance. It is in the interest of all producers to provide flooring and accommodation of a high quality.

*Stocking density and group size*
*The Welfare of Livestock Regulation 1994* specifies unobstructed pen floor areas for pigs of different weights. These range from $0.15m^2$/pig, where the average weight is 10kg or less, up to $1m^2$/pig, where the average weight is more than 110kg. Farmers should not stock growing piglets too densely; it may be helpful to provide barriers to enable subordinates to escape from aggressors. High stocking density may lead to an increase in interactions between individuals and, at its extreme, aberrant behaviour such as tail-biting and cannibalism; both result in a subsequent production loss.

The amount of land an outdoor herd needs will depend on soil type, local climate, the shape and location of the land available, and the age of weaning piglets. Generally, each hectare can support 14-19 sows. If stocking densities are too great it will cause

**Table 6.4    Guide to environmental requirements.**

| Pig type | Liveweight (kg) | Temperature °C | Space (m²/pig) | Trough length (cm) |
|---|---|---|---|---|
| Suckling pigs | up to 5 | 23-30 | 0.07 | |
| Weaners (flat deck) | 5-18 | 21-28 | 0.07-0.25 | 10 |
| Weaners (6 week +) | 10-20 | 17-24 | 0.25-0.37 | 15 |
| Early finishers | 20-45 | 13-21 | 0.37-0.6 | 20 |
| Late finishers | 45-90 | 13-17 | 0.6-1.33 | 23-28 |
| Sows in groups | 150+ | 10-15 | † | 30* |
| Dry sows (boars) in stalls | 150+ | 18-22 | 1.3 | 30* |
| Farrowing and lactating sows | | 15-21 | 1.6 | 30* |
| Average ventilation rate | | 1m³/h/kg of pig housed. Humidity 70-80% RH. | | |

†    No firm recommendations
*    30cm but related to housing and method of feeding.

excessive damage to grass cover. If the group size is too large there will be a greater competition for resources; this can lead to aggressive interactions. It is also more difficult to give sows individual attention in these conditions. For guidance on the space that should be given to pigs see MAFF 1997.

*Range management*

Rotation of land is vital to prevent excessive poaching of ground, and disease and parasite build-up; rotation cycles should therefore be less than the life cycles of parasites. Producers should attempt to maintain grass cover as it helps keep the animals clean and dry. Grass also helps prevent light soils blowing away and grass roots aid drainage over winter. Farmers should alter stocking rates and rotation intervals, and not resort to nose-ringing the pigs to prevent them rooting. Pigs must have shade and wallowing areas available to prevent heat stress and sunburn in hot weather.

*Bedding*

The UK *Codes of Recommendations* (MAFF 1983) encourage the use of bedding for all classes of breeding stock. They encourage the use of bedding in sow housing, because it can alleviate problems of lameness and skin abrasions. The sow can also manipulate and play with the staw during this enforced period of restriction. Adequate bedding is necessary in simple outdoor kennel-type housing to keep the animals warm and dry. To limit neonatal losses, it is important not to use too much straw in huts; an excess will hamper piglets as they try to move away as their dam goes to lie down. The decision to use bedding will depend largely on its availability, the degree of intensification, and the ease of disposal of either soiled bedding or slurry. Many units use solid manure systems for some classes of stock (eg sows, boars, retained gilts) and slurry systems for others (eg growing pigs). Absence of bedding may lead to a greater incidence of traumatic injuries to limbs, particularly if the finish of the slats and floors is too smooth. Straw or litter systems ease the burden of stockmanship.

*Effluent disposal*

The disposal of faeces, urine and washing water from any intensive unit can be a problem. Slurry-based systems ease the physical work in a unit, but increase the problems of storage and disposal. Straw or litter based systems increase the work of cleaning out, but disposal is easier because it is possible to stack and rot manure cheaply. Whatever the system, the final dispersal on to land is always a problem. Slurry has an obnoxious smell and requires a considerable acreage of land to avoid pollution. Untreated pig slurry has a biological oxygen demand some 20 times greater than that of crude town sewage. Farmers must take great care, therefore, to ensure that it does not enter rivers and other watercourses.

Producers have yet to find an environmentally acceptable method of effluent disposal which is also cost-efficient. This problem could become the greatest single limiting factor in the establishment of a pig unit in the future.

As with other intensive systems, the pollution potential is great, particularly in those units where effluent is disposed of as slurry. When spreading slurry continuously on a relatively small area of land, there is a considerable risk of contaminating the water table. There are also odour problems at spreading time. Planning regulations for establishing a new pig unit varies from one local authority to another. However, there is general legislation (see MAFF 1998a) that controls the pollution of water tables and water courses. The maximum application of total nitrogen from organic manure must not exceed 250kg/ha/year. Calculations suggest that a sow and litter weaned at four weeks need a manure area of 0.07ha, and a growing pig needs 0.04ha. In effect, statutory control of water pollution will limit pig numbers, particularly for those farmers keeping pigs outdoors in nitrogen-sensitive areas.

*Housing breeding females*

Sow housing ranges from simple outdoor kennel-type houses, through to yards and insulated force-ventilated buildings with individual pens.

Group-housing with individual feeding facilities is common between weaning and service. Such housing seems to encourage return to oestrus in case of an unsuccessful insemination, and helps in its behavioural detection.

*Stalls and tethers*

Until the late 1990s (but illegal from 1 January 1999)

producers housed most sows after serving, either individually in stalls, or in tethers. This system allows precise allocation of food, permits close observation and eliminates bullying. Individual accommodation also reduces space requirements and capital cost. Having higher stocking densities in an insulated building can provide sufficient heat to maintain the temperature in the house without further heating. Fan ventilation systems solely regulate temperature in these buildings. However, while stalls and tethers are efficient in production terms for housing pregnant sows, they deprive the sow of freedom of movement and deny her normal exercise. The *Codes of Recommendations* (MAFF 1983) emphasizes this point; many sows in these systems consequently show abnormal behaviour such as bar chewing, and it is probably right that these systems should now be banned.

*Group housing systems*

Producers converting from stalls and tethers should opt for a replacement system that is in keeping with the long term requirements of the business. They should consider making other changes at the same time, such as increasing sale weight or the overall herd size, which will strengthen the position of the business. Although the initial cost of a system is important, the overall expense of equipment and running costs should also be taken into account.

It is extremely advantageous to retain individual feeding for at least part of the pregnancy even though such feeding systems are more expensive to install. When existing dry sow accommodation is converted to loose housing systems it is likely to hold only 65 per cent of the original breeding stock. Farmers will therefore have to provide additional housing to accommodate the remaining 35 per cent or reduce their herd size. One solution is to invest in a purpose-built house with individual feeders for about a third of the herd and convert the existing system to a group-fed yard. This will make it possible to feed each sow the correct amount it requires during the initial part of pregnancy. Floor feeding, through dump or spin feeders, is usually the cheapest option but cannot cater for individual needs. Direct wastage of feed in the straw increases the running cost of this system. Individual feeding systems such as yards or kennels with scraped passages are more expensive in

terms of both space allowance and labour. Between the two are options such as electronic sow feeders (ESF) and trickle feeders.

*Outdoor housing*

Most outdoor huts for both dry and farrowing sows are simple half-round, timber-framed huts with solid backs and roofed to the ground with curved, corrugated iron. For pregnant sows, gilts or boars, producers often use large huts that hold up to five or six individuals. For lactating sows, they usually provide smaller huts, incorporating a low boarded extension to retain the piglets. It is important to insulate huts fully as partial insulation is of no benefit. Some producers fit their huts with farrowing rails in an attempt to prevent the sow squashing her piglets. Similarly, they restrict the amount of straw they use to limit neonatal losses. In the wild, sows choose to give birth in a nest away from the group. It may be advantageous to fence off periparturient outdoor sows for a few days. Such action may also encourage them to spend more time suckling their piglets.

*Farrowing crates*

Most intensive systems use farrowing crates. They restrain the sow so that she can stand or lie down, but not turn around. Producers group the crates in easy-to-clean, temperature controlled rooms. Room temperature needs to be 15-21°C; farmers also need to provide a further closed creep area at about 21-27°C, for the piglets, the higher temperature being preferable in the first few days of life. Farrowing crates come in a variety of designs, but are usually about 0.7m wide and 2.4m long. The creep area may be at the front or side of the crate. There has been little alteration to the basic crate over the years, although some designs have attempted to further reduce neonatal mortality. The extreme restless behaviour of the sow has stimulated the re-evaluation of farrowing accommodation, but has not led so far to any major changes on common commercial practice.

*Housing growing and fattening pigs*

Production systems use various forms of accommodation for growing pigs, but all aim to reduce housing costs while maintaining an adequate

**Table 6.5**     **Some types of housing for growing pigs.**

| Housing type | Controlled environment | Age of pig (weeks) | Group size | Litter/ straw | Comments |
|---|---|---|---|---|---|
| **Weaners** Multiplier pens | Yes | 2-5 | Whole litters | No | A very small number of producers use this system for early weaning. May have a future role for intensive care of small piglets. |
| Flat decks* | Yes | 3-11 | 10-20 | No | Popular system. allows good observation of piglets, easily cleaned; slatted floors used. |
| Weaner kennels* | Heating possible | 3-11 | 10-20 | No | Easily constructed, boxed units within an insulated building, with open dunging area discrete from lying area. Observation difficult. |
| Weaner verandahs* | Heating possible | 5-11 | 10-20 | No | A kennelled undercover area with an outside slatted dunging area. Success depends on stocking levels and the frequent movement of the kennel roof to control temperature. Low capital cost, but high level of stockmanship needed for continued good production. |
| Weaner pools* | No | 5-11 | Very variable | Yes | Any building can be used. Straw builds up for length of time pigs are penned. Fair level of observation possible. Disadvantages are increased work-load (mucking out) and high building costs as much space needed. |
| Open-front monopitch | No | 5-10 | Whole litters | Maybe | Form of housing popular for farrowing and follow on. Now used for weaners on some farms. Simple layout and good health control an advantage but high labour input and high level of stockmanship needed to maintain an optimum environment. |
| **Finisher pigs** Enclosed and insulated | Yes | 10 weeks to slaughter | 11-20 | No | Usually with central feed passage. Pens slatted or part slatted along outside walls. Good observation possible. |
| Kennels with covered dunging areas | No | 10 weeks to slaughter | 11-20 | Yes | Popular in arable areas. High capital costs because of space requirement. Easy to manage. |
| Open-fronted monopitch | No | 10 weeks to slaughter | 11-20 | May be | Low capital cost. Stockperson exposed to environment. High level of skill needed to maintain an optimum environment. |
| Kennel verandah | No | 10 weeks to slaughter | 11-20 | No | Kennel for pig with outside dunging over slats. Low capital cost but efficiency of use dependent on skill of stockperson. |

* Illustrated in Figure 6.3

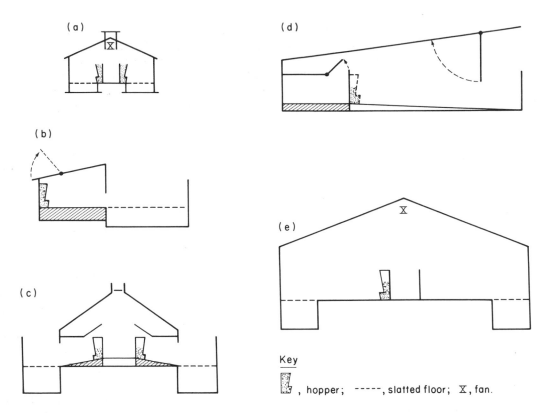

**Figure 6.3**      Diagrammatic representations of housing types (not to scale). (a) Flat deck rearing pens. Insulated house. Piglets on slatted or perforated floors. Extractor fan in roof. Feed hoppers. (b) Weaner kennel. Insulated pen with hinged roof and insulated floor. Feed hopper. (c) Kennel pens with verandah within insulated building. Kennels with insulated floor. Feed hopper. Open ridge to roof. (d) Weaner pool. May be straw based. Insulated roof, floor and walls. Includes kennel with hinged, removable lid that pigs enter through a pop-hole; usually hopper or trough fed. Of sufficient height to allow cleaning out from front with tractor loader. (e) Finisher accommodation. Insulated building. Complete environment control with fan and heater system. Pens with slatted or perforated floor dunging area. May be hopper, trough or floor fed.

environment for efficient food use. Table 6.5 and Figure 6.3 give details of some housing types that are available.

Pen shape depends considerably on the type of pig, feeding method and dunging provision. Rectangular pens (whose length is 2-3 times the width) with solid insulated floors encourage animals to be clean and dry. This is probably because they have room to define and maintain a dunging area at the one end.

If farmers feed from a trough, then the length of this controls the number of animals in a pen at a certain weight (Table 6.4).

**Welfare considerations**

Sainsbury (1984) summarizes the essential factors that influence the welfare of pigs (Table 6.6).

**Table 6.6**      **Essential factors influencing the good welfare of pigs.**

1. Consider limiting the total size of breeding units to 250-500 sows.
2. The housing of all young or growing pigs should allow depopulation, cleaning and disinfection after each batch.

3. Limit number of pigs in one common air space - about 16 sows and piglets, 200 weaners and 500 growing pigs, but smaller numbers are always advantageous.
4. Maintain temperature, humidity and ventilation rates at correct ranges for age.
5. Reduce the opportunity of a build-up of disease or cross-infections by separating pigs from their respiratory or intestinal excretions as quickly and completely as possible.
6. Ensure stocking rates are as generous as possible and the number of pigs in each pen are minimal to reduce competition and assist in the uniform distribution of pigs.
7. Whenever possible provide some bedding, preferably good quality straw.
8. Separation of various age groups is a great advantage in helping growth and reducing disease incidence.
9. There must be space for the essential behavioural needs of the pigs, namely freedom to stand up, lie down, stretch their limbs and groom themselves.
10. Where there is no bedding there is a real danger of discomfort and injury for the pigs. Good design and construction of the flooring is therefore essential.
11. The building must have ample lighting to ensure good management, hygiene and disease control.
12. Pigs should not need to compete for feeding, watering, resting, exercising and dunging areas.
13. There must be full provision to cope with mechanical breakdowns of essential equipment and emergencies such as fire.
14. Producers should not operate housing or management systems in which pigs are dependent on continuous drug administration.
15. Enough fibre and bulk are nutritionally essential for pigs to ensure the normal function and comfort of the digestive system. Housing design must provide uniform access to their feed for all the pigs.

*Reproduced from Sainsbury (1984)*

## NUTRITION AND FEEDING

The pig, like man, is monogastric and omnivorous; anatomically and physiologically both have similar digestive systems and therefore similar nutritional requirements. The major components of most pig diets are cereals, animal and vegetable proteins, and cereal by-products. Farmers may feed a wide range of vegetable products, the choice of which depends on the management system. For example, it is possible to feed adult stock on root crops or other arable by-products with appropriate supplementation, while

slaughter pigs require a more closely controlled ration. Increasing the density of the diet may reduce intake. Bulking, using inert diluents, increases intake but lowers digestive efficiency. As sows are unable to consume sufficient bulky forage to satisfy nutrient needs for maintenance and foetal growth, it is vital they have a balanced concentrate ration. Diets may be either bought-in compounds or home-mixed: quality control is as important for the home-mixed as for the factory product. Producers prefer compounded pelleted feeds for piglets and weaners, as they are more convenient and it is easier to maintain a small, regular supply of a fresh, palatable feedstuff.

Genetics, sex, behaviour, processing or nutrient density can influence appetite. In addition, an animal with free choice of feeds may choose those of which it has had previous experience. Reduced appetite is often an early sign of disease. When designing diets, it is necessary to take the system of production into consideration, together with these factors.

Different strains of pig may react in different ways to diet quality and pattern of feeding, but there are few major variations among UK-bred hybrid pigs. Given the same amount of feed of the same diet, boars grow faster and are leaner than castrated males; gilts have higher rates of lean tissue growth than castrates. Boars and gilts probably will respond, with extra lean growth, to higher levels of protein in the diet than will castrates. The superiority of the boar is only noticeable when the diet contains enough protein to support the high rate of gain of which it is capable. When offered feed *ad libitum*, castrated males will eat slightly more than will boars and gilts. Behavioural factors can affect feed intake and production performance. Processing food by grinding, soaking, heating, drying, or pelleting, improves digestibility. However, very fine grinding increases the risk of gastric ulcerations. Over-heating or over-drying also reduces digestibility. Pelleting sometimes improves performance, possibly because the steaming process improves digestibility, and because there is less wastage.

### Rationing
The amount of food a producer feeds a pig largely decides the performance and profitability of a system. The scale of feeding must be at such a level that animals grow as fast as possible or, with breeding

stock, maintain sufficient body reserves, but without becoming too fat. Pigs have a large appetite, therefore providing food *ad libitum* throughout the growing period can lead to over-fat carcases. The nutritional requirements of growing pigs need continuous revision as genetic progress improves the quality of the animals. Many hybrid pigs can now be fed *ad libitum* to high carcase weights.

To choose suitable diets and feed-scales for a particular system, the producer needs to know the energy and protein concentrations of the feedstuffs available (MAFF, 1980a, contains tables of feed composition), and the biological value of the protein. The concentration and the nature of the fibre content is also important, because pigs can only digest fibre to a limited extent. Giving diets rich in fibre and thus low in digestible energy concentration will restrict feed intake and may limit growth rate.

## Energy

Animals need energy for maintenance functions, and for lean tissue growth. Once the animal has satisfied these needs will use the remainder to lay down fat. Energy is quantified as digestible energy (DE) expressed as megajoules of digestible energy per kilogram of food (MJ DE/kg). It is easy to alter the nutrient density of a diet. The usual values are medium density (12.5MJ DE/kg), high density (13.5MJ DE/kg), and very high density (14.5MJ DE/kg). Including cassava and/or fat in a diet will increase its energy concentration. Table 6.7 gives recommendations for DE and crude protein (CP) allowances for different classes of stock.

Typical food conversion ratios (FCR) (kg feed/kg of liveweight gain) for a medium density diet are: weaner production, 1.5; porkers (20-63kg), 2.8; baconers (20-90kg), 3.1, and heavy hogs, 3.6. Using a diet of higher DE concentration will produce better ratios.

## Protein

Protein is expressed as crude protein (CP), or digestible crude protein. The nature and quality of dietary protein is important. Some 22 amino acids may be found in pig diets; of these nine are essential (ie the pig cannot synthesize these), and therefore the diet must include them. A shortage of essential amino- acids will limit the production of lean meat

and adversely affect growth. Pigs need greater amounts of lysine, histidine and methionine than valine and isoleucine. Unfortunately, the former are less abundant in common feedstuffs than the latter. Cereals supply about 66 per cent of the total protein in the diet of growing pigs, but they provide less than 50 per cent of the requirement for lysine. It is important to include, therefore, protein concentrates, such as fishmeal or soyabean meal, which are rich in lysine. In diets for growing pigs (15-90kg), lysine concentrations should be about 7g/100g CP. Since growing pigs use amino-acids for meat protein production, and sows for body maintenance and milk, the ideal protein in diets should mirror those in meat and milk.

**Table 6.7        Some typical diets.**

| Class of stock | MJ DE/kg | CP (g/kg) |
|---|---|---|
| Young growers* (5-20kg liveweight) | 14-16 | 200-220 |
| Growers (20-60kg liveweight) | 13-14 | 150 |
| Finishers (60-100kg liveweight) | 12-14 | 135-150 |
| Pregnant sows | 12.5 | 140 |

* The most concentrated diet necessary for piglets weaned at 2-4 weeks of age and would probably include milk products, such as dried skim milk. It is possible to lower nutrient density when piglets reach 10kg liveweight.

## Practical feeding

The energy to protein ratio is an important factor in diet formulation. Small piglets and growing pigs have lower maintenance needs (energy) and higher protein needs (lean tissue growth) than a pregnant sow. The ratio runs between 1:14 (DE:CP) for a young growing animal down to 1:11 for a pregnant breeding animal or a growing pig approaching slaughter at 100kg liveweight.

Farmers often keep gilts on the same feed or ration scales as bacon pigs until 14 days before the second or third oestrus, when they will weight about 90-115kg. After this point they change to an *ad libitum* feed containing 14 per cent CP. After service, producers restrict females to about 23MJ DE/day for a period. As pregnancy progresses, the sow will require extra nutrients for the growth of the foetuses and associated tissues, especially during the last 30 days. Feeding during pregnancy can affect litter size, birth-weight and sow body-weight. Stockpersons should feed gilts and sows so that they gain 20, 15 and 10kg in net body-weight per parity for the initial three parities. However, it will probably be necessary to adjust the amount to take account for the weight gain and condition of the individual animal.

Sows draw on their body tissues during lactation; a 10kg weight loss over this period is usually acceptable, but it is important to avoid excessive weight loss. A lactating sow will need about 50MJ DE/day for up to five piglets and 5MJ DE/piglet above that. If the farmer weans the sow after five weeks, she will require 15 per cent CP in her ration; if weaning takes place earlier she will require 17 per cent.

Feed allowances and diet calculations, as above, are only guidelines; it may be necessary to modify either or both according to the response of the pigs.

*Feed ingredients*
The major ingredient in all European pig diets is cereal. In the UK, barley has been the most popular cereal ingredient for many years, though this may in some part reflect the vast quantities farmers grow. Barley is a medium-energy ingredient, which it is possible to feed at high levels to growing and breeding stock. It contains a fair level of lysine in its protein content. Producers feed wheat and oats more rarely, probably because they are less readily available for use in animal rations. Wheat has a higher, and oats a lower, DE content than barley. The higher fibre level in oats limits the use of this cereal to older classes of stock where maximum feed conversion efficiency is not so important. Pig rations can utilize other cereals, such as maize, rye, sorghum and rice, depending on availability and price.

Traditionally, producers used fishmeal, meatmeal, meat-and-bonemeal, milk by-products, bloodmeals and feathermeals as sources of animal protein. These products are all dried and ground. It is vital to avoid over heating in the processing of these as it renders the lysine content unavailable by causing denaturation of the protein. Fears over disease transmission to humans, coupled with

legislation has reduced or eliminated the use of a number of these products in the UK.

Vegetable protein includes oilseeds, pulses and microbial protein. The usefulness of these products is dependent on their amino-acid composition and the presence or absence of toxic or anti-nutritional factors (plant products that may inhibit enzyme activity, or deter consumption). Their crude protein content ranges from 72 per cent for microbial protein down to perhaps 9 per cent for sunflower meal.

Pig rations frequently include cereal by-products from the milling of cereals for humans. These products often have a high fibre content as they are the outer husks of the grain. Their protein, energy and oil levels reflect the degree of milling.

Root crops are a useful source of energy but are low in dry matter and protein; producers mostly use them in rations for adult stock. It is possible to supply root crops through pipeline feed systems, but they will first need macerating. Pigs digest potatoes, particularly if cooked to make the starch more digestible, as efficiently as cereals.

Other feedstuffs, such as manioc, dried sugar beet pulp, molasses, dried green crops and distillery by-products, may all contribute to pig rations if at relatively low levels. Very high density diets for growing pigs use oils and tallows as energy sources in rations.

Liquid separated milk (about 9 per cent dry matter containing 37 per cent protein and 53 per cent lactose) is very digestible and suitable for liquid feeding systems. Such milk contains 0.1 per cent formalin, which preserves it for about seven days. Farmers can also feed other milk by-products, such as whey (the watery residue from cheese making). However, it is important to ensure an adequate water supply, because whey often has a high salt content.

Swill is a very variable product, which producers usually collect at source. Canteen wastes, for example, will contain meat and vegetable fragments, and confectionery residues and trimmings produced during food preparation. Swill manufacturers must by law boil it for at least an hour. This is important for disease control purposes and may also improve the quality of the product. The Ministry of Agriculture must licence the plant that treats the swill, and farmers of swill-fed pigs must transport them directly to a slaughterhouse. These stringent health and hygiene precautions are set out in the *Diseases of Animals (Waste Food) Order 1973* as amended in 1987 and 1996. These restrictions, coupled with the unpredictable composition which makes it difficult to feed with any accuracy, have resulted in swill being an unpopular feedstuff.

### Vitamins and minerals

Producers should include vitamins and minerals in all diets at or above the levels a pig requires. Textbooks, such as Macdonald *et al* (1995), give tables of these requirements. There is still much to learn about both minerals and vitamins, particularly vitamin deficiencies, and the interrelationships that occur between various minerals.

### Growth-promoting substances

Rations for growing pigs commonly include various non-hormone growth-promoting substances; the *Council Directives 70/524 EEC* and *96/51 EC* control their use. For example, it is legal to use some antibiotics, such as bambermycin, and virginiamycin, which are without any therapeutic value. Therapeutic antibiotics can only be incorporated in an animal food under a Medicated Feedstuffs Prescription issued by a veterinary surgeon.

Including copper sulphate in the diet will usually improve growth rate and food conversion ratio. *EEC Directives (83/615 and 85/520)* limit its inclusion rate to 175ppm for pigs up to 17 weeks of age, 100ppm for pigs from 17 weeks to six months, and 35ppm for pigs over 6 months and breeding stock. Nitrovin is another non-antibiotic growth promoter, but farmers cannot use it with antibiotics. Providing producers follow instructions for withdrawal, it is possible to include all these substances without causing unacceptable residues.

Research is also investigating other substances as growth-promoters. For example, some naturally occurring intestinal bacteria (probiotics), such as lactobacilli, may provide an alternative to antibiotics. They prevent colonization of the intestine by pathogens and enhance feed utilization. Researchers are also investigating porcine growth hormone (somatotrophin), which has no biological activity in humans, as a growth-promoting substance.

### Feeding systems

In most commercial situations, producers house pigs in groups. Individuals within groups show considerable differences in feed intake, and consequently in growth. The following are the most common methods of feeding growing pigs.

*Controlled feeding.* In this system the farmer controls the amount of feed a pig receives. The amount is restricted to less than the pig's maximum intake at each meal-time. Feeding is to a pre-determined feed scale that varies with either weight or age. It is the most common system of feeding.

*Ad libitum or semi-ad libitum.* Pigs have continuous access to feed, or access for a fixed period. This is invariably less efficient in carcass terms than controlled feeding. Up to pork weight the differences are slight. Above this, and particularly for bacon pigs, it is necessary to restrict the scale to achieve good food conversion and a lean carcass.

*Self choice.* This is common in the USA. Pigs have access to at least two types of feed and must select sufficient of each to satisfy their requirements. In comparison to any form of restricted feeding, this system is usually inefficient.

*Alternating feeding.* The stockperson provides access to, say, barley meal at one feed and a protein supplement at the next. If they offer both feeds within a 24-hour period, this does not affect performance. With longer intervals between feeds, differences in the weights of tissues and organs sometimes occur. There are obvious savings in the cost of mixing.

*Frequency of feeding.* There are no advantages in growth terms in feeding older pigs more than once per day, and to omit one feed in a seven day period is generally acceptable. However, behavioural problems may occur in some pigs. Baby pigs and young animals require either frequent feeding or continuous access to feed.

*Wet or dry feed.* If given a choice, pigs prefer wet feed and can eat it much more quickly than dry feed. In some piggeries, farmers pump meal suspended in water to troughs. Dry feed is used more commonly however, it is easier to handle, the risk of bacterial infection is less, and capital costs are lower.

*Trough or floor.* Feeding from a trough gives slightly better results for pellets and considerably better results with meal; this is because there is less wastage.

### Boars

Producers usually feed boars the same or a little more than sows; the exact level relates to housing conditions. Individually penned boars outdoors will need extra food to compensate for winter heat loss, particularly if the housing does not have insulation or much straw. Farmers should maintain mature boars at approximately 175kg. (AFRC 1990).

### Breeding females

It is vital to feed all breeding animals the correct amount they require. A sow that becomes overfat or too thin will have a lower fertility. Two to two and a half kilograms per day of a medium-energy ration is generally adequate for pregnant females. Farmers should increase rations if they feed sows in groups and competition occurs. They often reduce rations when feeding sows individually in an environmentally controlled house.

A sow's nutrition during gestation and the food she receives during lactation will influence her milk quantity. The lactation curve of the sow has only a slight peak at 3-5 weeks. It is important not to stimulate milk production too early, so that supply exceeds demand. Equally supply should not be so generous in the latter part of lactation as to cause drying-off difficulties at weaning. Producers should also avoid under-feeding because, if the sow continues to use her own body reserves for milk production, consequential problems, such as a delay in returning to oestrus, can result. Feeding during lactation should therefore take account of previous nutritional status, housing conditions, number of piglets in the litter, and intended weaning age. Many recommendations exist for feed allowances at this time, but are usually within a range of 4-7kg/day of a balanced concentrate ration.

### Piglets

It is usual to begin offering creep feed to piglets from about five days of age. Creep feed is a high density, highly palatable product, which the stockperson offers in small quantities at frequent intervals. Generally piglets eat very small amounts until they are 3-4 weeks old. Creep feed tempts the piglets to eat solids and aids gut enzyme development.

### Growing and fattening pigs

Producers may feed pigs close to their *ad libitum* intake in the early phases of growth, up to about 60kg liveweight. After this, and as pigs approach slaughter weight, they restrict feeding. Ration scales cover frequency of feeding and size of individual meal. At the lowest level they must satisfy the maintenance need of the animal. At their highest level they represent either the appetite of the pig as regulated by physical gut capacity, or total energy limit. If producers feed rations of low nutrient density, then they must take care to ensure adequate gut capacity for the bulk of feed an animal needs. High-density diets often satisfy the pig without using the full physical capacity of the gut. This results in a lower gut weight and a higher killing out percentage.

The ideal feeding system for the young pig would be *ad libitum* access. The newly weaned pig may be unable to cope with this, so farmers usually restrict their feed. At three week weaning a pig will eat 100-150g/day. Gradually increasing amounts of food in two daily feedings should result in the pigs eating almost to appetite. Animals should finish their ration and be keenly demanding food at the next meal. A breakdown in this situation suggests over-feeding, ill-health or an environmental fault. The good stockperson should be on the look-out for this.

As pigs grow older, their tendency to lay down fat increases. Although there has been significant genetic selection against this in recent years, there is still usually the need to restrict the level of feeding. In practice there are various ways to do this. Producers may allow pigs to eat to appetite up to a certain weight. They then restrict either on a weight-based or a time-based scale, before finally feeding a flat rate until slaughter weight. Time-based scales are useful in that they smooth out any fluctuations in

**Table 6.8**      **Daily feed requirements of growing pigs based on liveweight.**

| Liveweight (kg) | Approx daily gain 0.6kg | | Approx daily gain 0.7kg | |
|---|---|---|---|---|
| | Medium energy diet (kg) | High-energy diet (kg) | Medium energy diet (kg) | High-energy diet (kg) |
| 20 | 1.00 | 0.95 | 1.00 | 0.95 |
| 30 | 1.50 | 1.40 | 1.50 | 1.40 |
| 40 | 1.75 | 1.60 | 1.95 | 1.80 |
| 50 | 1.95 | 1.75 | 2.15 | 1.95 |
| 60 | 2.05 | 1.85 | 2.35 | 2.05 |
| 70 | 2.15 | 2.00 | 2.40 | 2.20 |
| 80 | 2.30 | 2.10 | 2.50 | 2.30 |
| 90 | 2.40 | 2.20 | 2.60 | 2.40 |

*Source: MAFF (1980a)*

**Table 6.9**      **Daily feed requirements of growing pigs based on age (approximate daily gain of 0.6kg).**

| Age (weeks) | Approx liveweight (kg) | Feed requirements | |
|---|---|---|---|
| | | Medium-energy (kg) | High-energy (kg) |
| 9 | 20 | 1.00 | 0.95 |
| 11 | 27 | 1.35 | 1.25 |
| 13 | 34 | 1.60 | 1.50 |
| 15 | 42 | 1.80 | 1.65 |
| 17 | 51 | 1.95 | 1.75 |
| 19 | 59 | 2.05 | 1.85 |
| 21 | 68 | 2.15 | 1.95 |
| 23 | 76 | 2.25 | 2.05 |
| 25 | 85 | 2.35 | 2.15 |
| 27 | 93 | 2.45 | 2.25 |

*Source: MAFF (1980a)*

growth. Also, altering a scale more frequently means it will match the pigs' requirements more closely. Tables 6.8 and 6.9 give examples of ration scales. Individual pig producers manipulate the form of feed scale to take account of the genetic merit, sex and market outlet for their stock. Better strains of pig can produce good carcasses at high levels.

*Outdoor pigs*
A medium energy compound (12.5MJ DE/kg) with a crude protein content of 16 per cent should be adequate. Farmers feed the ration as a roll or cob; the relatively large size of these reduces wastage through loss in the mud and to birds. Scattering the food or feeding it in long lines, and not in heaps or piles, will help to limit hoarding and aggression.

Under poor weather conditions it is important to feed generously to allow for the needs of gestation. Producers should regularly condition-score sows to ensure that the animals are receiving adequate food. In cold weather, farmers may need to feed up to 4.5kg/day to a sow at service, rising to 8kg during lactation. This quantity will include an element for litter size. A target feed figure for an outdoor sow is 1.25 tonne/year.

In-pig sows can also eat arable products such as potatoes, carrots and other root crops. As these are bulky foods, it is important to only offer them in proportion to the total appetite of the sow. To consume a nutritionally balanced ration, it may be necessary therefore to also feed a concentrate supplement.

Outdoor pigs often have high levels of minerals and vitamins, such as biotin, in their diet to ensure sound skeletal development, so limiting lameness.

### Water

It is vital to provide a fresh clean supply of drinking water for all animals. New-born pigs consume 0.25-1 litre/day. Growing animals need at least 2-6 litres, and adult stock 8-12 litres, daily. Environmental conditions such as hot weather or poor heat control in buildings may increase this requirement.

A clean source of water should be available from about seven days of age. Whatever the chosen system – nipple drinker, straw drinker or trough – it should be the same type as that in use in the weaner house and be at a suitable height for the piglets. If the piglets learn to use a drinker in the farrowing pen, there is less of a risk at weaning of dehydration resulting from inadequate water consumption.

When giving dry food, it is important to ensure that there are sufficient water points within a pen, as animals will take in water several times during a meal. An inadequate supply can lead to bullying and over-consumption of feed by dominant animals. If water is not available ad libitum, then farmers should provide it at a ratio of 2.5 units per unit of dry feed.

## HEALTH AND DISEASE

Producers must keep accurate records of drugs they use and follow instructions regarding withdrawal periods and slaughter. The National Office of Animal Health (NOAH) has included a *Code of Practice for the Safe Use of Animal Medicines on Farms* in their *Animal Medicine Record Book* (1993) and the Health and Safety Executive (1998) has issued a leaflet on the safe use of veterinary medicines.

### General disease prevention

Ideally, a pig unit should be totally closed to the entry of animals from other sources. Feed, transport and personnel should undergo a full hygienic preparation before entry, and the pigs themselves should live in a pathogen-free environment. However, even under experimental conditions, this is difficult to sustain. The commercial pig producer can achieve an acceptable compromise by preventing infectious organisms from entering the unit as far as possible, and washing out and sterilizing buildings thoroughly. Routine vaccination and medication programmes will also help to maintain animals in good health. In practical terms, the producer needs to isolate the unit from all but essential services. It is important, therefore, to site a unit, and design the layout, to limit the risk of disease transmission.

To limit airborne infection the farmer must consider the prevailing weather conditions. It is equally vital to reduce, if not eliminate, small mammals and birds which are potential disease vectors. Accidental introduction of a major infectious disease into a modern intensive unit can result in great economic loss. The effects of many subclinical infections interacting with an inadequate environment can be equally loss-causing, since affected animals fail to perform at their optimum biological levels.

To avoid production losses, it is essential to detect disease early and rapidly. This will depend on efficient observation of all pigs; the competent stockperson will therefore inspect the animals several times a day. Healthy pigs have a keen interest in feed, they are bright, lively and alert, and look clean and sleek. Stockpersons should note variations from the normal quickly, and search for an explanation. Besides examining the pig in detail, the search will involve scrutinizing management actions and records, since it might be possible to trace the cause to some past event.

Unfortunately, as herds become larger, so the effects of infection increase. This is because animals often share the same air space, or come from the same susceptible population. Large units normally have a continuous through-put of pigs and a low labour input. This can lead to a delay in detecting infection, with consequential production loss. Paradoxically, separating pigs from their excreta has probably been detrimental to young pigs in their development of adequate immunity.

The commercial producer aims to reduce disease levels so that they are not a drag on production, by using preventive hygiene and medicine. Preventive hygiene relies on the thorough cleaning of buildings between batches of pigs. This involves using high-power pressure washers, with Ministry of Agriculture approved disinfectants, sometimes with fumigation to follow. A regularly updated list of products suitable for disinfection and sterilizing procedures is available from the Animal Health sections of MAFF. The timing of such a stringent hygiene programme will depend on the pattern of production, namely when periodic depopulation of buildings between successive batches of animals takes place. It is comparatively simple to apply such a programme to farrowing accommodation and housing for weaned pigs and finishers. The problem becomes more difficult with adult stock because their housing is seldom empty. Here, preventive medicine using routine vaccinations and other treatments (see Table 6.10) is important in keeping infection down. Such treatments would include vaccination against swine erysipelas, atrophic rhinitis, parvovirus and *E. coli* and *Pasteurella* infections. They also involve incorporating products for the control of swine dysentery and intestinal parasites in the diet. A veterinary prescription is necessary to include some of these products routinely. It is important to follow instructions regarding withholding time before slaughter if using these products. It is vital, therefore, to keep records of the drugs used for preventative medicine and to treat sick animals.

**Health schemes**

Besides individual health maintenance, which producers operate in liaison with their veterinary surgeons, other nationally based schemes exist. These strive to raise the health status of pigs countrywide. Schemes that merit especial mention are those run by the Pig Health Control Association. In addition to other measures, they seek to register and advertise herds that are free of specific infections. These schemes use independent inspection and sampling of blood, etc. to regularly monitor their participating herds.

**Common disease conditions**

*Notifiable diseases*

Notifiable diseases that may occur in pigs include anthrax, Aujesky's disease, foot-and-mouth disease, swine fever, swine vesicular disease, Teschen disease and rabies (see Table 6.11 for details of those most likely to occur in the UK). A producer is legally obliged to report the presence, or even the suspicion, of any of these immediately to the police or to the appropriate government department. Usually only the most stringent measures (including slaughter, possibly of whole herds) are effective; without these such diseases are liable to put whole populations at risk.

*Zoonoses*

Anthrax and rabies have already been mentioned under notifiable diseases. Possibly the main danger to humans from pigs in the UK is *Salmonella* infection. Helminths such as *Trichinella spiralis* and *Trichuris suis* affect humans and pigs, but are not a major health problem in the UK. Other possible zoonoses include erysipelas, leptospirosis, streptococcal meningitis and influenza. The occurrence of these in humans via pig contact is rare. Care in the use of sharp implements and good personal hygiene is most important.

*Management related problems*

Slats, concrete floors and confined accommodation for pigs often provides little opportunity for movement, and causes various arthritic and rheumatoid conditions in some older animals. Often a more generous form of housing such as a straw yard will improve the condition.

**Routine husbandry procedures**

The public and those involved in pig production are increasingly scrutinizing those routine procedures and mutilations which inflict damage and/or pain and distress to the animals. Legislation now exists to control most of them.

With all routine procedures and mutilations, stockperson competence, together with a high standard of equipment maintenance and general smoothness of the operation. All of these procedures cause some stress to the animals and it is usual to carry out as many as possible simultaneously. Therefore, ear-notching, teeth clipping, iron injection, tail docking and castration may be done together within 48 hours of birth. This blocking-up of procedures reduces handling and seems to have the least unsettling effect on the piglets.

*Teeth clipping*

Most breeders clip the incisor teeth of newborn piglets, since these might lacerate the sow's udder and abrade the faces of siblings during competition

**Table 6.10**      **Common disease conditions: their prevention and treatment.**

| Class of pig | Housing | Common conditions | Routine control/treatment |
|---|---|---|---|
| Gilts | Yards/stalls | Endemic virus infections | Acclimatization and vaccination before exposure |
| | | Helminths | Anthelmintics ar regular intervals |
| | | Lice and mange | Regular lice and mange washes |
| | | Coliforms | Vaccinate animals |
| Pregnant sows | Yards/stalls | As for gilts | As for gilts, with rational use of booster vaccines |
| Lactating sows | Farrowing house | Lice and mange | Suggests inadequate late pregnancy treatment. Repeat wash if necessary |
| | | Mastitis/metritis | Rapid treatment with antibiotics plus feeding attention for piglets. |
| | | Agalactia | Improve farrowing house conditions. Pay attention to feeding |
| | | Hip and foot problems and lameness | Suggests poor pregnant sow management. Check records and accommodation. Cull severe cases |
| Piglets | Farrowing house | Chilling/anaemia | Correct the housing conditions. Iron injections |
| | | Coliform scours | Routine use of vaccines. Treat with antibiotics |
| | | Joint ill | Correct flooring defects if necessary. Query cleanliness of stockpersons equipment. |
| | | Other enteric problems | Antibiotic treatment |
| Weaners | Follower housing | Coliform scours (*E. coli*) | Vaccination should have been carried out earlier. Treatment with antibiotics |
| | | Salmonella<br>Pasteurella<br>Atrophic rhinitis<br>Pneumonic conditions | Improve environment. Use vaccines and antibiotics if necessary |
| | | Swine dysentery<br>Erysipelas<br>Helminths | Medicate feed<br>Vaccinate at weaning<br>Anthelmintics usually administered in feed |
| Fatteners | Various house types | Problems as for weaners | Prior preventive measures, and the use of clean buildings should be adequate for this stage |
| Boars | | | Treat as for sows. Routine vaccines and regular use of lice and mange washes and anthelmintics |

**Table 6.11**  **Main symptoms of the most common notifiable pig diseases.**

| Disease | Causal organism | Species affected | Symptoms |
|---|---|---|---|
| Anthrax | *Bacillus anthracis* | All animals including man | Listlessness, throat swelling, deaths from choking in 8-16 hours |
| Aujeszky's disease | Virus | Mainly pigs occasionally cats, rats, dogs, cattle or sheep | Sneezing, coughing, high temperature trembling and/or convulsions |
| Foot-and-mouth disease | Virus | All cloven-hoofed animals | Blisters on mouth and feet, lameness |
| Swine fever | Virus | Pigs | Acute: high temperature, arched back, greenish scours, reddish tongue Chronic: ill-defined symptoms, hind-quarter weakness, high temperature |
| Swine vesicular disease | Virus | Pigs | Similar to foot-and-mouth disease |

for teats. Bacteria can easily infect such lacerations. However, a statutory instrument - *The Welfare of Livestock Regulations 1994* - bans the routine practice of tooth clipping except where injury to the sow or other piglets is likely. The *Codes of Recommendations* (MAFF 1983) advises that only a veterinary surgeon or a competent trained operator should tooth clip an animal; it makes no comment on the age. It is therefore better not to perform this procedure as a routine task but to confine it to those litters where damage actually occurs. In outdoor systems, some research suggests that tooth clipping is largely unnecessary.

*Docking*

Tail-docking of piglets, leaving only a short stump, is routine on some units; it usually takes place before the piglets are seven days old. Producers use this procedure where there is a history of tail-biting in growing pigs. In outdoor systems it should be unnecessary to tail dock animals. *The Codes of Recommendations* (MAFF 1983) discourage the technique and stress the need for only trained operators to carry it out. *The Docking of Pigs (Use of*

*Anaesthetics) Order 1974* and *The Welfare of Livestock (Prohibited Operations) Regulations 1982* requires that after seven days of age only a veterinary surgeon, using an anaesthetic, dock a pig. A statutory instrument - *The Welfare of Livestock Regulations 1994* - bans the routine practice of tail docking except where injury to the sow or other piglets is likely.

It is important to consider all management and environmental factors if tail-biting occurs on a unit. There is no single factor that causes this condition. Many factors have been put forward, ranging from the genetic characteristics of the animal to stocking densities, sudden temperature fluctuations, ration composition and iron deficiency. Despite many investigations, the situation is still unresolved. The removal of a piglet's tail does not necessarily cure the problem; docked pigs often increase the level of neck and shoulder biting. It is vital that the producer attempts to correct husbandry defects rather than mutilate the piglets.

*Castration*

This is an increasingly uncommon practice on both indoor and outdoor pig units. Producers frequently used to castrate male piglets at one day or so of age when they were small enough for a single operator to handle them easily. Castration usually involves removing the testicles through low-placed scrotal skin incisions; the operator slowly pulls each testicle until the arteriovenous plexus breaks. The *Protection of Animals (Anaesthetics) Acts 1954* and *1964* and the *Veterinary Surgeons Act 1966* forbids the castration of pigs with elastrators (rubber rings) after the first week of life. This legislation also requires that a veterinary surgeon, using anaesthesia, conducts this operation after the age of two months.

*Iron injections*
Piglets usually receive an intramuscular iron injection within 48 hours of birth. Some farmers are now eliminating this procedure and instead increasing the iron content of the sow's diet and so raising the level of iron in her milk. This practice may help to reduce production costs. For any injection, it is essential to either restrain the individual animal or pack the group of animals tightly together. Poor injection technique can lead to abscesses, carcass staining and broken needles becoming embedded; all these are causes of carcass rejection and down-grading.

*Identification*
To keep accurate records of performance, pigs need to be individually identifiable. To aid identification in later life, farmers can ear-notch or tattoo and/or tag new-born piglets with a range of metal or plastic tags. Producers should slap-mark over the shoulders those animals destined for slaughter. This tattoo makes carcass identification possible, which enables the tracing of an individual pig's complete performance record.

*Nose-ringing*
Many farmers nose-ring their free-range sows and boars to prevent them from rooting and destroying the grass cover. As it affects the pig's ability to perform this highly motivated behaviour, producers should avoid it wherever possible. This is particularly so if stocking rate is generous and the sows do not appear to have caused much soil surface damage.

**General care and handling**
*Condition scoring*
An experienced stockperson may condition-score sows visually (for details, see Whittemore 1987). Animals should have good cover over the pin bones each side of the tail, and not have hollow loins or prominent pelvic bones, or prominent spine and ribs. An animal that has a poor cover with such prominent bones is emaciated and scores a grade zero. Equally, sows should not be so fat that further cover seems impossible at any point (Grade 5). A moderate to good animal (Grade 2/3) has some cover at all points, with bone structure distinguishable with firm pressure.

*Handling*
Quiet handling, using alleyways that are well designed and constructed and that the farmer keeps in good maintenance, makes moving large numbers of animals easier. With such facilities force is unnecessary. If pigs do need encouragement to move, the occasional light tap from a stick is sufficient. Handling boards are also useful to help move pigs, especially if they need to change direction. Most units regularly use fixed weighing, handling and loading facilities; therefore, time spent in planning the layout of a unit will save effort later (Figures 6.4 and 6.5). Handlers should always remember that pigs are social animals and are more at ease in a fairly densely packed space than when separate.

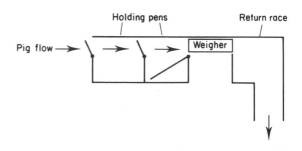

**Figure 6.4**    **Weighing and handling facilities.**

The traditional ways of handling individual pigs relates to their size. It is easy to cradle piglets in the hands or arms. To restrain a small to medium sized pig, a person needs to stand astride the animal while perching it on its hams and holding the front legs. To control large pigs, the handler needs to use a snout noose, which fits around the upper jaw behind the canine teeth. Stockpersons need to be aware that pigs are always noisy when they restrain them; they should therefore try to restrain animals quickly and quietly.

## Transport

Producers transport most meat pigs directly to the abattoir. European Agriculture Ministers have recently agreed minimum standards for transport time and associated issues. For details of the new complex rules and regulations see the guidance notes issued by the UK's Ministry of Agriculture (MAFF 1998b).

It is vital to handle, transport and slaughter animals as humanely and quickly as possible to safeguard the animals' welfare and produce meat of a high quality. Handlers must avoid causing pigs stress and physical damage, such as bruising, so well designed races and ramps with non-slip floors are essential. It is important to avoid mixing pigs, as they are liable to fight. The loading area should be at the periphery of the unit, with sufficient pens to avoid mixing groups. The entrance should also be separate from the exit to the loading ramp (Figure 6.5). To keep weight loss to a minimum, the slaughterhouse should hold pigs in the lairage for the minimum length of time necessary. Poor pig handling, particularly of very lean pigs, results in meat that may be either pale, soft and exudative (PSE) or dark, firm and dry (DFD). In both cases the meat is unpleasant in appearance and lacking in palatability. Some strains of pig are susceptible to PSE, but more problems result from poor pre-slaughter treatment.

## Casualty slaughter

On all units there is sometimes a need to kill animals. These are most often small, runt or deformed piglets and, less frequently, older pigs that are poor doers, or which have some traumatic injury such as a broken limb. Often these animals are of no commercial value, and the farmer will need to organize the disposal of the carcasses. The larger pigs do sometimes present real problems. For example, if an animal is in obvious pain there may be no recourse

but to send it to a slaughterhouse immediately; this is if it is still able to walk. If the condition of the pig is such that it requires slaughtering on the farm, the producer must obtain a veterinary certificate before an abattoir will accept the carcass. If the animal is acute pain it may have to be killed immediately. A good stockperson will usually be aware of the need to dispose of animals for humane reasons before emergency slaughter becomes necessary.

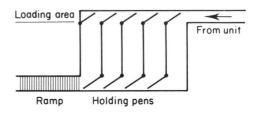

**Figure 6.5** **Loading area.**

Slaughter should be as humane as possible; the pig should not feel menaced or mishandled. A clean sharp blow to the back of the skull with a hammer will kill a piglet. It is best first to stun growing pigs and adults with a captive-bolt pistol before pithing. The stockperson carrying out these tasks needs training as much for this as for the more usual activities. For killing pigs in an emergency, it may be necessary to use a twelve-bore shotgun, or a heavy-bore rifle. The operator should aim the gun behind the ear of the animal such that the shot passes forward through the brain.

The Pig Veterinary Society (Blackburn 1996) has issued guidance notes which detail the procedures necessary for the transport and emergency on-farm slaughter of casualty pigs. It stresses that well run production units must have mechanisms in place to deal with emergencies. It highlights the critical role of stockmanship in dealing humanely with animal suffering.

### Disposal of carcases

The disposal of dead pigs and foetal membranes is an ongoing problem on pig units. On larger units there are two main methods. Incineration (usually fired by propane gas) is suitable for smaller pigs, but prior dismemberment is necessary for larger animals. Running costs are high and there is some odour. The other method involves microbial digestion of the carcase in a special pit. Farmers can construct these from a large-diameter concrete pipe, with an earth floor and an airtight lid (see MAFF 1980b). The Local and Water Authorities usually need to give planning consent.

## THE WAY AHEAD

The pig farmer, in common with other livestock producers, has responded to consumer demand for cheap foodstuffs by intensifying his production methods. As consumers have become more detached from animal agriculture, and as they become more affluent, so they have become more critical of farming methods and have begun to re-evaluate the product. The pig industry needs to respond to the consumer in several ways: it should be more open about how it produces pigs; it must explain that traditional methods of production are not automatically the best; it must be responsible in its use of drugs and, wherever possible, avoid the use of other feed additives. Farmers need to produce products the consumer wants. This implies that production methods must change, but not necessarily regress to more traditional technology, and that it must be seen to lose its apparently inhumane aspects. The area of pig behaviour and welfare needs considerably more research and producers need to take up the findings of experimental work. Interaction between stockperson and pigs is a difficult area that needs close analysis if we are to improve their welfare. Transport times and standards are another area that will continue to be debated heavily.

Outdoor production can have positive implications for animal welfare, but such systems are not always synonymous with good welfare. There is a growing market for pig meat products that are from a humane, drug-free system. Several producers are attempting to fulfill this requirement, and the number farming in this way will probably rise. It remains to be seen what system(s) replace sow stalls and tethers after the impending ban in 1999.

If it were not for financial constraints, many producers would possibly abandon farrowing crates for a generous litter area in which the sow could make her own farrowing nest. For such a system to work efficiently it needs a high level of stockmanship. Such a system is also counterproductive in terms of the efficient use of expensive buildings. An increase in farrowing space might lead, under certain conditions, to a reduction in piglet mortality. However, producers are wary of change because of the costs involved in providing efficient and humane farrowing accommodation.

The increasing interest in pig AI suggests that more producers are now realizing the benefits. Mounting pressure for efficient lean meat production will probably lead to an increase in the use of superior genetic material via AI. Another area of technology that the industry is making strenuous efforts to improve is methods of incorporating slurry into soil. It is evaluating other possible uses, such as methane production.

## REFERENCES AND FURTHER READING

**Agricultural Research Council** 1981 *The Nutrient Requirements of Pigs*. CAB: Farnham Royal, UK

**Agricultural and Food Research Council (Technical Committee on Responses to Nutrients)** 1990 *Advisory Booklet, Nutrient Requirements of Sows and Boars*. HGM Publications: Bakewell, UK

**Baxter S** 1984 *Intensive Pig Production: Environmental Management and Design*. Granada: London, UK

**Blackburn P W** 1996 *The Casualty Pig*. The Pig Veterinary Society: UK

**Brent G** 1986 *Housing the Pig*. Farming Press: Ipswich, UK

**Brent G** 1987 *The Pigman's Handbook, 2nd edition*. Farming Press: Ipswich, UK

**Bourne F J** 1986 Pigs. In: Cole D J A and Brander G C (eds) *Bioindustrial Ecosystems*, pp 183-196. Elsevier: Amsterdam

**English P R and MacDonald D C** 1986 Animal behaviour and welfare. In: Cole D J A and Brander G C (eds) *Bioindustrial Ecosystems*, pp 89-105. Elsevier: Amsterdam

**Fox M W** 1984 *Farm Animals: Husbandry, Behavior and Veterinary Practice*. University Park Press: Baltimore, USA

**Health and Safety Executive** 1998 *Veterinary Medicines. Safe Use by Farmers and Other Animal Handlers. Leaflet A531*. HSE: London, UK

**Hill J R and Sainsbury D W B** (eds) 1995 *The Health of Pigs*. Longman Scientific and Technical: Harlow, UK

**Hughes P and Varley M** 1980 *Reproduction in the Pig*. Butterworth: London, UK

Lawrie R A 1991 *Meat Science, 5th edition.* Pergamon: Oxford, UK

McDonald P, Edwards R A, Greenhalgh J F D and Morgan C A 1995 *Animal Nutrition, 5th Edition.* Longman Scientific & Technical: Harlow, UK

Meat and Livestock Commission 1997 *Pig Yearbook.* MLC: Milton Keynes, UK

Ministry of Agriculture Fisheries and Food 1980a *Nutrient Allowances for Pigs. B 2089.* HMSO: London, UK

Ministry of Agriculture Fisheries and Food 1980b *A Simple Disposal Pit for Foetal Material and Small Carcasses. L 648.* HMSO: London, UK

Ministry of Agriculture Fisheries and Food 1983 *Codes of Recommendations for the Welfare of Livestock: Pigs (reprinted 1991). PB 0075.* HMSO: London, UK

Ministry of Agriculture Fisheries and Food 1984 *Pig Production and Welfare. B 2483.* HMSO: London, UK

Ministry of Agriculture Fisheries and Food 1997 *Pig Space Requirements, Guidance on Schedule 3 of the Welfare of Livestock Regulations 1994. PB 3225.* MAFF Publications: London, UK

Ministry of Agriculture Fisheries and Food 1998a *Code of Good Agricultural Practice for the Protection of Water. PB 0587.* MAFF Publications: London, UK

Ministry of Agriculture Fisheries and Food 1998b *Guidance on The Welfare of Animals (Transport) Order 1997.* MAFF: London, UK

Muirhead M R 1983 Pig housing and environment. *Veterinary Record 113:* 587-593

Muirhead M R and Alexander T J L 1997 *Managing Pig Health and the Treatment of Diseases: a reference for the farm.* 5 M Enterprises Ltd: Sheffield, UK

National Office of Animal Health 1993 *Animal Medicine Record Book.* NOAH: London, UK

Pond W G and Houpt K A 1978 *The Biology of the Pig.* Cornell University Press: Ithaca, USA

Ridgeon B 1993 *The Economics of Pig Production.* Farming Press: Ipswich, UK

Sainsbury D W B 1984 Pig housing and welfare. *Pig News and Information 5(4):* 377

Sainsbury D W B 1986 *Farm Animal Welfare: Cattle, Pigs and Poultry.* Collins: London, UK

Taylor D J 1995 *Pig Diseases, 6th edition.* Published by the author. Distributed by Farming Press: Ipswich, UK

Walton J R 1993 *A Handbook of Pig Diseases, 3rd edition.* University of Liverpool Press: Liverpool, UK

Whittemore C T 1987 *Elements of Pig Science.* Longman: London, UK

Whittemore C T 1998 *The Science and Practice of Pig Production, 2nd edition.* Blackwell Science: Oxford, UK

# 7 Rabbits

J O L King

## THE UK INDUSTRY

In the past rabbit farming had a risky reputation with many enthusiastic enterprises starting well, but after a time collapsing because of uneconomic production costs due, at least in part, to high mortality rates. Now, however, efficient farmers are making steady profits throughout the year.

Rabbit farms vary in size from specialist units with over 200 breeding does, to enterprises with as few as 30 does providing a part-time occupation. Most farms breed and fatten all their own rabbits and send them to specialist packing centres, where they are killed and their carcases marketed.

Rabbit meat is white, fine grained and has a high protein, low fat content.

The production and consumption of farmed rabbit in the UK is below that of most European countries and yet the UK imports large numbers of carcases from China.

### Breeds

The rabbit industry has developed three pure breeds for meat production.

*The New Zealand White*. This albino breed was the most popular and many producers still keep them. The bucks weigh about 4.5kg and the does about 5kg. The bone structure is rather heavy and the coat is dense. The meat is pale in colour. The young grow rapidly and are efficient converters of food into meat.

*The Californian*. These rabbits have white bodies with black markings on the nose, ears, feet and tail; parts of the pelt that furriers do not use. Adult animals weigh from 3.5 to 4.5kg and are blocky in type. The bone structure is fine and the fur is dense, but not long. The flesh has a fine texture and is of excellent quality. The best strains have a good dressing out percentage.

*The Commercial White*. This is really a strain evolved from the New Zealand White and has a lighter bone structure. It is becoming the most popular type.

*Hybrids*. Farmers can breed quality rabbits by crossing individuals from three or more, unrelated, inbred breeds or strains and then crossing the best progeny from these lines to produce a generation of superior hybrid animals of a uniformly high standard. Only large scale producers develop these hybridization schemes as they require both a scientific input and the deployment of substantial physical resources.

Continental producers use a number of white breeds for meat production, including the Géant Blanc de Bouscat and the Blanc de Vendée. Overseas producers, particularly those in China, tend to keep coloured breeds which have darker flesh; a characteristic the British market does not favour.

## NATURAL HISTORY AND BEHAVIOUR

The wild rabbit, *Oryctolagus cuniculus*, is the ancestor of all domestic ones. Wild rabbits are social, colony animals which live in burrows which they tunnel into slopes. Adults weigh about 1.5kg and exist on a diet of grass and weeds. Their alimentary canals are well adapted for the digestion of large quantities of forage. Rabbit colonies have a social hierarchy with a small number of dominant bucks fighting for the best grazing areas.

The does usually mate for the first time at about 4.5 months and produce from four to six litters a year consisting of four to seven young. They give birth in short tunnels some distance away from the main site, in nests they make from leaves and grass and which they line with fur they pluck from their breasts.

## REPRODUCTIVE PHYSIOLOGY

Farmers should mate does before they reach much beyond the age of five months, otherwise they may lay down internal fat, with a resultant failure to conceive.

Does may not always show signs of oestrus (see Table 7.1 for descriptions), some will mate successfully when the vulva appears normal. The action of mounting by the buck induces ovulation. Farmers must keep does under good environmental conditions to maintain oestrus throughout the year. Cold houses which do not provide artificial lighting to extend the day length are likely to cause anoestrus. To avoid reducing fertility in bucks, it is important that they do not overwork. To maintain fertility, bucks should not be allowed to mate more than 12 times in any week, and then mating should be so arranged that the buck never mates more than 4 times in one day.

## NUTRITION

The rabbit is, whenever possible, a frequent feeder, and its digestive system works most efficiently when there is a steady progression of food material through the alimentary canal. Compared with other mammals, the digestion is remarkably rapid. The rabbit has a large caecum in which there is microbial digestion of starch and cellulose. The stomach wall of the rabbit is thin. In a healthy animal the stomach is never completely empty. Caged rabbits need continually available pelleted food to eat in order to facilitate a steady gut flow.

The practice of coprophagy also helps to ensure that the digestive system always has partially digested contents passing through. The rabbit forms hard pellets during the day, which they excrete, and soft pellets during the night, which they eat directly from the anus. Rabbits swallow soft pellets in larger numbers after a period of fasting; animals in cages perform this behaviour during the night. The pellets then move through the alimentary canal as the rabbit starts eating again. Another advantage of coprophagy is that previously undigested components of the diet, which have been broken down by the microorganisms in the caecum, pass through the alimentary canal again thus enabling the rabbit to digest them.

**Table 7.1        Reproductive and breeding data.**

|  | Buck | Doe |
|---|---|---|
| Age at onset of puberty | 5 months | 4 months |
| Age at which first bred | 6 months | 5 months |
| Duration of oestrus | - | Long periods. No definite cycle |
| Signs of oestrus | - | Vulva purplish in colour. Slightly enlarged |
| Gestation period | - | 31 days ± 2 days |
| Age of natural weaning | - | 6 weeks |
| Age of commercial weaning | - | Usually 4 weeks |
| Normal breeding life | 3 years | 2-3 years |
| Average number of young | - | 8-10 |
| Seasonality of breeding | All year | All year |
| Ratio of males to females for natural mating | 1: 10-15 | |

Commercial farmers almost always exclusively feed their animals all-purpose pelleted diets. Such rations generally contain 60 to 70 per cent cereals or cereal by-products (barley meal, ground oats, maize meal, weatings and bran), 10 to 20 per cent grass meal and about 20 per cent of a high protein food, such as fish meal and soya bean meal, together with a mineral and vitamin supplement. It is important that this includes calcium, phosphorus and sodium chloride, and vitamins A and D. Commercial diets rarely contain growth promoters and antibiotics, but they almost always have a coccidiostat which reduces the incidence of coccidiosis in growing animals. Unlike other types of livestock, rabbits are fastidious and will sometimes refuse to eat a new batch of food for no apparent reason. Thus, it may be necessary to introduce a new type of pellet gradually over a period of days. Feeding a small quantity of hay or straw in addition to the pellets is beneficial and may assist digestion and help to prevent rabbits pulling hair from others.

Rabbits need a constant supply of clean water; the quantities they drink will vary with the individual, its size and age, and the temperature and humidity of the surroundings.

## ENVIRONMENT AND HOUSING
### Site
Prospective farmers should obtain official approval before starting to erect any buildings. It is important that the site has approach roads wide enough for the lorries which will deliver food and collect rabbits. An electricity supply should be near to keep the cost of installation low and an ample quantity of clean water, preferably from mains, must be available. Producers who do not kill their own rabbits should select a site near a packing centre willing to purchase all the fattened animals as soon as they have reached slaughter weight, throughout the year.

### Effluent disposal
If a sewer system is not available for the removal of the urine and waste water from washing down floors and cages, the farmer must install proper septic tanks to prevent these liquids contaminating local ditches and streams. Rabbit faeces are an excellent fertilizer for spreading on farm land. If the farm itself does not cover a sufficient area on which these can be spread, the producer will need to make arrangements with another local farmer to remove and utilize the faeces. Although rabbit manure can be heaped, it is advisable to remove it regularly to minimize fly problems.

Some rabbit farmers breed earthworms of a large, productive strain in the manure pits under the cages, claiming that this can reduce offensive odours. They sell the worms to fishermen, fish hatcheries, aquarium keepers and zoos, which can be quite profitable. Before embarking on such a project, it is important to locate a merchant willing to accept regular supplies.

### Type of housing
There are no optimum environmental standards for rabbits, but the data in Table 7.2 appear satisfactory in practice. It is important to avoid extremes of temperatures. Farmers will need to take appropriate measures to prevent heat stress – recognisable when the rabbits pant for prolonged periods. At temperatures below those in the table, food consumption increases to supply the additional energy the rabbit requires to maintain body temperature. Many does will not come into oestrus during cold weather and, if litters are born, the losses from chilling are likely to be high. Thus, although it is possible to farm rabbits in outdoor hutches with solid walls and floors, it is an inefficient system and in practice it is rare. Producers, therefore, normally provide substantial buildings. These need good insulation to retain the body heat coming from the rabbits during the winter months, because the provision of artificial heat is uneconomic. A regular supply of fresh air is essential, but it is important to avoid draughts. A natural ventilation system, with outlets in or near the roof which allow the warmer and lighter exhaled air to escape as it rises and so draw in heavier fresh air through inlets sited low down in the external walls, is both adequate and economic. The building doors may need to be left open in very hot weather; however it will be necessary to prevent vermin and predators entering. Poor ventilation leads to an accumulation of water vapour, carbon dioxide and ammonia which would render the rabbits more susceptible to disease. Expert advice may be necessary to ensure correct temperature, airflow and humidity.

**Table 7.2**  **Environmental data.**

| Air temperatures | |
|---|---|
| breeding stock | 10-15˚C |
| weaned young | 16-18˚C |
| young in nest | 30˚C |
| variation over 24h | Not more than 5˚C |
| Relative humidity | Under 80 per cent |
| Light intensity | About 10 lux |
| Lighted period in 24h | 14h |

The building should be well lit. Farmers should utilize fully the natural lighting coming through the windows, but they will still need sources of artificial light so that they can see the rabbits clearly at all times. Light plays an important part in regulating the activity of the pituitary gland. Under natural lighting conditions sexual activity falls during the autumn and winter when the hours of daylight lessen and rises in the spring as the days lengthen. The use of supplementary lighting to provide a uniform light period of 14 hours a day throughout the year has great breeding advantages. It encourages the regular production of litters and avoids the reduction in reproductive activity, and the resultant shortage of young rabbits, which occurs during the winter if relying on daylight alone.

Producers nearly always divide large buildings into several small units so that they can periodically clean and disinfect them more easily. Farmers normally house all classes of stock in similar buildings, although weaned young grow faster and utilize their food more efficiently at temperatures slightly higher than those suitable for breeding stock. By providing nest boxes in which the does can make warm nests, farmers are able to ensure that newly born rabbits have the high temperatures they require.

A separate food store for keeping pellets and hay in dry conditions is necessary. Although expensive it is advisable to build a strong fence round the premises to keep out dogs and foxes.

## Cages

On commercial farms producers keep rabbits in wire mesh cages (see Figure 7.1). These are made of galvanised wire mesh and the joints must be smooth. An even floor surface is essential to avoid causing abrasions on the under surfaces of the hind feet. The square mesh should not exceed 19x19mm, as this allows faecal pellets to pass through, but does not normally cause foot injuries. Farmers can arrange individual cages in a single tier, but to make better use of the space in a building, they stack them in two or three tiers. The upper tiers are set back so that no cage is directly above another; this allows urine and faeces to fall into the pit below. Whatever the system, staff must be able to observe, handle and feed all the rabbits easily.

The floor area must be sufficient to allow all the rabbits in the cage to lie comfortably and move around without disturbing the others, and eat and drink without difficulty. For rabbits over 12 weeks of age the Ministry of Agriculture Fisheries and Food (MAFF) recommend a minimum cage height of 45cm. A cage of this height with a floor space of $0.56m^2$ can accommodate one breeding buck or doe, a doe and litter up to four weeks of age, or eight weaners until 10 weeks old, when they go for slaughter. Many farmers now only use cages of this standard size. The *Codes of Recommendations* (MAFF 1987) set out minimum space allowances for different stock.

Each cage has a pellet hopper which hooks on to the front and possibly a hay rack which also fits outside the cage. An automatic watering system with a nipple drinker in each cage is essential. Supply pipes should connect to a header tank fixed inside the building so that the heat warms the water before it reaches the rabbits. This reduces the risk of drinking freezing water in cold weather.

## MANAGEMENT
### Breeding stock

Producers always house stud bucks and breeding does individually. They also carefully ration breeding stock to prevent over-fatness. Most bucks, and does which are not heavily pregnant nor lactating, require about 110g of pellets a day, but stockkeepers will need to feed more to any which fail to maintain good bodily condition. They should also feed pregnant does *ad lib*. During the last week of pregnancy consumption of food pellets will rise to 200-225g per day. Water must always be available; an adult will drink about 0.28 litres and a pregnant doe about 0.57 litres daily. The exact amounts will depend on the temperature and humidity of their surroundings.

**Figure 7.1**     **Welded wire mesh cage with food hopper and automatic watering. (Courtesy of J C Sandford).**

**Mating**

It is important to place the doe in the buck's cage, and not the other way round, otherwise serious fighting can occur. The handler should place the doe with her back to the buck because some keen bucks mate at once. A doe anxious to mate will raise her hind quarters ready for the buck to mount, but some only adopt this position after the buck has begun to mount. It may be necessary to hold a young doe during her first mating. The stockperson should place one hand over the ears and shoulders and use the other to support the body, raising the hind quarters so that the doe is in a suitable mating position.

After mating the buck falls sideways, often emitting a slight scream. Many breeders allow a buck to mate twice before removing the doe from the cage. If there is no mating within a few minutes the stockperson should remove the doe and re-introduce her again about six hours later.

Although it is possible to artificially inseminate (AI) does successfully, its use in commercial units is difficult to justify economically because of the high labour requirements. Farmers should seek expert instruction if they are going to attempt to use AI.

**Pregnancy diagnosis**

It is possible to diagnose pregnancy at between 14 and 16 days after mating, by palpating the abdomen. To do this the handler will need to gently place the doe on a non-slippery surface so that she is relaxed – if she is frightened she will tense her abdominal muscles making palpation difficult. The handler will need to place a hand under the body slightly in front of the hind legs, he/she will then be able to feel the embryos as marble-sized ovals slipping between the thumb and fingers on either side of the mid-line of the abdomen. The operator must learn to distinguish between the foetuses and the faecal pellets. Routine palpation allows the stockperson to detect non-pregnant does and remate them on day 18 or 19 after the unsuccessful mating. These days are selected because a barren mating may result in a pseudo-pregnancy which lasts for 16 or 17 days, at the end of which there is usually a period of greater fertility. Also it gives the stockperson the opportunity to feed the doe an adequate, but not over-fattening, ration until the doe becomes pregnant.

## The parturient animal

About three days before a doe is due to give birth it is important to provide her with a nest box (see Figure 7.2). The nest box should contain some absorbent material, such as shavings, under a bed of

hay. The doe will supplement this with fur she plucks from her breast. Most breeders use nest boxes made of wood which they disinfect before re-use, but others favour stout cardboard boxes which they use only once. The great disadvantage of these is that some does eat away the walls and so reduce their efficiency. Metal boxes are easy to sterilize, but in cool buildings they tend to lose heat rapidly, thereby chilling the young. They are also noisy and restrict ventilation. A nest box must be large enough to keep the young together for warmth. They should be about 40x25cm, with three walls each 25cm high and a front of 15cm to provide a sill. This enables the doe to enter without difficulty, but prevents the young from being drawn out on a teat while sucking if she is suddenly disturbed. Most nest boxes have open tops to allow farmers to inspect the young easily and it is usual to place them on the cage floor. Sunken nest boxes which fit below the cage floor have the

**Figure 7.2**     **Californian doe with young in nestbox. (Courtesy of J C Sandford).**

advantage that very small rabbits straying from the nest can find their way back without difficulty.

As soon as the female has given birth, it is important to record the number of young born dead and alive and to check the state of the nesting material. Some does will urinate in the nest box, therefore it may be necessary to replace any wet bedding. If the doe has not pulled out much fur the stockperson may pluck some more from her belly (which is quite loose at this time) or add fur which they have preserved from other nest boxes containing healthy young. As the most critical age for a rabbit is soon after it is born it is important to pay attention to detail at this time. After the initial inspection, it is usual to disturb litters under a week of age as little as possible.

Farmers remove nest boxes as soon as the young leave the nest, which is usually when they are about three weeks old, to allow more floor space. Leaving the nest boxes in the cage for a longer period leads to a build-up of moisture and heat which encourage bacterial growth and so enhance the risk of infection with pathogens.

### Fostering

By mating groups of does together, farmers synchronize females to give birth at about the same time so that when they first examine the litters they can foster some youngsters from large litters into smaller ones. Ideally each doe should have about 10 young to suckle. To foster new young, it is important to first remove the doe from the cage before carefully opening the nest from the top, so as not to disturb the fur lining from its original position. Some breeders rub the young they are fostering with faeces from their new mother before placing them in the nest, but with placid does this is not necessary. Attempts to foster animals more than four days old are almost always unsuccessful.

### Food and water

Food consumption rises markedly during lactation; a doe will consume 450g a day or even slightly more. Thus, it is essential to ensure that pellets are always available. A doe with a large litter may drink as much as 4.5 litres of water a day at the time of peak lactation. If she is unable to obtain all the water she requires her milk yield will suffer and the young will go hungry. Therefore on economic, as well as on welfare, grounds it is vital to check that the nipple drinkers in the breeding cages are working efficiently.

### Weaning

Farmers usually wean young at the age of four weeks, by moving them to another cage and leaving the doe in the one with which she is familiar. The doe has usually been re-mated when the young are three weeks old; at which time her milk yield is rapidly decreasing. This re-mating/weaning regime allows three weeks in which the doe can build up body reserves before the birth of the next litter. Such a regime enables females to produce six litters per year. Some producers re-mate does at two weeks after parturition but still wean the litter at four weeks of age. This enables the doe to give birth to seven or even eight litters per annum. The farmer can obtain satisfactory results providing he/she supervises the stock carefully and feeds them very well, but it is vital to pay great attention to the condition of each doe throughout the year. A mating between the third and seventh days after parturition generally leads to conception, but this practise is unacceptable on welfare grounds.

## MEAT PRODUCTION

Unlike many other farming systems, producers usually breed, rear and sell all their own animals. Some have tried to establish fattening units, collecting rabbits at weaning, but the outbreaks of disease resulting from mixing from several sources make this system uneconomic.

Before starting a farm it is essential for the prospective producer to enter into a provisional agreement with a meat marketer. It is also important to ascertain the weight of rabbit they require, because there is no standard grading system and different packers have slightly varied market needs to satisfy. Most, however, require weights from 2.2 to 2.5kg, which is normally when the rabbits are nine to ten weeks old. As rabbits may lose as much as 0.1kg during transit from the farm to the packing station, producers should dispatch the animals at a slightly heavier weight to account for this loss.

Meat rabbits should have a blocky conformation with well muscled hindquarters and a short chest area. The killing out percentage at slaughter is a good guide to body type in life. This is the measure of the amount of meat in a carcase of which about 70 per cent should consist of edible meat. The flesh must be free from bruising, necessitating careful handling when loading rabbits into crates for transport. As the liver is a valuable by-product, it is important to keep the herd free from hepatic coccidiosis, a disease which causes white spots on this organ leading to down-grading and rejection.

The rabbit producer must try to provide a regular supply throughout the year. It is more difficult to breed during the autumn and winter when the demand for carcases is highest. The industry can even out this discrepancy by freezing carcases during the summer and selling them in winter, but this adds to the cost of marketing and frozen meat is unpopular amongst some consumers.

Most producers fatten rabbits in groups of about eight in cages of the standard type, but it is possible to run larger groups in colonies on wire mesh floors. The additional exercise in the more spacious areas increases food consumption and there is a greater risk of bullying. It is also not possible to arrange colony type pens in tiers.

The amount of food a growing rabbit consumes in the period from weaning to sale varies between 110 and 115g a day. In view of the high cost of pelleted foods the food conversion ratio (FCR) is an important economic measure and should be about 3:1, although some producers achieve a narrower ratio. FCR is the quantity of food the stockperson feeds to the litter (which includes the doe until the litter is weaned) divided by the weight they gain.

## REPLACEMENT STOCK PRODUCTION

Although rabbits can live for up to nine years most commercial farmers cull breeding stock when they are about three years old because prolificacy falls. Breeders of New Zealand White, Commercial White or Californian rabbits usually breed their own replacement stock, occasionally buying in a new buck to avoid excessive inbreeding. Large producers with hybrid stock retain animals which they have carefully selected into inbred parent lines, for as long as it is economic.

The breed and genetic characteristics a producer uses are very important because the future success of the enterprise depends, to a large extent, on the quality of the breeding stock. Farmers should select breeding stock from the progeny of mature does which have produced several litters. It is important that they have successfully re-mated at the required times, particularly during the winter months. The parents should have produced uniformly high litters with few deaths between birth and weaning. At weaning all the young should have been uniform in size and should have then grown steadily to attain the killing weight at the earliest possible age for the strain, having efficiently utilized their food (as shown by the food conversion ratio). The producer will need to check the opinion of the packer to ascertain whether or not the carcase quality of rabbits from earlier litters was good. Prospective replacements should have been sired by bucks which had mated keenly and produced good sized litters with a high carcase quality.

A careful examination of each individual is essential; the breeder should not use any animal which shows any weakness in conformation. A blocky appearance is most important in a buck and a doe must have at least eight teats. The producers should individually cage those rabbits he/she has selected until they reach breeding age; it is important to reject any which do not grow evenly. Maturing rabbits will consume an increasing quantity of pellets; they should not consume more than about 140g a day, otherwise they will grow fat and unfit.

## PELT PRODUCTION

The young age at which the meat trade slaughters rabbits reduces the value of their pelts because these are not sufficiently mature for manufacturing into garments. Manufacturers do use pelts from older rabbits, after tanning, to make coats, coat collars or gloves. They prefer white furs to coloured because they can dye them. In the UK, however, farmers keep rabbits primarily for their meat because the low value of rabbit skins means that pelt production alone is uneconomic.

## WOOL PRODUCTION

In the UK, Angora rabbit wool production is only small scale and many breeders spin and knit the wool from their own stock into a variety of fashion garments. The world price for Angora fibre is lower than that which would be economical for British growers and so Angora farmers wishing to sell the fibre they produce in bulk to knitwear firms must ensure that they have an outlet for a quality product before they start an Angora rabbit farm. The largest Angora fibre producer is China.

Some breeders still use British strains of Angoras, which produce high quality fibre, but most are now keeping the much higher-yielding European strains from Denmark and Germany. The coats of British Angoras need grooming periodically, but those of the European strains are comparatively free from matting and so grooming is rarely necessary. Shearing white adults every 90 days, when fibre lengths reach 60-80mm, will yield up to 1400g per rabbit per year, while coloured animals producing golden, chocolate, smoke and blue fibre, yield about 1000g per year each. Economically, rabbits produce the best yields between the age of nine months and four years, after which the rate of wool growth slows considerably.

Angora rabbit management is labour-intensive and farmers need to cage animals individually to avoid damage to the wool. The does are not very productive, producing an average of 24 young per year from four litters.

## HANDLING

To lift a rabbit the handler needs to place one hand over the ears and firmly grasp the skin over the shoulders, at the same time slipping the other hand under the body to bear its weight. It is possible to carry them in a similar manner. Rabbits must never be picked up or held by their ears. A handler may lift young animals by grasping the loose skin about half way down their backs.

## GENERAL CARE

### Sexing

Whilst it is possible to sex rabbits soon after birth, it is more common to do it at weaning. The handler needs to hold the rabbit with one hand grasping the ears and underlying skin, while using the other to apply gentle pressure to the sides of the reproductive orifice. The male is identifiable by an emerging immature penis, while in the female a V-shaped slit appears.

### Toenail trimming

Stockpersons must periodically shorten the toenails of adult rabbits to avoid overgrown nails catching on the cage floors. The handler will need to place the rabbit on a flat surface and hold each foot firmly while cutting the claw end with strong nail clippers. In white footed breeds the red quick shows clearly and must not be cut. In black nails the quick is not obvious, so, to avoid causing bleeding, the operator should only remove the sharp tip.

### Overgrown teeth

An overgrowth of the incisors can sometimes interfere seriously with feeding and cause damage to the lips. Only a veterinary surgeon or a trained operator should trim the teeth.

## IDENTIFICATION

It is important to be able to identify breeding rabbits easily so that the stockperson can keep records on each. The most satisfactory method is to tattoo an individual number in the ear. Whilst it is possible to use aluminium ear tags, rabbits may tear these off, causing tissue damage. It is inadvisable to use the closed leg-rings which fancy rabbit breeders use. They are expensive and, as the leg sizes of commercial rabbits are not standardized, some rings will be too small and so cause pain and swelling.

## STOCKMANSHIP

The success of a rabbit farm depends on the efficient management of highly productive animals. A skilled stockperson can rear rabbits with few losses in moderate surroundings, while one who is not properly trained or is inefficient will not obtain good results even in excellent buildings. A good stockperson must know what action to take in an emergency. Anyone considering working on a rabbit farm should attend a short course and then, if possible, work as an apprentice on a successful farm. Some farmers selling rabbits and equipment are willing to provide these services.

It is essential to inspect all the rabbits carefully at least once a day to make certain that they are in good bodily condition and that they have ample supplies of food and water. It is desirable to weigh stock at different stages of their development; this requires accurate scales with a deep pan for rabbits to sit in.

## HEALTH

A healthy rabbit is alert with bright, wide open eyes. It is active and has a good appetite. The body should be well fleshed and evenly covered with fur and the faecal pellets lying in the pen should be hard. Stockpersons should inspect each rabbit frequently during the day because, once ill, they deteriorate rapidly. Listlessness, eyes half-closed, a huddled, tucked up posture, grinding of the teeth and a loss of appetite are indications of ill-health. Sneezing is a sign of respiratory disease and head shaking may denote ear canker. These variations from normality are visible while the animal is in its cage, and the stockkeeper on his/her daily inspection should look out for these. To examine the rabbit in more detail, it is necessary to handle it. The nose should be dry, the inside of the ears clean, the fur sleek without bare patches and the undersides of the hocks free from sores. The fur round the anus must not be matted with faeces and the mammary glands should not feel hot. The normal body temperature of a rabbit varies between 39 and 40˚C (102.2 and 104˚F).

When establishing a new herd the prospective farmer must take care to acquire only disease-free rabbits. On large farms producers can maintain a closed herd into which they do not introduce new rabbits, but most will need to buy rabbits occasionally in order to maintain stamina and increase performance levels. Animals must come from a reliable source; it is important to isolate them for two or even three weeks, because the stress following the move may allow the development of disease conditions which were dormant on the farm of origin.

Visitors, food, water, wild rodents and birds, and flies may also introduce infectious agents. Visitors who have been in contact with rabbits should wear disposable overalls and shoe covers. Food, particularly hay, should not have come in contact with wild rabbits and water should come from a pure source. Stockpersons should take measures to prevent rats, mice and wild birds from entering the building and they should also use sprays to control fly numbers.

Some common diseases are liable to occur despite the care farmers take to guard against them. Table 7.3 lists these together with recommendations for their control. The best, and most widely used method for the administration of medicines to large numbers of animals is to include the drug in the food in the case of disease prevention (this can be done by the compounder), and in the drinking water for curative measures. The treatment of individual rabbits may be uneconomical; if a farmer does treat an individual, he/she tends to give the medication by injection. There are three ways of injecting a rabbit; subcutaneously behind the shoulders, intramuscularly into the muscles at the top of the hind leg or intravenously into one of the ear veins.

There are two serious viral diseases which only affect rabbits and which occur in the wild. The myxoma virus causes myxomatosis. The distinctive clinical signs are oedema of the head, eyelids and genitals with swelling of the eyelids leading to blindness. Appetite usually remains normal until shortly before death. The virus is spread mainly by the rabbit flea, but direct or indirect contact can also transmit the virus.

Viral haemorrhagic disease (VHD) was first reported in the UK in 1992 by fanciers breeding show rabbits; in 1994 it was recorded in wild rabbits. The causal agent is a calicivirus and the *Specified Diseases (Notification) Order 1991* made it a notifiable condition in Britain. The disease appears most commonly in adults, but may affect any rabbit over eight weeks of age. The symptoms are depression, loss of appetite, respiratory distress, incoordination, a blood-stained mucus discharge and death within one or two days. The virus is spread by direct contact with infected rabbits, contaminated food or water or by mechanical transmission on equipment, clothing, rodents, insects or birds.

Vaccines which protect rabbits against these two diseases are available and producers should seek veterinary advice on their use.

As a regular hygienic measure, and certainly after a disease outbreak, farmers should disinfect the buildings. It is important to first thoroughly clean all the surfaces with which the rabbits have come in contact before scrubbing the inside of the building with a chemical disinfectant approved for general farm use under the *Diseases of Animals (Approved Disinfectants) Order 1978 as amended.*

**Table 7.3    Common rabbit diseases.**

| Disease | Clinical signs | Prevention |
|---|---|---|
| Coccidiosis<br>    intestinal<br><br>    hepatic | Diarrhoea in young<br>Sudden death<br>Often no signs in life. At post-mortem examination white spots on liver | Wire cage floors. Coccidiostat in food |
| Mucoid enteritis | Mucoid material instead of faeces | Increase fibre in diet |
| Snuffles (chronic rhinitis) | Sneezing, bilateral nasal discharge | Eliminate diseased animals.  Improve ventilation |
| Sore hocks | Necrosis on underside of hind feet | Avoid rough cage floors.  Breed for thick fur on the soles of the hind feet |

**Transport**

It is important to move adult rabbits in individual containers to avoid fighting. Farmers normally transport young animals in wooden or metal crates, preferably in batches of about 10. These crates must be strong and have adequate ventilation and be spacious enough to avoid overcrowding. The doors must be sufficiently large to allow the easy insertion and extraction of the rabbits and the floors must be solid to prevent the soiling of the animals in the lower units by faeces and urine falling down from those higher in the stack of crates. As it is impractical to feed and water in the crates, those transporting rabbits should not confine them for more than eight hours.

**Slaughter**

The usual method of slaughtering meat rabbits is to administer a heavy blow to the back of the head and then to decapitate the animal. The operator holds the rabbit firmly by the hind legs just behind the hocks with the left hand or places it on a flat surface and lifts it by the ears until the front feet are just clear of the stand. It is important to check that the blow kills the rabbit, not merely stuns it. For the small specialist market which requires a carcase complete with the head, either dislocation of the neck, or electrical stunning after which the throat is cut, will ensure that the skull and brain remain intact. To manually dislocate the neck, the operator will need to hold the rabbit by the hind legs with the left hand as before while grasping the neck behind the ears with the right and giving it a sharp pull with a downward and backward twist.

**THE WAY AHEAD**

As the success of a rabbit farm, and the welfare of the rabbits, depends to such a large extent on the skill and dedication of the stockperson, colleges of agriculture should consider holding specialized staff training courses.

An investigation into the optimum environmental conditions for rabbits of different ages could lead to better housing systems which might help to increase the resistance of rabbits to disease, accelerate the growth of fattening animals and make winter breeding more efficient. Producers should only attempt to raise the number of young weaned per doe per annum if they can achieve it without adversely affecting the health of the mothers.

**FURTHER READING**

**Adams C E** 1987 The laboratory rabbit. In: Poole T (ed) *The UFAW Handbook on the Care and Management of Laboratory Animals, 6th edition pp 415-435*. Longman: Harlow, UK (new edition in preparation)

**Cheeke P R** 1987 *Rabbit Feeding and Nutrition*. Academic Press Inc: Orlando, USA

**Lebas F** 1986 Nutrition of rabbits. In: Wiseman J (ed) *Feeding of Non-ruminant Livestock*, pp 63-69. Butterworths: London, UK

**Ministry of Agriculture, Fisheries and Food** 1985 *Commercial Rabbit Production*. Bulletin No 50. HMSO: London, UK

**Ministry of Agriculture, Fisheries and Food** 1987 *Codes of Recommendations for the Welfare of Livestock: Rabbits (reprinted 1991)*. PB0080 MAFF: London, UK

**National Research Council** 1980 *Nutrient Requirements of Rabbits, 2nd edition*. National Academy of Sciences: Washington DC, USA

**Okerman L** 1994 *Diseases of Domestic Rabbits, 2nd edition*. Blackwell Scientific Publications: Oxford, UK

**Portsmouth J I** 1987 *Commercial Rabbit Keeping*. Iliffe: London, UK

**Sandford J C** 1971 *Reproduction and Breeding of Rabbits*. Watmoughs: Idle, Bradford, UK

**Sandford J C** 1995 *The Domestic Rabbit, 5th edition*. Collins: London, UK

# 8 Red deer

## A J Hanlon

## THE UK INDUSTRY

UK producers mainly farm red deer (*Cervus elaphus*) although some use fallow deer (*Dama dama*). As a livestock species, red deer possess many attractive traits; these include a long reproductive life (over 15 years), high quality meat production and adaptability to a range of nutritional environments (Hamilton 1994a).

### Development of the industry

Deer farming started in the UK in the early 1970s as part of an initiative to find an alternate land use for poor quality hill pasture and to exploit the large populations of wild red deer in the Scottish Highlands (Blaxter *et al* 1974, 1988). During the past 20 years, the number of deer farms has increased in Great Britain to nearly 300 (Table 8.1).

### Structure of the industry

Producers generally slaughter farmed red deer for venison at approximately 16 months of age; they do however keep some hinds as replacement stock or to sell for breeding (Table 8.1). The type of production

**Table 8.1  Deer farming statistics for 1998 - Scottish Office Agriculture, Environment and Fisheries Department (SOAFD) 1997 and Ministry of Agriculture Fisheries and Food (MAFF) 1998.**

|  | England & Wales | Scotland |
|---|---|---|
| Number of deer farms | 240 | 47 |
| Total number of deer | 18,517 (red deer)<br>5,326 (fallow deer) | 7,227 |
| **Breeding stock**<br>Breeding females<br>    Yearlings<br>    Adults<br>Breeding males | <br><br>1,795<br>10,512<br>10,897 | <br><br>744<br>2628<br>244 |
| **Number of deer sold alive**<br>Breeding<br>    Females<br>    Males<br>Fattening<br>    Females<br>    Males | <br><br>531<br>146<br><br>673<br>1063 | <br><br>-<br>-<br><br>485<br>1121 |
| **Number of deer slaughtered**<br>    Females<br>    Males | <br>2,776<br>3,868 | <br>633<br>1,102 |

system depends on the land resources available (Hamilton 1994a). Hill farmers produce weaned deer calves that they sell onto upland or lowland farms for further feeding or breeding. In contrast, upland and lowland farms concentrate on producing prime venison and breeding stock.

The UK government does not subsidize venison production. In the past New Zealand and some other countries, farmers have gained high revenues from velvet antler production in addition to venison. The velvet is used in some oriental medicines. In the UK *The Welfare of Livestock (Prohibited Operations) Regulations 1982* forbids the harvesting of antler velvet.

**Meat quality**

Farmers produce prime venison from deer they slaughter at 1–2 years of age (maximum 27 months). Venison has a lower fat content and a higher lean meat to bone ratio in comparison with other meats such as beef, lamb and chicken (Table 8.2). The proportion of fat is higher in hinds than stags (except stags pre-rut) and in older animals, irrespective of sex (Drew 1994). The handling of animals pre-slaughter and the processing of the carcass post-slaughter also affects meat quality. For further details see Drew (1994).

**NATURAL HISTORY AND BEHAVIOUR**

Wild red deer evolved as forest-dwellers inhabiting open woodlands and forest margins. They have adapted to survive on open moorland, but will have a lower liveweight and reproductive success; this is due to the poorer quality of forage and the lack of shelter (Kay and Staines 1981). Deer generally are extremely vigilant and agile animals, which are easy to excite or frighten. However, it is possible to tame or habituate wild deer to the sight and sound of humans.

For most of the year, wild red deer form single-sex groups, only coming together during the mating (rutting) season. Adult females (hinds) form family groups. They consist of related adult hinds (mothers, aunts and cousins) and their offspring from the current and previous years, including males (stags) of up to three years old (Clutton-Brock and Albon 1989). Group sizes vary from two to more than 50 hinds, depending on habitat and season (Mitchell *et al* 1977). Stags form smaller, less well structured, groups (sometimes known as the bachelor herd or stag party) containing adult and juvenile stags.

**Feeding**

In temperate climates red deer show seasonal variations in feeding behaviour. Diet varies depending on the location and season and includes grasses, herbs and woody browse (Kay and Staines 1981). Photoperiod and the quantity and quality of food available determines the seasonal feeding activity (Heydon *et al* 1993). Generally red deer have 5–11 feeding bouts per day, the majority occur during daylight, with peaks in activity at dawn and dusk (Kay and Staines 1981; Sibbald 1994). Feeding behaviour may also vary between the sexes; hinds select areas with a higher quality forage than stags. Stags and hinds in areas of poor quality forage compensate by increasing the quantity of forage they consume (Kay and Staines 1981; Heydon *et al* 1993).

**Table 8.2**     Characteristics of different types of meat (*modified from Drew 1994*).

| Species | Age at slaughter (months) | Carcase weight (kg) | Lean (%) | Fat (%) | Bone (%) |
|---|---|---|---|---|---|
| Beef | 24 | 239 | 59 | 23 | 18 |
| Lamb | 6 | 16 | 51 | 27 | 22 |
| Chicken | <2 | 1.2 | 59 | 15 | 24 |
| Red deer | 26 | 63 | 73 | 7 | 20 |

## Pre-rut behaviour

Aggressive behaviour intensifies in stags during the pre-rut period as levels of testosterone increase. Stags perform dominance displays to establish or re-assert their place in the hierarchy. Typical behaviours include roaring, sparring and scent marking (Clutton-Brock *et al* 1982). Mating strategies vary with species of deer and between different populations of the same species. Red deer stags commonly compete for a territory, or rutting site, which are areas the hinds frequently use in the autumn. By mid-September stags leave the bachelor herd, moving to rutting sites, where mature stags (normally over five years old) compete for individual sites. Stags in possession of a site are extremely territorial, instigating fights against trespassing stags and those on neighbouring sites.

## Rut behaviour

By the end of September stags start to gather a harem of hinds on to their territory. Apart from mating, stags have no other investment in their offspring. To maximize their reproductive success it is important for them to mate with as many hinds as possible (Clutton-Brock *et al* 1982). Stags therefore devote most of their time and energy defending, maintaining and increasing their harem. Consequently they reduce the time they spend feeding by 39 per cent (Clutton-Brock *et al* 1982). Stags may lose up to 20 per cent in liveweight and 80 per cent of their fat reserves during the rut (Clutton-Brock and Albon 1989). Stags maintain their harems for approximately three weeks, abruptly stopping rutting behaviour once they are at a lower critical liveweight (Clutton-Brock *et al* 1982).

Interactions between stags and hinds intensify in early October, coinciding with the onset of oestrus in hinds. Hinds show no overt changes in appearance to signify their reproductive state, although in late oestrus hinds may mount each other or the stag. To increase their chances of successful matings, stags periodically lick and smell the perineal area, urine and faeces of hinds. This behaviour has the effect of stimulating oestrus and determining their reproductive state. Stags will chase and mount hinds in oestrus, but unless receptive, hinds will move away. Mating follows when a hind stands still as a stag approaches. Copulation is rapid, lasting less than 30 seconds. It is easy to differentiate from other mounts because during ejaculation the stag lifts himself off the

ground, pushing the hind forward. Once the stag has dismounted the hind squats to urinate. Each hind mates up to five times during each rut (Guinness *et al* 1971).

## Calving

In May, shortly before calving, hinds move away from the herd to give birth in isolation. As parturition approaches the udder begins to swell and shortly before calving the vulva becomes red and swollen. As labour commences, hinds show characteristic behaviours such as repeatedly touching and grooming their flanks and running in a high step gait with neck outstretched. The duration of labour generally depends on the size of the calf and the ease with which it passes through the pelvis. As labour progresses, hinds become increasingly restless, alternating between lying and standing. At parturition, when the calf has almost fully emerged, the hind stands up to ease the calf out and to sever the umbilical cord. Calves are normally born in the anterior-dorsal position, although breech births may also occur (Blaxter *et al* 1974).

## Mother-infant interactions

Following parturition, the hind consumes all traces of foetal membranes from the calf and the birth site. In farmed deer, newborn calves can normally stand within 30 minutes and suckle within 40 minutes (Arman *et al* 1974). After the hind has suckled, she moves away from the birth site with the calf slowly following. At this stage the calf is not fully mobile and so after a short distance it lies down; it remains in the same place – known as a lie-out – until the hind returns to suckle. As the calf grows stronger it follows the hind for longer distances until, by the end of the first week, it can run with the herd.

Hinds suckle their calves for the first few days at 2–3 hour intervals. While the hind suckles, she licks the calf's perineum to stimulate defecation and urination; she then consumes the faeces and urine as the calf produces them. Young and old hinds tend to suckle their calves for longer. Peak lactation occurs at approximately 40 days after birth (Loudon *et al* 1984).

Suckling bouts decrease as lactation progresses and by three months of age calves suckle, on average, 3–4 times a day. The hind rejects up to 50 per cent of sucklings the calf attempts by the time they are six months' old. At this stage the average bout lasts for only 30 seconds. The timing of weaning depends on the reproductive state of the hind. Pregnant hinds will gradually stop lactating in the winter, although yeld (barren) hinds may continue to nurse their young until the following summer (Clutton-Brock *et al* 1982).

## BREEDING
### Reproductive biology
Weight largely determines the onset of puberty in both males and females. Stags normally reach puberty at about 16 months of age (yearlings). The presence of mature stags, however, will preclude them from mating; yearlings may therefore not become fully sexually active until they are sufficiently large and strong to compete for territories and gain access to hinds.

In the wild, Scottish hinds reach puberty at 2–3 years of age (Kay and Staines 1981). Oestrus coincides with shortening day length in the autumn, but the presence of a rutting stag will also act as a stimulant (Lincoln 1985). Red deer are seasonally polyoestrous and come into oestrus every 19 days for 12–24 hours until they conceive. Hinds normally conceive in their first oestrous cycle, although some may continue to cycle 2 or 3 times before conceiving (Guinness *et al* 1971). Gestation lasts for approximately 231 days (range 226–238 days). Red deer normally give birth to singleton calves, although very occasionally twins.

Antler cleaning in mid-August precedes the mating season (the rut). An increase in the hormone testosterone triggers antler cleaning. Levels continue to rise in September and October, only returning to basal concentrations in late November (Fennessy and Suttie 1985). Other physiological changes also occur before the rut, affecting the appearance of stags, such as an increase in the size of the neck girth, growth of a mane and the enlargement of the testes.

### Antler development
At puberty, stag calves develop pedicles, circular bony stalks, on the frontal lobes of the skull, which later act as a platform for antlers. Antlers begin to grow between April and May when pedicles are 5–6cm long (Fennessy and Suttie 1985). During growth, velvet covers the antlers, forming a thick furry protective skin, containing a network of blood vessels and nerve endings. The rich supply of nutrients from blood vessels inside the velvet and the soft developing bone supports growth. By August antler development is complete. In the final stages of growth a ring, the coronet, forms around the base, constricting the blood and nutrient supply to the velvet. The antler velvet shrivels and begins to peel off and the stag aids its removal by actively rubbing its antlers against trees or through bushes, exposing clean, hard antlers. Hard antlers consist of dead bone. Stags retain their antlers during the rut and over winter, but cast them in early spring. Before casting, the stags leave the herd, antlers loosen and each is separately cast within a day or two of the other. New antler growth begins shortly after casting. In yearlings, antlers consist of simple spikes. The size and complexity of antlers increases during successive years, except in old stags, when the process begins to reverse (Putman 1988).

### Mating
The deer farming year begins in the autumn, with the rut. One month before the start of the rut, stockpersons de-antler (disbudding or castration prevents antler growth; see Goddard 1994a), weigh the stags and treat them with anthelmintics.

The farmer introduces stags to rutting groups in mid to late September or immediately following weaning (Hamilton 1989). Many farms operate single sire mating systems. The number of hinds per group (stag to hind ratio) depends on the age of the stag (Table 8.3) (Haigh and Hudson 1993; Fletcher 1994a). Producers leave stags in their mating groups for one to two months, but may replace them during the last few weeks to ensure that all hinds have been covered. Yearling hinds have a lower weight than adult hinds at the start of the rut and consequently come into oestrus up to three weeks later. Similarly, yearling stags become sexually active later in the rut than adult stags and some may fail to show rutting behaviour. This is particularly so if they are in a paddock next to a mature rutting stag (Haigh and Hudson 1993). Therefore, to maximize the reproductive success of yearlings, three or four yearling stags are put with a group of yearling hinds.

**Table 8.3    Recommended stag to hind ratios.**
*(modified from Haigh and Hudson 1993)*

| Age of stag | Number of hinds |
|-------------|-----------------|
| 16 months | 10 |
| 2 years | 20 |
| 3 years | > 30 |
| 4 years | > 40 |
| 5–8 years | 50 |
| 9 years | < 50 |

## Calving

On farms, calving normally occurs between the end of May and the beginning of July. The condition of hinds during the rut and the date of introducing the stag to the mating group partly determines the time. Calves born early in the season have higher weight-gains than calves born later (Adam and Moir 1987). In addition, late born calves are more vulnerable to a build-up of bacterial infections. Management during the mating period should therefore promote early calving and reduce the duration of the calving period to less than six weeks. This maximizes weight-gain, minimizes infection and advances weaning of calves, which enables hinds to recover body condition before the rut. To achieve these, producers must ensure that hinds enter the rut at their target weights (ie over 80kg), and remove stags in mid-November (Fletcher 1994a).

In early May, stockpersons allocate pregnant hinds to calving groups. It is common to separate yearling and adult hinds. Further segregation may be necessary according to the stage of pregnancy. Most hinds calve easily and, compared with other livestock, relatively few births need assistance. By monitoring the behaviour of hinds and the duration of calving, it is possible to identify those in difficulty. Labour normally lasts about two hours, but may take longer in yearlings or primiparous hinds (Hanlon 1996a).

Stockpersons should tranquillize, and move to handling facilities, those hinds in calving difficulty (Griffiths 1994). Once in the handling area it is important to disinfect the perineal area of the hind and hands and arms of the person assisting the calving. If there is any risk of brucellosis, leptospirosis or other infections, handlers should wear gloves and arm protection (Haigh and Hudson 1993). Using ample lubricant, the handler should check the position of the calf to see if its neck or limbs have become twisted and to assess the size. It may be necessary to use calf ropes (Haigh and Hudson 1993).

After delivery, the stockperson may need to remove mucus from the calf's nostrils, which they achieve by tickling its nose with straw to stimulate sneezing (Haigh and Hudson 1993). Since many hinds abandon calves following assisted calving, it is important to avoid further human contact. The calf should be left alone with its mother for at least one hour. After assisting a calving in the field, it is important to move the hind and calf to a confined space to allow bonding. If the calf is still wet after one or two hours alone with its mother, then she has probably rejected it (Haigh and Hudson 1993).

Farmers normally tag and weigh calves 12–36 hours after birth (Haigh and Hudson 1993). Tagging calves within 12 hours of birth can lead to abandonment and mismothering. However, tagging calves later in the season may be equally effective and avoid the risk of abandonment. Stockpersons can handle calves with their mothers in July (Gould 1989). Once in the handling area, it is possible to separate off calves and return hinds to the calving paddock. The calves are then weighed and tagged before being reunited with their dams. Mother and offspring will reunite after a period of separation if the handler releases the calves back into the paddock one at a time.

## Weaning

Producers usually wean calves at approximately 100 days after birth (Hanlon 1996b). They normally wean before the onset of winter, immediately before the start of the mating season, but the exact timing will vary with location and year. However, it is possible to wean late-born calves after the rut. At weaning, farmers weigh the calves, vaccinate them against clostridial diseases and transfer them to housing or a new paddock as far from their mothers as possible. Apart from tagging, calves usually have had no experience of handling and may be unfamiliar with the raceway and handling system; this makes them difficult to muster and prone to escape. Separation from the maternal herd, handling and changes in diet and environment impose a severe acute stressor on calves (Hanlon 1997); this may influence the effectiveness of prophylactic treatments at separation (Hanlon *et al* 1994).

## ENVIRONMENT AND HOUSING

### At pasture

Deer farmers generally maintain adults on pasture throughout the year. The number of paddocks depends on the production system. Location and access to paddocks are important considerations. Calving paddocks are often close to the farm buildings and handling facilities; this enables farmers to monitor calving from a distance without disturbing the deer. Also, it enables the stockperson to treat hinds experiencing calving difficulties with minimum disturbance due to the proximity of the handling facilities. Trees providing natural shelter, although not essential, will improve calving success (Hamilton *et al* 1994).

### Housing

In temperate climates producers usually house weaned calves over winter. This is because they possess proportionally less body fat than adults and are therefore less able to cope with climatic stress (Blaxter *et al* 1988). The minimum space allowance is 1.5m²/head; however recent research suggests that a higher rate of 2.25m²/head may be more appropriate (Hamilton and Soanes 1995).

## HANDLING

Farmed deer may be handled less than four times a year. At other times interactions with humans are restricted to routine treatment or to stockpersons delivering supplementary feed. The behaviour of deer during handling will depend on their sex, age and experience. New stock and calves at weaning are generally difficult to handle because they are unfamiliar with the stockperson and the layout of the handling facilities. At weaning, calves may try to escape through fencing or underneath gates. Pollard *et al* (1992) suggest that including a tame hind in the weaning group can reduce panic among calves. The same principle applies to new stock.

Individual deer may respond aggressively to stockpersons during handling or routine feeding. It is therefore important to be able to recognize threatening behaviour that precedes an attack, such as prolonged direct eye contact, ears flattened back, grinding teeth and snorting. Stags during the rut, and to a lesser degree, hinds during calving may also behave aggressively (Blaxter *et al* 1974). Stockpersons should generally avoid handling stags before or during the rut as it is dangerous. However, it may be unavoidable; for example it may be necessary to de-antler stags before the rut or move them into a new group during the rut. It is important to take special precautions when handling rutting stags. Stags may need sedating and handlers should muster and pen them individually, especially when in hard antler, because they will fight.

When mustered (gathered together), deer should be able to express their normal behaviour as far as possible. Except for adult stags, it is advisable to always muster deer in groups. Segregation from the group may cause individuals to panic and try to escape; in extreme cases this may cause capture myopathy (Haigh and Hudson 1993). In these conditions, it is better to allow the separated animal to regroup by mustering a small group of deer to join it. Deer will move better if the stockperson is familiar to them and musters them at their flight distance.

### Handling Facilities

Besides good stockmanship, it is possible to reduce stress during handling by incorporating specific features into the design and layout of handling facilities. The system should enable animals to move continuously through and prevent self-injury.

Paddock design and access to the handling facilities are important considerations. To increase the ease of handling, paddocks should be long and narrow or taper towards the gate. This will allow the handler to muster deer out of the paddock as a group (Haigh and Hudson 1993). Ideally each paddock should connect to a raceway leading to the handling facilities. Moving deer may be difficult in situations where access is via one or more adjoining paddocks (Blaxter *et al* 1988). Farmers can move deer with minimal handling by leaving the adjoining gate open overnight and using feeding stations to attract deer into the adjacent paddock. Alternatively, they can muster deer along 'guide' fences (Blaxter *et al* 1988).

Most accidents occur between the home paddocks and handling facilities (Burdge and Burdge 1989). Deer are extremely agile, therefore it is important to muster them at a walking pace, to prevent panic and pile-ups. Raceways should be sufficiently wide to allow handling in groups and to allow access for farm machinery.

Deer, like other livestock, move faster along curved, rather than straight, raceways. For long raceways, Haigh and Hudson (1993) recommend that straight sections should not exceed 100m. Inexperienced deer, or those reaching the end of the raceway, may try to escape over or through the fencing or by breaking-back along the raceway. The walls must therefore be solid, acting as a visual barrier and sufficiently high to deter deer from trying to escape. It is inadvisable to try to physically stop deer, particularly adults, from breaking-back. Fitting long raceways with gates, enables the stockperson to limit break-back by closing gates as deer move along.

All deer farms should have handling facilities so that they can test for tuberculosis if it is necessary. The complexity of the facilities depends on the size of the farm and number of deer. Most consist of a crush with a drop floor which in effect lowers the animal so that it is held firmly between the sloping sides of the crush, a weigh-crate and at least one holding pen. Some procedures, such as the application of pour-on drugs, do not need physical restraint and so the stockperson can conduct these in holding pens on groups of deer. It is advantageous to have holding pens of different sizes, to hold different numbers and sizes of deer (Blaxter *et al* 1974). Mobile handling facilities can offer advantages over permanent facilities in some circumstances (Hamilton 1994c).

### Fencing

Deer farmers use three main types of fencing: strained wire (horizontal wires), mesh and wire and baton. They also use electric fencing for controlling grazing or to prevent fighting between rutting stags in adjacent paddocks.

Fencing requirements differ for perimeter and internal use. Perimeter fences need to be approximately 2m high to prevent deer from escaping and other wildlife from entering (FAWC 1985). Some

areas of internal fencing, such as raceways, need to be stronger as they will be under greater pressure, particularly during mustering. At such pressure zones fencing needs to prevent escape, entanglement or injury. Fencing needs to be visible at deer pressure zones. Farmers often achieve this by covering the wire with hessian.

With all fencing, wire spacing needs to be sufficiently small to prevent deer from pushing their heads through, otherwise they can damage ear tags or antler buttons. Entanglement in fencing is one cause of neonatal mortality in calves; using a small wire-spacing or small-mesh netting on the lower portion of the fence will reduce this. It is inadvisable to use electric fencing in calving paddocks because of the risk of electrocution to calves.

### NUTRITION

Farmed red deer have a higher basal metabolic rate than sheep and require approximately one third more energy. Digestion is similar to that of traditional ruminant livestock given the same diet (Kay and Staines 1981).

### Nutrition of breeding hinds

Nutrition can dramatically increase the reproductive success of red deer hinds by reducing the age of first calving, increasing the rate of conception and calf survival. Liveweight primarily determines the age of first-calving (Blaxter *et al* 1988). By improving the nutrition of farmed hinds, producers have advanced the onset of puberty to 16 months; this is at least one year before puberty in wild hinds. Nutrition also influences the subsequent reproductive success of hinds. In the wild, lactating hinds have a 50 per cent chance of calving in the following season (Kay and Staines 1981). Conception rate positively correlates with the pre-rut weight of hinds: those over 70kg have a 90 per cent rate of conception compared with only 50 per cent of hinds less than 65kg (Hamilton and Blaxter 1980). Liveweight and nutrition influences conception date. For example, yearlings on heather-dominant pasture conceive approximately 17 days later than those on good quality pasture; similar trends occur in adult hinds (Hamilton and Blaxter 1980).

Adult hinds have a maintenance metabolizable energy (ME) requirement, in relation to their metabolic liveweight, of approximately $0.57MJ/kg^{0.75}$. Their requirement varies with live weight and reproductive status. For example, 80kg hinds require approximately 15MJ ME/day for maintenance; this increases to about 17-20MJ ME/day during late pregnancy, and to 32MJ ME/day at peak lactation (Adam 1994).

To provide adequate forage to grazing hinds during lactation, farmers need to maintain sward height of grass at 6-8cm (Hamilton 1989). Stocking rates depend on the quality and growth of pasture. For example, one hectare of heather moorland can only maintain 0.66 lactating hinds; on cultivated upland and lowland pasture it is possible to keep 10-12 hinds/ha and 12-16 hinds/ha respectively (Hamilton 1989).

Maintaining body condition during pregnancy is critical to ensure prenatal development and neonatal survival. Low levels of nutrition leading to a reduction in liveweight of hinds during pregnancy may lower the birth-weight of their calves and therefore reduce calf survival rate. However, over-feeding hinds, particularly during late pregnancy, may also have detrimental effects, causing excess calf growth, resulting in calving difficulties (dystocia). The constant provision of supplementary feed may also lower the muscle tone and fitness of hinds through lack of exercise associated with the reduced need to forage (Haigh and Hudson 1993).

### Nutrition of calves

The weight and body condition of hinds during pregnancy determines the birth-weight of calves and their subsequent growth rate during lactation (Blaxter and Hamilton 1980). Large hinds tend to produce heavier calves. Stag calves are approximately 1kg heavier at birth than hind calves and have a higher growth rate from birth to weaning (Blaxter *et al* 1988). Birth-weight is linked to mortality; for example, 4kg calves have a 100 per cent mortality compared with five per cent in calves of 6-7kg (Blaxter and Hamilton 1980).

Calves are totally dependent on milk for their energy requirements during the first 30 days (Loudon *et al* 1984). They consume between 200-600g of milk at each suckling bout and a maximum of 2000g/day (Arman *et al* 1974). As lactation progresses dependency on milk decreases. By the time calves are approximately three months, milk accounts for only 10-20 per cent of energy intake; grass and other forage provide the remainder.

The ME requirement for maintenance is approximately $0.45 MJME/kg^{0.75}$ and remains stable for calves between 3 and 16 months of age. However the ME requirement for growth varies with season and live-weight gain (Table 8.4). The weight of calves at weaning depends on the production system and may vary from 35-45kg (Loudon *et al* 1984). Live-weight gain falls as the quality of pasture decreases in the late summer. Seasonal inappetance over winter also greatly affects liveweight-gain, although skeletal growth continues at a slower rate. In many circumstances, particularly on hill and upland farms, supplementary feeding and housing may be necessary to achieve target liveweights at 15 months of age. Weaned calves that farmers house and feed maintenance rations over winter have a lower live-weight gain than those on *ad libitum* diets. However, the rapid growth of calves following turn-out on to pasture in late spring partly compensates for this growth differential (Milne *et al* 1987). To maximize growth potential and to reduce feeding costs, producers should provide weaned calves with maintenance requirements over winter. They should then follow it by *ad libitum* feeding at least one month before turn-out (Milne *et al* 1987).

**Table 8.4** **Seasonal estimates of the metabolizable energy (ME) requirements for calves of 50kg with different growth rates (g/day).**

*(modified from Adam 1994)*

| Season | Maintenance requirements in MJ ME for various growth rates (g/day) | | | |
| --- | --- | --- | --- | --- |
| | 50 | 100 | 150 | 200 |
| Autumn | 11.3 | 14.0 | 16.8 | 19.5 |
| Winter | 12.9 | 17.2 | 21.6 | 25.9 |
| Spring-Summer | 10.9 | 13.4 | 15.8 | 18.2 |

At weaning, the target liveweights for hind and stag calves on good quality pasture are 43 and 46kg respectively (Adam 1989). Farmers need to maintain a sward height of 8–10cm for calves they keep at pasture following weaning to maximize growth before winter. Growth rate can vary by up to 100g/day, depending on the quantity and quality of pasture and weather conditions (Loudon *et al* 1984). Supplementary feeding of calves on poor quality pasture is necessary to achieve mid-winter liveweights of 56 and 60kg for hind and stag calves respectively (Adam 1989).

## Nutrition of stags

Farmed red deer stags of 150 and 250kg have maintenance requirements of 24.4 and 35.8MJME respectively (Adam 1994). Stags have similar maintenance requirements in summer and winter (Blaxter *et al* 1988). However during the rut, stags may lose up to 30 per cent of weight due to an increase in activity and a decrease in feeding. The rut ends shortly before the onset of winter inappetance. It is important to maintain an adequate diet as energy requirements increase after the rut and stags may continue to lose body condition (Adam 1994).

## Mineral requirements

Calcium, phosphorus and magnesium are essential components in the diet. Deer can to some extent counteract dietary deficiency in the major minerals by mobilizing body reserves, although this may impair skeletal growth and muscle development in calves. Mineral requirements are normally high during growth and lactation. Hinds lose approximately 2.2g calcium and 1.9g phosphorus per kg of milk (Adam 1989). Phosphorus deficiency may reduce fertility. Deer require other minerals, such as copper, cobalt, selenium and iodine, in trace quantities. Deficiency in the first three of these can produce subclinical symptoms, which may only become clinical when compounded by other factors.

## HEALTH AND DISEASE

This section discusses some of the most common diseases, but for a more comprehensive review see Alexander and Buxton (1994).

## Tuberculosis

*Mycobacterium bovis* and *M. avium* cause tuberculosis (TB). Other strains of mycobacteria also exist but most are non-pathogenic and occur naturally in the environment. TB is a notifiable disease. It is particularly hazardous because it is transmissible to other livestock and humans. *M. bovis* is extremely infectious and can therefore have devastating effects in confined groups of animals such as calves housed over winter. The symptoms vary depending on the infectious dose, route of infection and the overall immunocompetence of the animal. A low dose of *M. bovis* may produce a localised infection, causing no clinical symptoms. The immune system controls localised infections, but subsequent challenges to the immune system may result in the spread of the infection, causing abscesses and eventually death (Reid 1994a).

## Johne's Disease

Unlike most other livestock species, Johne's disease is more common in young rather than old deer, and is transmissible in-utero. Another species of Mycobacteria, *M. avium paratuberculosis* causes it. The symptoms include loss of condition, retention of the winter coat and, as the disease progresses, diarrhoea. It is advisable to isolate infected animals because they shed many bacteria. There are no effective treatments available and death will occur in weeks or months. Vaccines for cattle are available and are effective in deer although they may confound future tests for *M bovis*; they are thus not routinely used (Reid 1994a).

## Malignant Catarrhal Fever

Malignant catarrhal fever (MCF) has been a major cause of mortality in farmed deer in New Zealand; together with TB and lungworm it is a most important deer disease (Blaxter *et al* 1988). Two viruses, *Alcelaphine herpesvirus-1* (AHV-1) and *Ovine herpesvirus-2* (OHV-2), can cause MCF (Reid 1994b).

Wildebeest and sheep carry AHV-1 and OHV-2 respectively, although normally neither develop a clinical infection (Reid 1994b). The viruses become pathogenic once they spread to another species. Stress exacerbates clinical disease in deer and some species are more susceptible than others (Mackintosh 1992). There are various symptoms, including inappetance, dysentery, enlargement of the lymph nodes, bilateral corneal opacity, eye and nose secretions and skin ulcers (Reid 1994b). There are no vaccines available;

to reduce the chances of an outbreak of MCF, producers should therefore maintain farmed deer in different areas from sheep and allocate separate stockpersons for each species (Blaxter *et al* 1988).

## Yersiniosis

The bacterium *Yersinia pseudotuberculosis* causes Yersiniosis. Both wild and domestic animals are carriers and it can occur in the faeces of healthy deer. Infected deer shed large quantities of the organism in their faeces and so infection can rapidly spread within a group. Deer commonly encounter *Yersinia* in their first winter, but do not show any clinical signs of infection. Clinical disease may follow exposure to stress such as poor nutrition, handling and transportation. The symptoms include weight loss, green, watery diarrhoea and prolonged inactivity before death (Buxton and Mackintosh 1994). If diagnosis occurs in time it is possible to treat with antibiotics; animals with diarrhoea also need treatment to prevent dehydration.

## Clostridial Diseases

The *Clostridium* species causes a variety of diseases. *C. perfringens* (also called *C. welchii*) causes enterotoxaemia. Clostridial infection can occur when the diet changes suddenly; it results from carbohydrates seeping into the gut, where the bacteria multiply, producing a lethal toxin. Other examples include blackleg and malignant oedema, which results from the infection of open wounds by Clostridial species. Both result in blood poisoning and are fatal unless diagnosed in time. Prophylactic treatment at weaning followed by re-vaccination at yearly intervals using multivalent vaccines can reduce clostridial infections (Buxton 1994).

## Parasites

*Ostertagia* species are the main gastro-intestinal parasites, although others may also be present in the gut, but at such low numbers that they are effectively non-pathogenic (Munro 1994a). Clinical infestation with *Ostertagia* causes the same symptoms as in other livestock, namely weight loss and poor coat condition, soft faeces and a soiled tail (Munro 1994a). The type of treatment will depend on the degree and spread of infestation and grassland management. The most effective systems combine anthelmintic treatment with pasture rotation (Munro 1994a).

*Cryptosporidium parvum*, which causes cryptosporidiosis, is a coccidian parasite that infects the small intestine. Red deer calves pick up the infection by ingesting *Cryptosporidium* eggs, shed by other infected animals. Clinical symptoms occur 5–6 days after ingestion and include inappetance and diarrhoea. Cryptosporidiosis is highly contagious and once one animal becomes infected it sheds vast quantities of eggs in its faeces; this contaminates the pasture or pen and leads to the disease spreading rapidly. It is therefore important to isolate infected calves to prevent the infection spreading. Anticoccidial vaccines have little effect and, since there is no specific treatment, farmers normally just treat calves for dehydration (Blaxter *et al* 1988; Angus 1994). *Cryptosporidium* is also a zoonosis and can cause acute gastro-enteritis in humans.

Lungworm, *Dictyocaulus viviparus*, is the most important internal parasite. Cattle strains of *Dictyocaulus* can cross-infect deer, although the deer strain is less virulent and produces milder symptoms (Blaxter *et al* 1988). Heavy infestations can dramatically reduce productivity in farmed red deer. Clinical symptoms include inappetance and gradual weight loss leading to sudden death (Munro 1994b). Deer may also show signs of respiratory difficulty, culminating in suffocation if farmers do not treat the condition.

Calves are the most vulnerable age group to *Dictyocaulus* infestation, particularly in the autumn and winter. Producers should consider prophylactic treatment of hinds with anthelmintic if infection among their calves is common. It is important to monitor regularly faecal egg count. This involves collecting 5–10 fresh faecal samples from different age classes of deer at 6–8 week intervals (Munro 1994b).

Nasal bot flies, *Cephenemyia auribarbis*, may cause respiratory problems in red deer. From May to July female bot flies inject larvae into the nostrils of deer. Initial larval growth occurs in the nasal cavities, surviving on the deer's blood; by April or May the larvae migrate to the pharyngeal pockets until they develop fully. Before pupating, larvae drop out of the nose and onto the ground. Heavy infestation with bot fly larvae may cause suffocation or secondary infections at the site of attachment following inhalation of larvae (Munro 1994c). Warbles (*Hypoderma diana*) and headfly (*Hydrotaea irritans*) can also cause problems (Goddard 1994b,c).

Although red deer are susceptible to liver fluke, *Fasciola hepatica* infection, it normally produces a subclinical condition. Liver fluke, however, is an important cause of liver condemnation post-slaughter (Munro 1994d). Treatment with flukicides is more effective for adult (90–100%) than immature (60%) flukes. Therefore it is important to re-treat infected animals approximately one month after initial treatment (Munro 1994d).

## Management-related problems

In the past poor management caused many of the injuries and diseases in farmed deer. Inappropriate mustering, handling and transportation can cause trauma and capture myopathy. Inexperienced deer such as wild or newly introduced stock are particularly prone to panic, resulting in a rapid rise in body temperature (ie hyperthermia). High environmental temperatures or when deer are in their winter coats, and certain restraining drugs that effect homeostasis, exacerbates capture myopathy. When working with deer it is important to remember to apply the '3T' rule: Training, Taming and Tempo (Haigh and Hudson 1993).

Metabolic problems such as acidosis or rumenal overload may arise from inappropriate supplementary feeding. Acidosis results from the production of excess lactic acid from fermentation in the rumen, following the consumption of large quantities of concentrated feed. The resultant increase in acidity can destroy the tissue lining and the natural microflora of the rumen. Once the lactic acid enters the circulation it can also damage other tissues and organs. The quantity and the type of feed the deer consumes determines the severity of acidosis. Grains such as wheat, barley and corn are worse than oats. Processing grain by crushing or rolling can also increase acidosis by breaking down the cell walls and making more starch available for fermentation.

Dominant animals are more likely to suffer from acidosis because they have priority of access to feed. Deer suffering from acidosis show inappetance, increasing inactivity eventually resulting in an inability to stand and in severe cases leading to death. The treatment for acidosis is to wash out the rumen, a procedure which must be carried out by a veterinary surgeon. However, farmers can easily prevent it by gradually introducing the ration of supplementary feed over several days.

Other diet related problems include mineral deficiencies (see Adam 1994) and dystocia. Allowing hinds to overeat in late pregnancy can cause dystocia. Producers can prevent it by maintaining the sward height between 6 and 8cm, which they achieve by altering the stocking rate or by mowing. Using large sires of different breeds and congenital deformities may cause dystocia (Haigh and Hudson 1993).

## TRANSPORTATION

*The Welfare of Animals (Transport) Order 1997* covers the legal requirements on the transportation of farmed deer. For guidance on these complex matters see MAFF (1998).

Farmers should only transport stags in velvet, hinds in late pregnancy, sick or injured deer, or young calves in cases of emergency for veterinary treatment. Transportation of entire stags (over 24 months old) during the rut is dangerous and inadvisable. However, if transportation is necessary, stags will need sedating and penning individually. Similarly, to avoid injury, it is advisable to de-antler stags in hard-antler several days before transit (MAFF 1989a).

Before transit, farmers should group deer according to sex, age and body-weight. Unfamiliar animals need penning together several days before transport so that the stockperson can remove problem animals and pen them separately. Penning before transport may also benefit the deer as it allows them to adjust to greater confinement. Stocking rates depend on the sex and age of deer (Table 8.5). Recommendations in relation to height of headroom also apply such that deer should have sufficient space to stand normally. Group sizes vary with sex, age and reproductive status. For example, MAFF (1989a) recommends 10 and 20 as the largest group sizes for yearlings and calves respectively.

**Table 8.5** **Recommended minimum space allowance during transit (MAFF 1989a).**

| Adult stags | 1.0m² |
|---|---|
| Adult hinds | 0.5–0.6m² |
| Yearling stags | 0.5–0.6m² |
| Yearling hinds | 0.3–0.5m² |
| Hind and stag calves | 0.3–0.5m² |

It is important to provide bedding, such as hay or straw, for long journeys (Fletcher 1994b). Hyperthermia is a potential problem in groups of animals during transit. Ventilators must be sufficiently small to prevent deer from pushing their heads through, and should be above or below eye level, to avoid eye irritation leading to conjunctivitis (Haigh and Hudson 1993). Provision of water and food is necessary on long journeys. Hauliers often supply fresh root vegetables or cut grass because they have a high moisture content.

During loading, handlers should muster animals in their groups along a purpose-built chute or a temporary hessian wall from the holding pens to the transporter, to deter escape. It is preferable to muster deer at a walking pace, without the use of sticks and goads. An experienced handler should accompany deer on long journeys (MAFF 1989a). Similar principles apply to unloading. Following transportation, it is important to rest deer before further handling. At unloading, fencing should be visible to stop deer from running into it. An alternative method is to unload deer into the middle of a field, or at night, to reduce the risk of them running into fencing (Fletcher 1994b).

## SLAUGHTER

Producers can either slaughter their deer on the farm (field-slaughter) or transport them to an abattoir. Field-slaughter is most common on farms with retail outlets (Hamilton 1994b). Legislation requires a veterinary surgeon to inspect the animals sometime within the 72 hours before slaughter. Giving a supplementary feed, so that the deer are standing motionless at the time of being shot, helps to increase the accuracy of field-slaughter. The type of firearms and ammunition must conform to the *Deer Act 1963* and *Deer (Scotland) Act 1959, The Firearms Act 1968* as amended in 1988 and in 1997 (twice). The code of welfare practice (MAFF 1989b) covers recommendations for abattoir slaughter of farmed deer.

After slaughter, producers can either process the carcasses on the farm, in Ministry approved premises, or transport them to a local abattoir. Processing (dressing) must occur within one hour of slaughter, or three hours if the carcass is refrigerated between 0–4˚C. A meat inspector will need to examine the carcasses within 24 hours of slaughter (Drew 1994). The *Farmed Game Directive 91/495/EEC* and *Fresh Meat (Hygiene and Inspection) Regulations 1992* list the requirements for on-farm processing facilities.

## THE WAY AHEAD

Public perception is very important. Most consumers view venison as a natural, wholesome food which has been produced under animal welfare-friendly conditions. Maintaining the good stockmanship and management of farmed deer is essential if incidence of disease and injury and the need for drug therapy is to be kept to a minimum.

## REFERENCES AND FURTHER READING

**Adam C L** 1989 Nutrition on the deer farm. In: Gould J (Ed) *Deer Farming: a handbook for the 1990s*, pp 24-29. British Deer Farmers Association

**Adam C L** 1994 Feeding. In: Alexander T L and Buxton D (Eds) *Management and Diseases of Deer: a handbook for the veterinary surgeon*, pp 44-54. Veterinary Deer Society: London, UK

**Adam C L and Moir C E** 1987 A note on the effect of birth date on the performance of suckled red deer calves and their dams on low-ground pasture. *Animal Production 44:* 330-332

**Alexander T L and Buxton D** (Eds) 1994 *Management and Diseases of Deer: a handbook for the veterinary surgeon,* 2nd edition. Veterinary Deer Society: London, UK

**Angus K W** 1994 Neonatal enteritis. In: Alexander T L and Buxton D (Eds) *Management and Diseases of Deer,* pp 136-139. Veterinary Deer Society: London, UK

**Arman P, Kay R N B, Goodall E D and Sharman G A M** 1974 The composition and yield of milk from captive red deer (*Cervus elaphus L*). *Journal of Reproduction and Fertility 37:* 67-84

**Blaxter K L and Hamilton W J** 1980 Reproduction in farmed red deer. (2) Calf growth and mortality. *Journal of Agricultural Science, Cambridge 95:* 275-284

**Blaxter K, Kay R N B, Sharman G A M, Cunningham J M M and Hamilton W J** 1974 *Farming the Red Deer.* HMSO: Edinburgh, UK

**Blaxter K, Kay R N B, Sharman G A M, Cunningham J M M, Eadie J and Hamilton W J** 1988 *Farming the Red Deer,* 2nd edition. HMSO: Edinburgh, UK

**Burdge J and Burdge J** 1989 Facilities for the deer farm. In: Gould J (Ed) *Deer Farming: a Handbook for the 1990s*, pp 35-37. British Deer Farmers Association:

**Buxton D** 1994 Clostridial diseases. In: Alexander T L and Buxton D (Eds) *Management and Diseases of Deer*, pp 121-123. Veterinary Deer Society: London, UK

**Buxton D and Mackintosh C G** 1994 Yersiniosis. In: Alexander T L and Buxton D (Eds) *Management and Diseases of Deer*, pp 117-119. Veterinary Deer Society: London, UK

**Clutton-Brock T H and Albon S D** 1989 *Red Deer in the Highlands*. BSP Professional Books: Oxford, UK

**Clutton-Brock T H, Guinness F E and Albon S D** 1982 *Red Deer: Behavior and Ecology of Two Sexes*. The University of Chicago Press

**Drew K R** 1994 Venison. In: Alexander T L and Buxton D (Eds) *Management and Diseases of Deer*, pp 20-23. Veterinary Deer Society: London, UK

**Farm Animal Welfare Council** 1985 *Report on the Welfare of Farmed Deer*. FAWC: Surbiton, UK

**Fennessy P F and Suttie J M** 1985 Antler growth: nutritional and endocrine factors. In: Fennessy P F and Drew K R (Ed) *The Biology of Deer Production*, pp 239-250. The Royal Society of New Zealand: Wellington, New Zealand

**Fletcher T J** 1994a Mating. In: Alexander T L and Buxton D (Eds) *Management and Diseases of Deer*, pp 74-79. Veterinary Deer Society: London, UK

**Fletcher T J** 1994b Transport. In: Alexander T L and Buxton D (Eds) *Management and Diseases of Deer*, pp 35-37. Veterinary Deer Society: London, UK

**Goddard P J** 1994a Deantlering. In: Alexander T L and Buxton D (Eds) *Management and Diseases of Deer*, pp59-60. Veterinary Deer Society: London, UK

**Goddard P J** 1994b Warbles. In: Alexander T L and Buxton D (Eds) *Management and Diseases of Deer*, pp 171-172. Veterinary Deer Society: London, UK

**Goddard P J** 1994c Headfly. In: Alexander T L and Buxton D (Eds) *Management and Diseases of Deer*, pp 172-173. Veterinary Deer Society: London, UK

**Gould J** 1989 Deer farming in the lowlands: the perfect marriage with arable farming. In: Gould J (Ed) *Deer Farming: a handbook for the 1990s*, pp 13-15. British Deer Farmers Association

**Griffiths L M** 1994 Dystokia. In: Alexander T L and Buxton D (Eds) *Management and Diseases of Deer*, pp 86-87. Veterinary Deer Society: London, UK

**Guinness F E, Lincoln G A and Short R V** 1971 The reproductive cycle of the female red deer (*Cervus elaphus L*). *Journal of Reproduction and Fertility 27:* 427-438

**Haigh J C and Hudson R J** 1993 *Farming Wapiti and Red Deer*. Mosby: St Louis, Missouri, USA

**Hamilton W J** 1989 Farming red deer- the keys to good management. In: Gould J (Ed) *Deer farming: a handbook for the 1990s*, pp 21-23. British Deer Farmers Association

**Hamilton W J** 1994a Land resources and related systems of deer farming. In: Alexander T L and Buxton D (Eds) *Management and Diseases of Deer*, pp 17-18. Veterinary Deer Society: London, UK

**Hamilton W J** 1994b Field and abattoir slaughter. In: Alexander T L and Buxton D (Eds) *Management and Diseases of Deer*, pp 17-18. Veterinary Deer Society: London, UK

**Hamilton W J** 1994c. Portable handling facilities to improve the welfare of farmed red deer (*Cervus elaphus*). *Animal Welfare 3:* 227-233

**Hamilton W J and Blaxter K L** 1980 Reproduction in farmed red deer. (1) Hind and stag fertility. *Journal of Agricultural Science, Cambridge 95:* 261-273

**Hamilton W J and Soanes C A** 1995 Effect of management practices on the welfare of weaned red deer calves. In: *Proceedings of the Third International Congress on the Biology of Deer, August 1994, Edinburgh*. Paper 165

**Hamilton W J, Littlewood C and Murray D** 1994 Shelter on an upland deer farm. *Deer Farming 46:* 20-22

**Hanlon A J** 1996a Farming deer: calving. *Irish Veterinary Journal 49:* 298-301

**Hanlon A J** 1996b Weaning farmed red deer calves. *Irish Veterinary Journal 49:* 611-613

**Hanlon A J** 1997 The welfare of farmed red deer. *Deer Farming 53:* 14-16

**Hanlon A J, Rhind S M, Reid H W, Burrells C, Lawrence A B, Milne J A and McMillen S R** 1994 Relationship between immune response, live-weight gain, behaviour and adrenal function in red deer (*Cervus elaphus*) calves derived from wild and farmed stock, maintained at two stocking densities. *Applied Animal Behaviour Science 41:* 243-255

**Heydon M J, Sibbald A M, Milne J A, Brinklow B R and Loudon A S I** 1993. The interaction of food availability and endogenous physiological cycles on the grazing ecology of red deer hinds (*Cervus elaphus*). *Functional Ecology 7:* 216-222

**Kay R N B and Staines B W** 1981 The nutrition of the red deer (*Cervus elaphus*). *Nutrition Abstracts and Reviews (B) 51:* 601-622

**Lincoln G A** 1985 Seasonal breeding in deer. In: Fennessy P F and Drew K R (Eds) *The Biology of Deer Production,* pp 165-179. The Royal Society of New Zealand: Wellington, New Zealand

**Loudon A S I, Darroch A D and Milne J A** 1984 The lactation performance of red deer on hill and improved species pastures. *Journal of Agricultural Science, Cambridge 102:* 149-158

**Mackintosh C G** 1992 Observations on the relative susceptibility to disease of different species of deer farmed in New Zealand. In: Brown R D (Ed) *The Biology of Deer,* pp 113-199. Springer-Verlag: New York, USA

**Milne J A, Sibbald A M, McCormack A and Loudon A S I** 1987 The influences of nutrition and management on the growth of red deer calves from weaning to 16 months of age. *Animal Production 45:* 511-522

**Ministry of Agriculture Fisheries and Food** 1989a *Guidelines for the Transport of Farmed Deer.* MAFF Publications: Alnwick, UK

**Ministry of Agriculture Fisheries and Food** 1989b *Codes of Recommendations for the Welfare of Livestock: Farmed Deer.* (reprinted 1990) PB 0055. MAFF Publications: London, UK

**Ministry of Agriculture Fisheries and Food** 1994 *Census of Farmed Deer September 1993: England and Wales.* MAFF Publications: London, UK

**Ministry of Agriculture Fisheries and Food** 1998 *Guidance on The Welfare of Animals (Transport) Order 1977.* PB 3766. MAFF Publications: London, UK

**Mitchell B, Staines B W and Welch D** 1977 *Ecology of Red Deer: a research review relevant to their management.* Institute of Terrestrial Ecology: Cambridge, UK

**Munro R** 1994a Gastro-intestinal parasites. In: Alexander T L and Buxton D (Eds) *Management and Diseases of Deer,* pp 126-128. Veterinary Deer Society: London, UK

**Munro R** 1994b Dictyocaulus. In: Alexander T L and Buxton D (Eds) *Management and Diseases of Deer,* pp 111-113. Veterinary Deer Society: London, UK

**Munro R** 1994c Nasal Bots. In: Alexander T L and Buxton D (Eds) *Management and Diseases of Deer,* pp 110-111. Veterinary Deer Society: London, UK

**Munro R** 1994d Live Fluke. In: Alexander T L and Buxton D (Eds) *Management and Diseases of Deer,* pp 129-130. Veterinary Deer Society: London, UK

**Pollard J C, Littlejohn R P and Suttie J M** 1992 Behaviour and weight change of red deer calves during different weaning procedures. *Applied Animal Behaviour Science 35:* 23-33

**Putman R J** 1988 *The Natural History of Deer.* Christopher Helm: London, UK

**Reid H W** 1994a Mycobacterial infections. In: Alexander T L and Buxton D (Eds) *Management and Diseases of Deer,* pp 93-101. Veterinary Deer Society: London, UK

**Reid H W** 1994b Malignant Catarrhal Fever. In: Alexander T L and Buxton D (Eds) *Management and Diseases of Deer,* pp 101-106. Veterinary Deer Society: London, UK

**Scottish Office Agriculture and Fisheries Department (SOAFD)** 1993 *Census on Farmed Deer in Scotland*

**Sibbald A M** 1994 Effect of changing daylength on the diurnal pattern of intake and feeding behaviour in penned red deer (*Cervus elaphus*). *Appetite 22:* 197-203

# 9 Laying hens

## R G Wells

### DEVELOPMENT OF THE UK EGG INDUSTRY

The marked changes that have taken place in the UK egg industry over the decades since World War 2 are illustrated in Table 9.1. The data show an initial rapid expansion in the size of the national laying flock, followed by a marked reversal in the post-British Egg Marketing Board free market situation. The decline in hen numbers has been due to a fall in egg consumption, largely due to a sharp decline in the eating of cooked breakfasts, combined with a consistent increase in egg production per bird.

A particularly striking development has been the swing to intensive production systems. A major impetus for this development was the prospect of financial gain from producing more eggs in the autumn and winter months when traditionally they were scarce and expensive. The better control of light with intensive systems gave more uniform seasonal production. Also, improved control of temperature, which raised house temperatures overall, meant that the birds ate less food, while the avoidance of temperature extremes resulted in less mortality. The latter also arose with the greater protection from predators and the reduced contact with disease-carrying wild birds. Another reason for the expansion of intensive production was that it permitted closer supervision and control of flocks, which saved on labour.

There has, however, been a slight revival in extensive or free-range egg production since about 1980. This has arisen because some consumers are prepared to pay the premium necessary to offset the extra costs of producing free-range eggs. They perceive the more natural system, where birds have access to outside range, as being more humane and may feel that free-range eggs are more nutritious than intensively produced eggs.

Another very pronounced trend has been the concentration of egg production into fewer, much larger units. Although there are still about 30,000 holdings with layers, three-quarters of the country's eggs come from under 300 units each with 20,000 or more layers. Some layer farms today accommodate over 500,000 birds. This development, which provided the normal economies of scale, was facilitated by the concomitant development of intensive systems and also by the introduction of

**Table 9.1    Trends in the development of the UK egg industry\*.**

|  | 1950 | 1960 | 1970 | 1980 | 1990 | 1995 |
|---|---|---|---|---|---|---|
| *Number of layers (millions)* | 25 | 40 | 65 | 55 | 40 | 30 |
| *Holdings with layers (thousands)* | 300 | 250 | 150 | 60 | 40 | 30 |
| *Layers in flocks > 1000 birds (%)* | 5 | 15 | 80 | 95 | 96 | 97 |
| *Eggs produced intensively (%)* | 1 | 60 | 95 | 98 | 92 | 88 |

\* Approximate data based on statistics published by the Ministry of Agriculture Fisheries and Food and industry organisations, those of the earlier years collated by Coles (1960).

modern methods of disease control which removed the traditional precautionary restrictions on the number of birds per site.

## Trends in intensive systems

Initially, as the controls on animal feed supplies were relaxed, there was a resumption in the use of battery cages. Interest then grew in the deep-litter system which had been successfully developed in the USA and was more adaptable to use on the general farm. In this system birds are penned in loose flocks on a floor covered with litter material. The birds' droppings mix in with the litter and the mixture accumulates over the entire production period. By 1960 about 40 per cent of UK egg production was in deep-litter units, but there was a huge swing back to cages during the 1960's and by 1980 the move to cages was almost total (see Table 9.2). One reason for this was the adoption of much higher stocking rates in battery houses. In the late 1950's, smaller bodied 'hybrid' stocks came on the market which could be housed at more than one per cage without adversely affecting their health or productivity. This practice effectively eliminated the previous disadvantage in capital cost per bird of the battery system. Modern cages are typically designed to hold 4 or 5 layers and are often stacked in tiers four or more high, giving stocking densities in excess of 30 birds/m² of house floor area.

Producers were also keen to benefit from the other advantages of the battery system. Birds in batteries tend to lay more eggs, eat less food and require less labour than those housed on litter. These economic benefits result largely from simplified stock management and disease control within a system that is also conducive to mechanisation. The main contributory factors are as follows:

1. Separation of the flock into small groups of some 4 or 5 birds gives social stability with a big reduction in aggressive behaviour and the attendant risk of cannibalism. Also, dividing up the flock in this way eliminates the risk of large pile-ups of birds and deaths from smothering in a panic situation

2. The sloping wiremesh flooring to cages through which the birds' droppings fall and down which their eggs roll simplifies egg collection and makes it more hygienic. Eggs remain cleaner in the absence of litter and nest boxes

3. Keeping birds from contact with their droppings breaks the life cycle of intestinal parasites and prevents related diseases, notably coccidiosis

4. The absence of litter dispenses with the difficult problem of maintaining it in a suitably dry, friable condition

5. The absence of nests stops all birds from going broody. Broodiness still occurs in a small proportion of modern hybrids when floor-housed

6. High stocking rates ease the job of maintaining optimum house temperatures during the winter months which, apart from saving on food consumption, leads to an increase in air changes and a less damp, stuffy atmosphere.

Four out of the above six factors contribute towards a reduced rate of death and disease in caged flocks and therefore, as well as providing economic benefits, confer welfare advantages on the birds. However, many people object to battery cages because they keep hens very closely confined in a barren environment; they greatly restrict movement and deprive the birds of much of their natural behaviour, such as running, wing flapping, scratching, dust bathing and nesting. The more recently developed perchery and aviary systems, for

**Table 9.2** **Trends in the use of housing systems for laying hens in the UK. (approximately percentage of eggs produced).**

| System of housing | 1960 | 1970 | 1980 | 1990 | 1995 |
|---|---|---|---|---|---|
| *Free-range* | 40 | 5 | 2 | 8 | 12 |
| *Battery* | 20 | 80 | 96 | 90 | 84 |
| *Deep-litter** | 40 | 15 | 2 | 2 | 4 |

\* Taken to include all non-cage intensive systems.

the production of so-called 'barn' eggs, are an attempt at a compromise. These are modifications of the deep-litter system which incorporate multi-tiered banks of perches or platforms. By making use of vertical space they permit an increase in stocking densities. The resultant higher temperatures give the savings in food consumption and improvement in atmospheric conditions during the colder weather, but birds remain in large comparatively unstable groups and still have access to their droppings. Part of the floor of a perchery may remain littered, which enables birds to scratch and dust bathe. Alternatively, the entire floor may be constructed of wooden slats or wire mesh over a droppings pit to avoid the complication of litter management and the associated disease hazards and dirty egg problems.

## THE FULL CYCLE OF BREEDING AND PRODUCTION ACTIVITIES

The complete chain of events in the provision of eggs for consumption is outlined diagrammatically in Figure 9.1. It begins with the primary breeders who hold the elite stocks and undertake the complex genetic selection process. Their overhead costs are very high so they have merged over the years to form a comparatively few large international companies with world-wide markets. They pass the parent stock, usually as day-old chicks, to secondary breeders or flock multipliers for rearing to sexual maturity and then mating them in the ratio of one cockerel to about 12 hens for the production of fertile hatching eggs. These operations are usually performed under intensive conditions in loose flock systems incorporating deep litter and varying proportions of raised slatted floor area. Typically the production period runs from 23/24 to 68 weeks of age when each breeder hen produces about 255 eggs of which almost 90 per cent are suitable for incubation.

Secondary breeder units are organised around a hatchery to which the eggs are transported and set in incubators. After three weeks of very precise temperature and relative humidity control, the chicks hatch at the rate of about 80 per 100 eggs set (80% hatchability). The male chicks of modern small-bodied egg producing hybrids have no commercial value for meat production, so they are killed at this stage, usually with carbon dioxide gas. The centralised, large-scale operations of the hatchery

present an enormous potential for disease spread around the industry. Hence, great emphasis is placed on hygiene at the hatchery and disease control measures on the supply farms.

The egg production chain

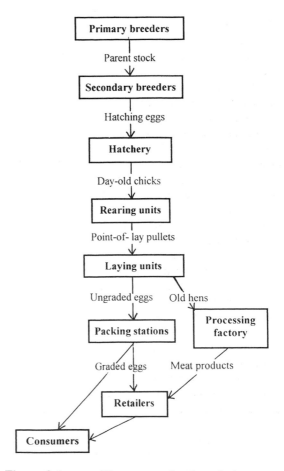

**Figure 9.1**     **The egg production chain**

The day-old pullet chicks are delivered in specially ventilated and temperature controlled vehicles to rearing units where they are grown to the point-of-lay stage (16–18 weeks of age). Both cage and deep-litter systems, in roughly equal proportions, are used for the rearing of pullets. Most rearing units, accommodating approximately 70 per cent of the national pullet flock, are in the hands of egg producers who therefore rear their own layer replacements. The remaining pullets are reared by specialists who sell them on to egg producers at point-of-lay.

The transport of point-of-lay pullets to producers from rearers takes place in specially equipped and ventilated lorries. It is subject to the Community Directive 91/628/EEC on the protection of animals during transport, which stipulates, among many other things, a maximum journey time without food and water of 12 hours (excluding the loading and unloading time). The most important stressor to consider in pullet transport is high ambient temperature leading to dehydration, making the need for adequate ventilation paramount. For a useful scientific review of the subject of the transportation of poultry see Freeman (1984).

The point-of-lay pullets commence egg production a week or two after being transferred to laying quarters. They are then kept in lay for some 50–56 weeks when each will produce in the region of 300 eggs. As noted earlier, most layers are housed in battery cages, but others are kept in free-range and in non-cage intensive systems (see Table 9.2).

The eggs produced are graded to European standards for weight and quality at registered packing stations. Many of these stations are on-farm where producers grade and pack their own eggs for sale to local shops and catering establishments and/or to consumers direct. These producer/marketers account for almost half of the shell egg market. The remainder of producers sell their eggs ungraded to large specialist packers who are the main suppliers of supermarkets. Eggs downgraded as unsuitable for sale in the shell, but fit for human consumption, go to breakers for processing into liquid, frozen and dried products for use predominantly by bakers and confectioners.

Hens at the completion of their productive lives are transported for slaughter at specialist processing factories. The carcases are usually de-boned, in the raw or cooked state, and the meat supplied to manufacturers of pastes, pies, soups and canned meat products. The bone and meat scrap can be used for pet food.

Sometimes transportation to the processing factory has not been carried out with due care and attention and has resulted in many birds suffering bone fractures. This has been particularly true of birds taken from battery cages, as they have comparatively weak bones from lack of exercise. It is very important that all personnel involved in the catching and transporting of hens are familiar with, and adopt, the measures set out in the *Joint Industry Welfare Guide to the Handling of Spent Hens*[1] (obtainable from NFU, Agriculture House, 164 Shaftesbury Avenue, London WC2H 8HL) originally prepared by the National Farmers Union and the then British Poultry Federation. In particular, handlers should catch the birds by both legs and remove them from the cage singly, while supporting the breast manually or with the aid of a special breast support slide over the food trough. It clearly helps if the cage front can be fully opened. Broken bones in hen carcases create problems of bone splinters in the meat products, so their prevention is good business as well as good welfare.

More comprehensive textbooks on poultry production, such as those by Austic and Nesheim (1990) and North and Bell (1990) give details on breeding stock management and hatchery operations. The emphasis here will be on the rearing and laying unit stages. Relevant aspects of the biology and social behaviour of the fowl will be incorporated as appropriate. More specific reference to these subjects is made by Appleby *et al* (1992) and Rose (1997).

### Breeding and choice of breeds

Hybrid stocks, first developed in the USA, became available on the UK market in the late 1950's and were taken up rapidly in place of the traditional breeds of poultry for commercial egg production. They are the outcome of scientific breeding methods and have probably contributed more than anything else to the consistent improvement in laying hen productivity over the past three or four decades. Modern hybrids are practically all products of four-way crosses, (see Figure 9.2). The nucleus stocks or 'gene pool' comprise closed pure-bred lines of different traditional breeds and/or strains within breeds. A primary breeder may have 20 or more different pure-bred lines in the gene pool for a layer hybrid, giving a vast number of possible permutations

---

[1] New edition now being prepared to reflect changes in the industry

of four-way crossing programmes. The four grandparent lines are selected so that the resultant genetic mix is optimal in terms of the combination of desirable performance characteristics and the expression of hybrid vigour.

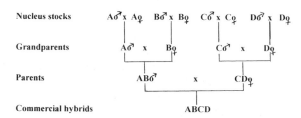

| Nucleus stocks | A♂ x A♀   B♂ x B♀   C♂ x C♀   D♂ x D♀ |
| Grandparents | A♂   x   B♀       C♂   x   D♀ |
| Parents | AB♂       x       CD♀ |
| Commercial hybrids | ABCD |

**Figure 9.2**    **Basic structure of a four-way cross breeding programme.**

Two main types of egg producing hybrid are available internationally: (1) small-bodied white birds, of about 1.8kg mature weight, which lay white eggs and have been bred exclusively from different strains of the White Leghorn breed; and (2) larger brown-feathered birds, of about 2.2kg mature bodyweight, which lay brown eggs and have been developed predominantly from different strains of Rhode Island Reds, but usually with the infusion of one or more other pure breeds. The smaller white birds tend to convert food into eggs more efficiently because of their reduced body maintenance requirement. They produce smaller eggs than brown birds, but usually a few more of them. However, the gap between white and brown hybrids has narrowed considerably over recent years. In the UK there is a strong preference for brown eggs and consumers have been prepared to pay a premium to offset the higher cost of their production. The market throughout the EU is roughly 75 per cent brown eggs and 25 per cent white, while world-wide the ratio is about 50:50.

When choosing from the available hybrids it is important that, apart from laying well and showing efficient food conversion under the prevailing environmental conditions, they should also produce eggs within a range of weights appropriate for the egg price differentials of the local market. Reliable comparative data on stock performance under UK conditions are not readily available. There are the results from European random sample tests in which the rearing and laying performance of representative samples of birds are recorded under standardised conditions, but the majority of these tests are undertaken in Germany (Heil and Hartmann 1994). Most producers base their choice of stocks on their own experience and that of others in their area. Many will run their own 'tests' on new strains of bird, which may be of particular value in assessing specific local disease resistance. The standard of the supply hatchery in terms of chick quality and back-up service will influence the final decision.

The bird that has been the most popular with UK egg producers over the past 20 years or so is the ISA Brown from Institut de Selection Animale in France. Its main rival has been the Hisex Brown from Euribrid in the Netherlands. Some other strains of layer of current interest to UK producers are the Lohmann from Germany, Hyline from America and Shaver from Canada. All these hybrids are used in both intensive and free-range systems, although bred exclusively under intensive conditions.

Producers purchase their replacement pullets at day-old for home rearing, or at point-of-lay. Home rearing can give some cost savings, at least in relation to transport. Also, by giving complete control over the rearing process, it is more conducive to the adoption of vaccination programmes appropriate for the specific disease challenges of the home site and to the correct linking of rearing and laying programmes in such things as lighting and nutrition. However, buying in point-of-lay pullets eliminates the need for capital investment in rearing facilities and simplifies management. Also, it provides more flexibility in the timing of flock replacements and avoids the risk of disease transfer to the growing pullets from adult hens where both types of stock are housed on the same site.

### Rearing and laying stages
This outline of production cycle lengths and performance standards relates to brown egg producing hybrids kept under UK conditions.

In the rearing period pullet chicks grow from about 40g at day-old to 1.4–1.5 kg at 17–18 weeks of age (point-of-lay) and consume some 6–7kg of food per bird. Mortality is low, at 2–4 per cent. Selective breeding over the years has gradually increased the rate of sexual maturity, resulting in birds coming into lay earlier. The move to laying quarters, therefore, is taking place increasingly nearer 17 than 18 weeks of age to ensure that pullets have sufficient time to settle in before egg production actually commences.

Egg production starts at 18–20 weeks of age and the birds are usually kept in lay till 70–76 weeks old. A graph of typical egg production from a flock is shown in Figure 9.3 where the per cent egg production, based on the number of birds housed at point-of-lay, has been plotted against the age of the birds in weeks. In practice, of course, the slope of the production curve is far less smooth than depicted here. The age at 50 per cent egg production, 22 weeks in this example, is the normal expression used for the whole flock's age at sexual maturity. The term 'hen housed' or 'bird housed' is used to describe egg production calculated on the basis of the initial number of birds housed at point-of-lay. The alternative expression of hen-day production accounts for mortality and is based on the number of surviving birds.

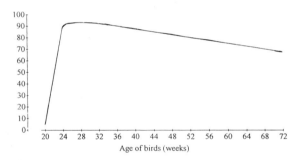

**Figure 9.3    Typical flock egg production curve.**

The normal mortality rate is about 0.5 per cent per month among caged layers. It tends to be more variable and, on average, higher in free-range flocks.

Total production from caged flocks is usually in excess of 25 dozen eggs per bird housed for the 52 weeks between 20 and 72 weeks of age. Mean egg weight over this period is about 64–65 g. Food intake over the same period is typically 42–43 kg per bird, giving a food conversion ratio of 2.1-2.2 kg food/kg eggs. In comparison, free-range flocks tend to

produce one or two dozen less eggs per bird, although mean egg weight is usually higher, and they eat about 10 per cent more food, mainly due to the lower environmental temperatures. In striking comparison, the wild Jungle Fowl, from which all domestic types of chicken have originated, will usually produce two clutches per year of about 12–15 eggs, each weighing about 25g.

The period from 20 to 40 weeks of age is a particularly critical one. Apart from the initial rapid onset of lay, egg weight is increasing markedly over this period from about 45g to 63g and the birds are still growing, each adding about 0.5kg in body weight. The maximum egg mass output from the flock at 56–58 g per bird housed per day occurs at about 36 weeks of age. Failure to meet the birds' exacting nutrient requirements in these early stages of lay will show as a post-peak dip in the egg production curve.

Most flocks are culled at the end of one production period of 50–56 weeks and replaced with new point-of-lay pullets. The exact timing of the replacement is determined by the rate of decline in the number and quality of the eggs produced. With modern hybrid layers it tends to be the deterioration in egg quality that is the more limiting factor. Egg albumen becomes more 'watery' as the laying period progresses, but the main problem with advancing age of bird is a decline in shell thickness. This is largely responsible for the rise in overall downgrading from 3–4 per cent in the early stages of lay to 10–12 per cent or more after a year or so (the overall average equals approximately 7%). 'Cracks' and other shell defects account for approximately 85 per cent of all downgrading of eggs.

Approximately 5–10 per cent of the national flock is kept for two laying cycles. This spreads the cost of pullets over an extended laying life and may increase economic margins in times of low egg prices. It also results in larger eggs, by avoiding the small eggs in the early stages of the first cycle, which is of advantage in certain markets, such as for free-range eggs where worthwhile premia only apply to the heavier weight grades.

The hen's laying life is extended by inducing a short period of rest from egg production through a moult. As well as the loss of feathers, the hen's reproductive tract regresses and then rejuvenates along with the regrowth of feathers. Birds are typically moulted by 60 weeks of age and then kept in lay till 96–100 weeks old. Production post-moult is likely to be between 150 and 170 eggs per bird housed. Egg quality also recovers but not quite to the original pullet level.

Moulting is usually induced by giving birds an environmental shock sufficient to disrupt the formation of hormones that stimulate and sustain egg production. It is important that the cessation of egg production is uniformly rapid and complete. However, compliance with the current welfare codes (MAFF 1990) restricts producers somewhat in this objective, particularly the requirements to provide at least eight hours light per day and not to withhold food. A technique that complies, but one that is not well tested, is to replace the layer diet with whole grain barley, oats or some other low-calcium diet for four weeks while providing only eight hours of dim light per day. It is interesting to note in this context that in the USA, where induced-moulting is more prevalent, the normal practice is to starve birds of food for about 10 days!

The completion of one or more production cycles is followed by a non-productive period of cleaning out between flocks, which typically takes two weeks. It is a time of heightened activity to meet demanding deadlines for bird handling, cleaning and disinfection, maintenance, repairs and reassembling. However, it does not normally extend to the whole site, except for specialist pullet rearers, but involves individual houses in sequence. Most layer sites, therefore, are not operated on an 'all-in, all-out' basis, but usually accommodate three or more flocks of different ages to maintain a more uniform supply of eggs, both in number and weight. This is of particular importance to producers who do all their own marketing.

The clean-out operation often includes disposal of manure accumulated over the entire rearing or laying period. For example, at the completion of rearing pullets on litter there will be approximately three tonnes of manure to remove per 1000 birds. A deep-pit layer unit, in which the droppings have accumulated for a 52-week production period, will yield some 30 to 40 tonnes of manure per 1000 birds, depending on moisture content (50–60%). The question of manure disposal has assumed much greater importance with moves to reduce nitrate pollution of water. Compliance with the *Code of Good Agricultural Practice for the Protection of Water* (MAFF 1998) has serious implications for many specialist producers. It may be necessary to have firm agreements with neighbouring farmers to use their land for manure spreading. There is certainly an encouragement to develop alternative methods of manure disposal, treatment and use.

### Housing and stocking densities

The great majority of replacement pullets and laying hens are housed under so-called controlled environment conditions, that is, in windowless, insulated and mechanically ventilated buildings.

High standards of insulation and ventilation are required for effective temperature control within the building. The 'U' value of the overall structure should not exceed $0.4 W/m^2/^{\circ}C$. That is, for every $m^2$ of the structure no more heat than 0.4 watt should flow through from the hot to the cold side for each $1^{\circ}C$ difference between inside and outside temperature. The inlets and outlets for ventilation, which may be located in any combination of positions in the sides, ends or ridge of the building, need to be carefully linked, preferably by automatic means, to ensure the matching of air throughputs. Fans are incorporated either in the inlets, which provide a pressurised system of ventilation, or more usually in the outlets. Both inlets and outlets should be well baffled to minimise the interference with temperature and light control from gusts of wind and seepage of light, respectively. The details on temperature requirements, ventilation and lighting appear later.

The rate at which a building is stocked is important both to economic performance and bird welfare. Excessive rates of stocking, apart from subjecting the birds to undue stress, impair pullet development and reduce hen-housed egg production. At times of high economic margins (high egg prices and/or low food costs), it generally pays producers to stock layer houses comparatively densely to increase total egg output from the flock, but this is achieved at the expense of individual bird performance and welfare. At times of economic squeeze, contrary to popular belief, the reverse is true and the emphasis should be on high production per bird which improves welfare (Roush 1986).

The maximum acceptable stocking density for bird welfare, irrespective of the economic climate, is debatable. What may be acceptable in one situation can be completely unacceptable in another. It depends much on the quality of the housing, in particular on the effectiveness of temperature and light intensity control. High temperatures and bright lights predispose birds to aggressive pecking, and exacerbate the effects of dense stocking. Also, stocking density, or more specifically floor space allowance per bird, cannot be considered in isolation from other space provisions, such as for feeding, drinking and, in the case of non-caged layers, nesting. Behavioural problems in a flock indicative of over-crowding may not be the result of a high stocking density in itself. Moreover, some strains of bird are more aggressive than others and all strains appear to overcome the stress of high stocking densities better if group or colony size is small.

Stocking rate, therefore, is a complex issue; observance of any particular rate cannot ensure acceptable welfare of the birds. This is acknowledged as qualification to the recommended maximal stocking densities for different housing systems presented in the Codes of Welfare (MAFF 1987). The UK code for replacement pullets reared under the deep-litter system is a maximum of 17kg liveweight per m² of floor area. It is logical to express stocking density in terms of bodyweight with birds of varying size, but in practice producers must calculate the equivalent bird numbers. With brown pullets typically weighing 1.4kg at the completion of the rearing period, the equivalent is 12 birds (16.8kg) per m². Colony sizes are not mentioned in the code, but it is advisable to divide large flocks of pullets into pens of no more than about 4000 birds to reduce the risk of pile-ups. The recommended maximum stocking rate for laying hens kept on deep litter is also 17kg per m² of available floor area, but in this case a maximum of 7 birds per m² is specified. Stocking rates in deep-litter units may be increased in compliance with the codes up to a maximum of 25 birds per m² of available floor area where at least 15cm of suitable perching is provided per bird.

The code for replacement pullets reared in cages is a minimum of 250 cm² of cage floor area per kg liveweight (350 cm² per 1.4kg point-of-lay pullet). A typical example of a commercial rearing cage is one measuring 0.6m deep x 1.2m wide for 20 birds. There is no UK code for the stocking of laying hens in battery cages. This is subject to legislation (MAFF 1994) which stipulates a minimum cage floor area of 450cm² per bird where four or more birds are kept in the cage. The regulations also prescribe minimal feeding and drinking space provisions, that is, at least 10cm per bird of food trough and of drinking channel length, or access to two drinkers (nipples or cups) from within each cage. Specifications for cage height (minimum of 35cm with at least 65 per cent of the area 40cm or more) and floor slope (maximum of 8 degrees for rectangular wire mesh, 12 degrees for other floors) are also included. All of these legal requirements apply irrespective of bird size. A popular choice of cage for housing 2.2kg brown egg layers, which meets the regulations, has a floor area of 50cm x 50cm and holds five birds.

In addition, the regulations set out in principle certain requirements for cage design, construction, maintenance and cleaning, as well as general conditions relating to the management of battery units. These will be cited, as appropriate, in subsequent sections.

## HUSBANDRY SYSTEMS

### Deep-litter systems

Very few producers keep laying hens on litter, but approximately half of the UK national replacement pullet flock is reared under this system. The most common litter material used is white woodshavings which, for rearing pullets, is spread initially to a depth of 5–7.5 cm; about double this depth is required for housing layers. The actual amount used varies with season, house condition and stocking density. A cheaper alternative to woodshavings is chopped wheat straw, which is of comparable quality in terms of water absorption and insulation, but it must be clean, that is, not dusty or damp and musty, which would put young chicks at risk from fungal disease organisms such as *Aspergilli*. Shredded newspaper is another material sometimes used as a litter. All litter is usually cleaned out at the end of each batch of birds, although in some countries litter is re-used after heaping it between batches to heat and 'sterilise'.

The main problem with a deep-litter unit is the management of the litter itself. It is critical to maintain the litter in a suitably dry, friable condition. If too dry (less than about 20% moisture in a litter based on woodshavings), the atmosphere will become unpleasantly dusty, which predisposes stock and attendants to respiratory disease. If too wet (more than about 40–45% moisture), the surface of the litter tends to cake, so that the droppings do not intermix but accumulate on the surface. This results in birds with dirty feet which, with adult layers, gives a particular problem of dirty eggs. Damp litter also puts the birds more at risk from coccidial infection, encourages the breeding of flies and increases the production of ammonia from bacterial breakdown of uric acid in the droppings.

In the UK climate over-damp litter is the more likely problem. It is multi-factorial and demands considerable skill to combat. The following is a summary of the main things to do in order to keep litter dry and friable:
1. Provide an adequate depth of good absorbent litter
2. Remove patches of litter that become badly caked or where drinkers accidentally flood
3. Provide ventilation designed to keep atmospheric humidity below about 70 per cent and to ensure that the incoming cold air is mixed adequately with internal warm air before falling to litter level
4. Incorporate a waterproof membrane in the floor to prevent rising damp, and insulate the building well to prevent condensation
5. Manage drinkers so that water spillage is minimal
6. Avoid excessive stocking densities, especially in the winter time, which increase humidity of the atmosphere as well as add to the load of damp faeces
7. Achieve even distribution of birds over the available floor area through uniform temperature, ventilation and lighting, and even distribution of feeders and drinkers
8. Avoid food ingredients or levels of them which cause birds to drink excessively and thereby produce wet droppings. For example, avoid excess sodium chloride, potassium (plentiful in soya bean) and protein, especially poor quality indigestible protein. Also, restrict the use of barley or feed it with a multi-enzyme additive to prevent the problem of sticky droppings

9. Prevent diseases that cause enteritis and/or malabsorption of water leading to diarrhoea, for example, coccidiosis and Gumboro disease.

A droppings pit in a deep-litter unit, that is, a raised platform of wooden slats or wire mesh about 0.75m high over one to two thirds of the floor area, will ease litter management by reducing the load of droppings on the litter area. The birds roost on the pit platform and they are encouraged up during the day by locating the drinkers and some feeders over the pit area. This also means that spillage from the drinkers is kept out of the litter, although it can result in the pit becoming a hotbed of ammonia production. The need to dismantle, clean and reassemble the pits between flocks creates additional work at a time of high labour demand. Pits are therefore a feature of layer units where clean-outs only take place about every 56-60 weeks, rather than pullet units which usually operate a 18-20-week growing/clean-out cycle.

**Perchery systems**
Percheries are a development of deep-litter units which make use of the space above floor level. This allows higher stocking densities, which raise ambient temperatures and increase ventilation throughputs. The development has been fostered by the search for a loose flock or 'colony' system for intensive egg production that competes more effectively with battery cages than the deep-litter system.

Essentially, percheries contain banks of tiered perches and/or platforms which incorporate feeders and drinkers at the different levels and alternate with tiered runs of nest boxes (see Figure 9.4). Many configurations of perchery have been developed, but no standard form has emerged. There is the problem of the concentration of droppings associated with the tiered perches. If droppings are simply allowed to fall through to floor level, they will soil birds and probably some feeding and drinking equipment on the way. They then accumulate on the floor, for example in shallow pits, at a rate that necessitates fairly frequent removal by some form of belt or scraper system. Incorporating wide plastic belts below each tier of perches to collect the manure for disposal at about weekly intervals also has its problems. Apart from increasing both capital and running costs, it makes it even more difficult for the stockperson to negotiate the tiered banks for cleaning and stock inspection, and to reach birds, equipment and 'floor' eggs.

Most producers of barn eggs in the UK have opted for the comparatively uncomplicated belt-less arrangements of perchery and simplified things even more by installing them on slatted and/or wire mesh floors over a deep pit. The manure then falls right through to the pit where it can accumulate for the entire production period before being cleaned out. The passageways between perch banks can have solid littered floors. Alternatively, the whole floor area may be of slats or mesh to avoid the problems of litter management. However, such percheries increase restriction on the birds' natural behaviour.

**Figure 9.4**      **Perchery system of housing laying hens**
Bank of perches (right), tiered nest boxes (left) and a wire mesh floor.
*(Courtesy of Poultry World)*

The correct configuration of percheries has still to be clearly established. There is evidence that birds can suffer severe knocks and many broken bones from flight and landing accidents in existing perchery installations (Gregory *et al* 1990). In contrast, caged layers, although their bones are weaker, break less of them during their lives. It seems that birds should not be required to fly more than about 1m for safe

negotiation between perches, but other important aspects of design and lay-out still need critical examination. Specifications for the actual perches are open to question. Traditional perches are constructed of 5x5 cm bevelled softwood timbers at 30cm centres and provide 15cm length per bird. This length is sufficient for hens weighing about 2.2kg, enabling all birds to perch simultaneously. The type of platform, whether wooden slats or mesh of wire or plastic, is also debatable. Wire mesh (usually 2–3 mm diameter wire of 25x75 mm mesh) remains comparatively clean, but some consider slats to be more comfortable for the birds. Traditional slatted platforms are constructed of 25–30mm square bevelled softwood with 25–30mm tapered gaps. Hardwood and plastic are more hygienic, as are narrower slats.

Attempts to stock percheries at the density of 25 birds per $m^2$ of available house floor area, the recommended maximum in the Welfare Codes (MAFF 1990), have often been unsuccessful. A high level of stockmanship is required in percheries and a stocking rate of 16–20 birds per $m^2$ may be more appropriate. The optimum colony size is also unclear, but large flocks need penning into colonies of no more than some 2000–3000 birds. Recommendations on stocking density and other aspects of the welfare of hens in percheries are included in a report by the Farm Animal Welfare Council (1991).

**Cage systems**
Roughly 50 per cent of replacement pullets are reared in cages. Cage rearing has the advantage that there is no change in housing system when point-of-lay pullets are moved to layer cages. Litter-reared birds may be checked and have age at sexual maturity delayed by a day or two while adjusting to a cage environment. On the other hand, cage-reared pullets are unsuitable as layers for non-cage systems since they do not develop immunity to coccidiosis or the ability to perch. Cage rearing involves a high proportion of capital expenditure on specialist equipment. The capital cost per bird space is about the same for cage and deep-litter rearing units, but the proportional split in expenditure for house and contents is approximately 40:60 for a cage unit and 70:30 for a deep-litter unit.

Traditionally, cage rearing was a two stage process involving the movement of birds from heated brooder units to larger grower units at about four weeks of age. Modern systems are usually single stage, accommodating birds from day-old to point-of-lay. Some cages are specially equipped for brooding purposes and birds are spread out from these to occupy all cages by five or six weeks of age. Then each bird has some 5–7 cm of food trough space and about eight birds share each nipple-type drinker. Rearing cages, like layer cages, may incorporate belt cleaning or a deep-pit system of manure disposal (see below).

Cages, whether for rearing or laying, are constructed of wood and wire netting or of metal and wire mesh. Most modern cage installations are of the more hygienic metal construction which is easier to clean and does not harbour so readily the troublesome red mite parasite. The main difference with layer cages is the sloping, forward extending floor for eggs to roll away. The actual slope is critical – too gentle and the eggs hold up in the cage to be pecked at and trodden on, too steep and they roll down rapidly to be

broken on impact at the collection point. It requires about 7.5 degrees of slope for conventional 25x50 mm wire mesh and 11.5 degrees for tightly stretched wire netting. It is important for the comfort of the birds that the slope is the minimal necessary. Also, it improves welfare if the mesh is smaller than conventional (for example, 25x25 mm) which increases the number of points at which the bird's foot is supported and reduces the incidence of foot lesions. The essential features of a conventional cage for laying hens are shown in Figure 9.5.

The configurations of laying cages are many, but basically there are two main types of arrangement: (1) vertically tiered and assembled back-to-back; and (2) stepped and assembled on an 'A' frame (see Figure 9.6). Stepped cages were developed in California and are particularly suitable for use in a hot climate. The stacks of cages run lengthwise down the building, separated by service gangways varying in width depending on whether egg collection is manual or mechanical.

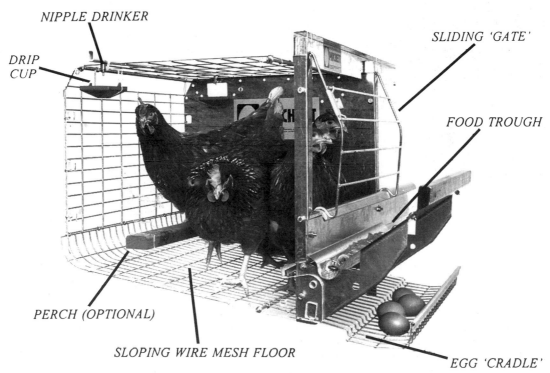

**Figure 9.5**     **Conventional battery cage**
(*Courtesy of R J Patchett Ltd*)

convenience and labour saving of cleaning out only once per cycle. In order to provide vehicle access for efficient manure removal, the pit needs to be about 2m deep, which in practice means that the cages are supported on walls and pillars about 2m above ground level. In other words, for a system described as deep-pit the buildings are surprisingly high rise!

A big disadvantage of stepped cages under UK climatic conditions is the large area that they occupy, which results in a comparatively low stocking rate per unit of house floor area. A popular compromise has been the use of semi-stepped cages which overlap front-to-back by about one-third (see Figure 9.6). The lower cages are protected from droppings by sloping the roof below the area of overlap and covering this slope with a plastic flap or metal sheet which deflects the droppings into the pit. The problem with this, under the current regulations for layer battery cages, is that it seriously restricts the available cage area with a minimum height of 35cm (MAFF 1994).

The deep-pit system can be used with vertically tiered cages if they are assembled with a gap down the middle of the stack so that they are not quite back-to-back. Mechanically operated scrapers at each tier are used to push the manure off solid boards down through the central gap into the pit. Here it tends to cone up more than under stepped units, which exposes a greater surface area for drying. This helps to keep down the production of ammonia and the growth of fly populations.

Most layer cage installations are now four or more tiers high and of necessity incorporate mechanical egg collection. It is a requirement under the EU layer battery regulations for all installations of more than three tiers to have special devices or measures to ensure effective daily inspection of the birds at all levels. The legislation also requires the stockperson to thoroughly inspect all automatic equipment on which the birds depend at least once a day (MAFF 1994).

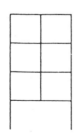

Vertically
tiered

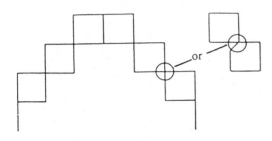

Stepped ( or semi-stepped)

**Figure 9.6** **Diagrammatic representation of the two main types of cage configuration.**

Vertically tiered cages usually have reinforced plastic belts between the tiers which collect the manure. These belts are wound down to one end of the stacks once or twice a week to deposit the manure on to a cross conveyor for removal from the building to a store or to an awaiting trailer or manure spreader. This is described as a belt-clean system.

Stepped cages adapt well to the deep-pit system of manure handling where, simply by force of gravity, the manure accumulates below the cages for the entire production period and affords the

### Brooding of pullets

It is necessary to provide supplementary heat in the first 3–6 weeks of a chick's life, depending on housing quality and weather conditions. This is known as the brooding period, when the chick is still developing effective measures of body temperature control and when, in the natural state, it is dependent on the broody hen for warmth.

*Brooding on litter*

The most common method of brooding under deep litter conditions is to suspend radiant or canopy-type gas fired heaters over the litter, usually at appropriate intervals down the centre of the house (see Figure 9. 7). As the heat is localised around individual brooders, the system is usually described as 'local' or 'spot' brooding. Gas is chosen as the normal source of heat on the combined grounds of convenience, reliability and cheapness. Gas combustion provides a moist heat which aids feathering, although the addition of moisture to the atmosphere is a further factor to note in the complex task of preventing damp litter.

The brooders should be turned on a day or two before the day-old chicks arrive in order to dry out the house and to achieve stable target temperatures. They should be suspended at a height that provides an initial temperature of 30-32˚C immediately below them at litter level. The temperature within the overall house should be 24–27˚C. Other preparations include:

1. levelling and compressing the litter for uniform floor insulation and ease of chick movement
2. covering 20–25 per cent of the brooding area with feeding stations (none directly under the brooders), supplementing the permanent feeders with strips of paper, pans or trays

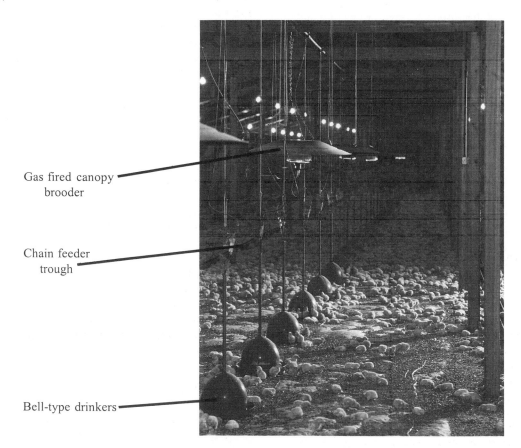

Gas fired canopy brooder

Chain feeder trough

Bell-type drinkers

Figure 9.7      **Brooding chicks in a deep-litter unit.**

*(Courtesy of Poultry World)*

3. setting up drinking points between the feeding stations, supplementing the permanent drinkers with mini-drinkers (filled in time to take the chill off the water)
4. providing maximum illumination over the feeding and drinking areas (usually in practice 20–30 lux at floor level)
5. setting up solid or netting surrounds around the brooders to confine the chicks near to the source of heat (chicks wandering off and becoming cold may huddle together and smother).

On arrival the chicks should be placed under the brooders without delay and left to settle down for an hour or two. They should then be uniformly distributed and eating and drinking readily, not cheeping noisily and bunched under the heaters, when too cold, or ominously quiet and spread out to the far regions of the brooding area, when too hot; nor predominantly down one end or to one side as when avoiding a draught.

Temperatures should be progressively reduced to provide a uniform 20–21°C by 4–5 weeks of age. The actual rate of decline depends on the quality/health of the chicks and their rate of feathering. After a few days lights remote from the initial brooding area are switched on to establish more even chick distribution and to bring more of the permanent feeders and drinkers into use.

Brooder surrounds are expanded, merged and eventually removed after 10 days or so. It is at this time that there is the main risk of smothering, especially when the lights go out. Installation of red pilot lights over the brooding area, which come on for about 30 minutes as the house lights go off, creates an artificial dusk and helps to attract chicks back to the heat. Alternatively, the house lights may be extinguished gradually by fitting an automatic dimmer.

Levels of food and water in the troughs, pans or bowls are high to begin with to encourage early feeding and drinking, but during the second week levels are gradually reduced, ideally down to about 1.5cm, to restrict food wastage and water spillage.

*Brooding in cages*
The most common method of brooding in cages is to use large gas- or oil-fired blower heaters which blow hot air, either directly or through some form of

ducting, into the house to create a uniformly high temperature throughout the building. It is often referred to, therefore, as 'whole house' brooding. It is not exclusive to cage rearing and can be used in deep-litter units.

Chicks confined to cages are very dependent on the correct temperature being provided within every cage. They cannot move out of uncomfortably hot zones into cooler areas and *vice versa*. The main risk in practice is over-heating so the recommended initial brooding temperature of 28–30°C is a few degrees below that usually provided under 'spot' brooders in deep-litter units. A high/low temperature alarm is particularly important as well as measures to prevent over-/under-heating. It is wise not to rely on just one big heater and to have back-up thermostats for all heaters.

Oil-fired heaters produce a very dry heat which increases the risk of dehydration of baby chicks. The relative humidity should be kept in the region of 70–80 per cent, if necessary by spraying water over the floor in the early stages.

Typically, brooding takes place in the top of three-tier or the middle two of four-tier cage installations. Special features are required here to contain the baby chicks and to enable them to reach food and water. The mesh floors are usually covered with non-slip corrugated or crimped paper for the first 2–3 weeks. Various adjustments to the cage fronts are necessary, usually for the initial 5–6 weeks, to enable chicks to feed without escaping (see Figure 9.8). One or two bowl-type drinkers are required per cage in the first 7–10 days to supplement permanent nipple-type drinkers. The nipple lines are height adjustable and raised as the birds grow.

Lights should shine brightly into the brooding cages till all chicks have established satisfactory feeding and drinking behaviour. Light intensity is then gradually reduced to about 5 lux at the food troughs to prevent aggressive pecking and general flightiness.

**Post brooding temperature requirements**
The optimum ambient temperature for pullets post-brooding and for hens throughout the laying period is about 21°C. In normal practice in the UK, no artificial heating or cooling is used at this stage. Reliance is placed on good house insulation coupled

(a)                                                                    (b)

**Figure 9.8**      **Cage rearing of pullet chicks**
   **(a)       Food trough entirely outside the cage**           Photographs showing 2 ways of
                (*Courtesy of Autonest Ltd*)                             containing baby chicks within the
   **(b)       Food trough partially inside cage**                 cage while enabling them to
                (*Courtesy of R J Patchett Ltd*)                      to reach the food.

with high stocking densities to keep temperatures up in cold weather and efficient ventilation to keep them down in hot weather.

Birds show a saving in food intake with rising temperature (approximately 2 per cent for each 1˚C rise in temperature over the range from 15˚C to 25˚C), but this economic benefit is offset by adverse effects on production factors. Egg weight starts to fall as temperature increases above about 18˚C (approximately 1g in average egg weight for each 3˚C rise in temperature up to 28˚C). Beyond about 21˚C shell thickness is in decline, and after a further degree or two rise egg numbers start to decrease. With growing pullets high temperatures slow down growth and development. Under most UK conditions a mean daily temperature of about 21˚C seems to

give the right balance between food consumption and other economic factors.

The quality of feathering influences the response of the birds to varying temperatures. In flocks which have considerable feather loss, the optimum temperatures are two or three degrees higher. Also, it should be noted that the recommended temperatures relate to the general house environment and not to the microclimate within cages or among birds squatting on a littered floor.

With intensive units it is generally more problematic to keep temperatures down in hot weather than to avoid sub-optimum temperatures in cold weather. Birds have no sweat glands and rely on evaporative cooling from the respiratory tract as the chief means of heat loss. Above about 28˚C they start to pant to increase evaporative cooling and prolonged panting is the main sign of heat stress. When this occurs producers should be in a position to take emergency action.

Special means of creating high air speeds over and around the birds are of particular value during hot weather. They increase sensible heat loss through convection, which occurs particularly from the unfeathered areas of the birds, such as the combs and wattles. In very hot conditions, air speeds should be sufficient to ruffle the birds' feathers (about 2m/s). This can be achieved by installing special circulatory fans within the house and/or directing incoming air straight on to the birds. It may cause birds that are sitting to stand up, which will facilitate their own main means of increasing sensible heat loss, which is to raise their wings and expose the poorly feathered areas underneath. Spraying water in the house is very effective at reducing heat stress in hot dry climates, but increasing the humidity in an already humid atmosphere will impair the birds' own evaporative cooling through panting and may do more harm than good.

Birds also require an ample supply of cool drinking water to help them combat high temperatures. For example, water consumption of laying hens almost doubles with a $10°C$ rise in ambient temperature above the recommended $21°C$. In contrast, they markedly reduce food consumption to restrict metabolic heat output. In the emergency of heat waves it may help to withdraw food completely during the hottest mid-day/afternoon periods. Apart from emergency measures, it is important that houses are not over-stocked, that insulation is kept in good condition and that the permanent ventilation system is correctly installed and regularly maintained.

## Ventilation

Food intake is commonly used as the fundamental basis for determining the requirement of poultry for ventilation, due to its close relationship to metabolic activity.

The minimum ventilation requirement is a throughput of about $2m^3$ of air per second for each tonne of food eaten by the birds per day ($2m^3$ /s/t/d). Thus, if birds in a building are eating two tonnes of food in a day (which approximates, for example, to the intake of some 50,000 five-week-old replacement pullets or 16,000 adult laying hens) their minimum requirement for ventilation is $4m^3$ /s. Apart from bird respiration, this is needed to remove volatile products from the manure, which include water vapour and obnoxious gases, notably ammonia that arises from bacterial breakdown of uric acid. The latter can be particularly problematical in deep-litter and deep-pit units during the winter months, creating an unpleasant atmosphere if ventilation is held down at minimal levels in order to maintain the optimum temperature of around $21°C$.

The requirement for maximum ventilation is ten times the minimum rate, that is, $20m^3$/s/t/d. This large increase is needed just to remove excess heat during periods of hot weather and is, of course, the rate that the ventilation system of a building must be designed to achieve. In mechanically ventilated units this involves the installation of fans of sufficient size and number. For example, if, as has been customary in the poultry industry, use is made of 610mm diameter, 940rpm fans which have a throughput of about $2.5m^3$/s, then one such fan is required per 1000 laying hens or per 1375 replacement pullets.

The operation of fans, apart from those required to provide the minimum ventilation rate, is under thermostatic control. They usually work by extracting foul air and thus drawing in clean air through the inlets but sometimes they are set to blow in fresh air and force foul air out through the outlets. It is important for efficient ventilation that inlet/outlet area varies in proportion to fan operation and throughput. In large modern units inlets/outlets usually open and close automatically with ventilation demand. This can be achieved through either some form of flap which billows open as the fan pressure builds up, or a more sophisticated system of hydraulically operated panels that open and close with variations in temperature and air pressure. Increasingly, modern electronic systems and computers are being used to control fan operations and related openings to achieve the correct temperature and air quality.

There are many possible ways of combining fans and inlets/outlets to create workable ventilation systems (see Figure 9.9). All systems, however, have the same objective of providing uniform air movement at bird level(s) without cold draughts or over-heating. This means that the system should be designed so that in cold weather incoming air is mixed well with warmer internal air before reaching the birds (best achieved with the high speed inlet jet), whereas during hot spells just the reverse applies. In all deep-pit houses it is important to reverse the natural upward flow of warmed air so that air is not drawn up over the manure and polluted with obnoxious gases before reaching the birds. Also, as it passes through the birds first and warms up in the process, a reverse air flow is more effective at drying the manure.

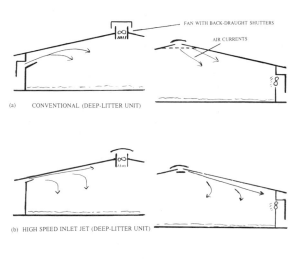

(a)   CONVENTIONAL (DEEP-LITTER UNIT)

(b)   HIGH SPEED INLET JET (DEEP-LITTER UNIT)

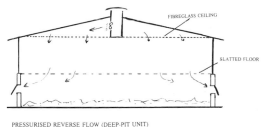

(c)   PRESSURISED REVERSE FLOW (DEEP-PIT UNIT)

**Figure 9.9      Diagramatic representation of the main ventilation systems**

Most ventilation systems need special fail-safe arrangements to provide an adequate level of natural ventilation in the event of a power failure. In fact, it is a legal requirement in the EU to provide such arrangements, or equivalent mechanisms, as well as an alarm to warn of a failure to the ventilation system, which must be thoroughly checked for defects at least once per week (MAFF 1994). Fail-safe devices, when activated, should allow air to enter at

a fairly low level and escape from the top of the building. The fan shafts provide useful openings, top or bottom depending on the system, provided that the back draught shutters, which normally close up the shafts automatically when the fans stop running, fall away when the stoppage is due to a power failure. This is achieved by holding the shutter panels in place with electromagnets. Inlet areas may need to be increased for effectiveness in an emergency and this can be achieved through the use of electromagnets which allow constrictive panels to fall away if the power fails.

Producers are increasingly installing auxiliary ventilation in belt-clean battery units to dry the manure as it collects on the belts. This involves blowing fresh or a mixture of fresh and recycled air along perforated ducting which runs centrally down the cage stacks below each tier of cages. It improves the atmospheric quality and facilitates the task of weekly disposal of the manure, especially if compliance with the *Code of Good Agricultural Practice for the Protection of Water* (MAFF 1998) necessitates transporting it over long distances to suitable spreading sites. Such in-house drying can also constitute the first stage of a manure composting process.

## Lighting
### Light duration
The sexual development of pullets and the egg production of hens are greatly influenced by light duration. Increasing daylength stimulates them and *vice versa*. It is the **change** in the pattern of lighting, not the absolute duration of light that is of prime influence. Wild birds lay their eggs in the spring and moult in the autumn because of the changes in daylength. The stimulus is received through the bird's eye and also through the cranium. An outline of the biological basis for the effect of light is presented in Figure 9.10.

There are basically two things to avoid in order to achieve a satisfactory pattern of lighting for egg production. The first is to prevent a decrease in light duration in the laying period and the second is to prevent an increase in light duration during rearing. If the latter is not done, pullets will come into lay too early when they are too small. This will result in unacceptably small eggs throughout the laying period.

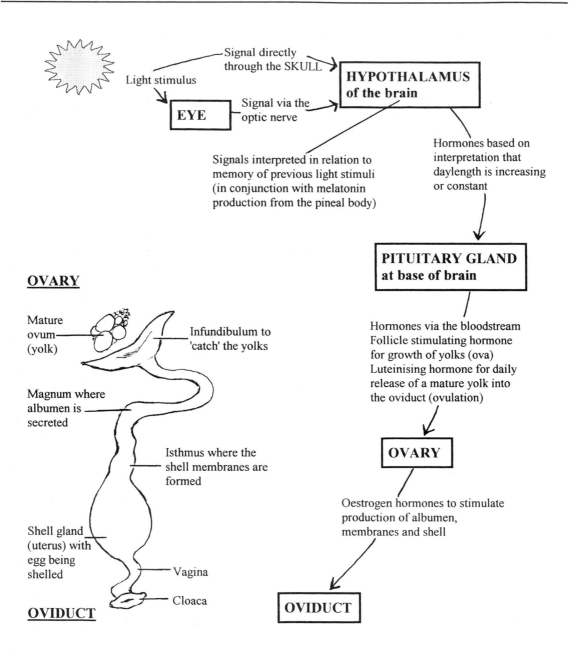

Light stimulus

Signal directly through the SKULL

**EYE**

Signal via the optic nerve

**HYPOTHALAMUS of the brain**

Signals interpreted in relation to memory of previous light stimuli (in conjunction with melatonin production from the pineal body)

Hormones based on interpretation that daylength is increasing or constant

**OVARY**

Mature ovum (yolk)

Infundibulum to 'catch' the yolks

Magnum where albumen is secreted

Isthmus where the shell membranes are formed

Shell gland (uterus) with egg being shelled

Vagina

Cloaca

**OVIDUCT**

**PITUITARY GLAND at base of brain**

Hormones via the bloodstream
Follicle stimulating hormone for growth of yolks (ova)
Luteinising hormone for daily release of a mature yolk into the oviduct (ovulation)

**OVARY**

Oestrogen hormones to stimulate production of albumen, membranes and shell

**OVIDUCT**

Note: Approximate timing of egg formation ( an example):-After an egg is laid (oviposition) 30 minutes elapse before the next yolk is ovulated and then 24.5 hours are involved in the formation of the albumen (3 hours), shell membrane (1 hour) and shell (20.5 hours).

**Figure 9.10      Biological basis of light stimulation of egg production**

Mean egg weight will fall by about one gramme for each week of premature egg production. Also, bringing pullets into lay too early will cause an increase in mortality from prolapse of the uterus. Early sexual maturity induced by lighting is therefore undesirable; this is not the same as genetically developed early maturity which, in contrast, is economically beneficial.

One possible satisfactory lighting programme for egg production is a constant daylength of, for example, 12–14 hours from the beginning of rearing to the end of lay. However, in practice there is usually some stepping up and/or down of light duration in an attempt to gain positive benefits as opposed to merely avoiding the negative effects. The following gives an outline of typical conventional practice in light-proofed buildings.

| | |
|---|---|
| 0 → 2d | almost continuous light |
| 3d → 2wk | abrupt fall or step down to 8 h/d |
| 3wk → 17/18wk | constant 8 h/d |
| 18/19wk → 22wk | increase by 1 h/wk |
| 23 wk → end of lay | increase by 20-30 min/wk to maximum of 17 h/d and then hold at this level |

This programme will bring the birds into lay at about 19 weeks. The main alternative is to make the step down to 6-8 hours light much more gradual by extending it over the first 8 weeks, which will delay sexual maturity by about a week and give an increase of approximately one gramme in average egg weight. In buildings with minimal light seepage, daylength in the laying period may be maximised at 14–16 hours to save on electricity.

Hens lay eggs in the light period. The lights are usually switched on early in the morning so that worthwhile numbers of eggs are available for collection at the start of the working day. A flock lays approximately half its eggs within 5 hours of the lights coming on, and 90% or more of them within 8-9 hours. Individual hens lay their eggs progressively later on succeeding days as egg formation takes more than 24 hours (see Figure 9.10). This continues until they have to break the sequence and miss a day before starting a new 'clutch' back at the beginning

of the light period. In high producing modern hybrids the period of egg formation is only a little over 24 hours and as a consequence the 'clutches' are very large.

A number of alternative lighting programmes can be used during the laying period. One possibility is to employ so-called ahemeral lighting which involves cycles of different length than the normal 24-hour day. Some producers use extended cycles of 28 hours, for example, 16 hours light and 12 hours dark (16L:12D) or 20L:8D which give a six-day week. Although this restricts egg production to six per week, because of the extra time available for egg formation, the resultant eggs are larger and total egg mass output is unaffected. Also, the shells of the eggs are thicker and, if brown, darker in colour. Such a programme can be of value at any time when price premia for the larger eggs are particularly high. It may also be useful towards the end of the laying period when it will arrest the normal decline in shell quality and when the adverse effect on egg numbers is far less than in the early stages of lay.

A very different lighting programme, but one that also produces less but larger eggs of improved shell quality, is intermittent patterns of symmetrical light: dark cycles. An example sometimes used in practice is continuous cycles of 3L:3D. As the hen can recognise only one 'dawn' and one 'dusk' in a 24-hour period and there is nothing in this symmetrical pattern to distinguish one cycle from another, the control of light over egg production is lost. In this situation individual hens are producing successive eggs in their own good time, which is in excess of 24 hours and does not restrict laying to the light period. Compared to ahemeral lighting, the 3L:3D programme is easier to apply in practice. It can be arranged to have at least six hours of light each working day, while the mid-day 3D period can be broken into without any adverse effects. Also, it is possible to change back and forth at will from 3L:3D to conventional lighting to suit requirements for egg size without the hens recognising any decline in daylength. A problem experienced sometimes with symmetrical intermittent lighting is a breakdown of response and return by the birds to a normal 24-hour production pattern. This might be due to light seepage into the building, or diurnal fluctuations in temperature or disturbance and noise by attendants on a 24-hour basis.

When subjected to intermittent lighting with asymmetrical cycles hens regard the longest dark period as the one 'night'. For example, under repeated cycles of 2L:4D:8L:10D the 10D period is night to the hen, while the 4D period is perceived as light. The programme provides the equivalent of 14 hours light per day with no electricity cost for four of those hours.

This characteristic of the hen to interpret all but the one longest dark period in a 24-hour cycle as night is further exploited in the 'Biomittent' lighting programme. This involves dividing each hour of the day into 0.25 hour light and 0.75 hour dark, for example, (0.25L + 0.75D)x16:8D which is as effective in terms of egg production response as 16L:8D, but uses less electricity and gives a saving of about 5 per cent in food intake. The business of lights going on and off throughout the working day is not a particular problem as the only consequence of putting the lights on now and then during the 0.75 hour dark spell is a slight reduction in the saving on electricity. However, the 'Biomittent' pattern goes against the recommendation within the Welfare Codes (MAFF 1987) that birds should be given at least eight hours lighting per day. This recommendation is not supported by the findings of Lewis *et al* (1992) that asymmetrical intermittent and 'Biomittent' lighting reduce mortality, vices and obesity.

When birds receive natural light the only manipulation possible is supplementation to prevent daylengths increasing during rearing or decreasing during lay. The appropriate programme of supplementation will depend on the natural daylength at point-of-lay. Pullets should be reared on a programme that either steps down to this, as would occur naturally in the autumn, or provides constant lighting at the point-of-lay level. Laying hens should be allowed to follow naturally increasing light patterns, but be held at constant high levels when natural daylengths fall. For example, if rearing starts in mid-winter and natural daylength is increasing to reach 14 hours at the point-of-lay stage, then natural light can be supplemented to provide a constant 14 hours per day throughout the rearing period, or a step down from 24 hours at day-old to 14 hours per day at some stage along the way. From then on, the laying birds can experience the rise in the natural

daylength to its maximum (approximately 17 hours in mid-June in the UK) and then be held at this level for the remainder of the laying period.

In modern practice natural lighting is restricted mainly to free-range egg production and to the laying stage only, as the majority of flocks are reared intensively in light-controlled housing and transferred to range accommodation at point-of-lay. Moreover, modern free-range layer units tend to be windowless so that artificial lights need to be switched on throughout the natural day as well as during the period of supplementation. This does mean, however, that natural daylength can be restricted by controlling the time that the birds are allowed out of the house. This can be of benefit in the very early stages of lay in mid-summer and for any period of induced moulting.

*Light intensity and type*
Young chicks require bright lights (at least 20–30 lux) initially to find food and water. After a few days in light-proofed units most producers reduce the intensity, usually in gradual steps using a dimmer, to about 5 lux, or less if necessary, to control feather pecking and cannibalism. Low light intensity during rearing appears to have no adverse effect on subsequent egg production. Lights can be turned up for bird inspection purposes.

Light intensity should not be reduced on moving to the laying quarters, nor should there be a big increase. It is a particularly critical period with regard to cannibalism as the birds come into lay. If there is evidence of a problem, light intensities should be kept down at this stage and increased later when there is much less risk of cannibalism developing. Light intensity needs to be kept up at 5–10 lux or egg production may be adversely affected.

Light should be evenly distributed and this may be difficult to achieve in cage units, especially if the tiers are stacked vertically. Maximum use needs to be made of reflected light from walls and ceiling, while shades can be used to reflect light down to the lower tiers and lights can be installed in the pits of deep-pit units.

In addition to providing sufficient light, it is important to ensure adequate darkness in the prescribed dark periods. Light seepage into so-called light-proofed buildings can negate carefully planned lighting programmes. Houses should be checked for darkness when the artificial lights are off and the sun is shining outside. Light seepage should not reach an intensity sufficient to register a response with the birds, which for many years has been regarded as approximately 0.4 lux. However, more recent studies indicate that the minimum light intensity needed for an effective response is a proportion of the maximum intensity provided, rather than an absolute value, and that this varies with the pattern of lighting used. The prevention of light seepage entails a gap-free construction and well sealed doorways, but the main problem is the ventilation inlets and outlets, which should be baffled.

Some poultry keepers have tried using red lights to reduce the incidence of cannibalism. Any benefit is probably because the change from white to red has reduced the intensity of the light. There is no evidence that red or any other coloured lights improve bird welfare or economic performance. Birds are unable to see in blue light which is sometimes used in deep-litter situations to prevent panic and pile-ups of birds at times of catching.

## NUTRITION AND FEEDING

### Egg formation and nutrition

Egg farming is an exploitation of the hen's reproduction process, involving essentially the conversion of animal feedstuffs into eggs for human consumption. At the onset of maturity in the pullet, hormones released from the anterior pituitary gland stimulate the Graafian follicles of the single ovary (unlike mammals, birds only fully develop the left ovary) to develop into the future yolks of the eggs (see Figure 9.10). The yolk is a single ovum vastly distended by phospholipids which are synthesized in the liver, and transported to the ovary in the blood. The numerous follicles are at different stages of development, the largest being the first to be shed from the ovary (ovulation). The released ova pass into the oviduct where the albumen (egg white), the two shell membranes and the calcareous shell are laid down. The hen's food intake increases markedly to meet the demands of egg production and a strong specific appetite for calcium develops.

The digestive system of the fowl is shown in Figure 9.11. The crop is a thin-walled storage compartment at the base of the oesophagus; the glandular proventriculus secretes hydrochloric acid and digestive enzymes; the thick-walled, muscular gizzard contains grit or stones to assist in its grinding function; the duodenum receives bile and pancreatic juices; the small intestine's upper and lower loops are the main sites of absorption; the paired, blind-ended caeca arise at the junction of the small and large intestine which opens at the cloaca. The ureters from the kidneys also open into the cloaca so that faeces and urine are mixed and voided together as the droppings.

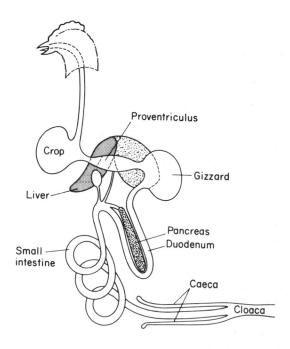

**Figure 9.11       The digestive system of the fowl.**

The fowl is omnivorous, but with an inability to digest cellulose and similar fibre. The caeca contain micro-organisms that can produce small amounts of cellulose digestion products, but these are of no nutritional significance. The alimentary canal is relatively short and food passes through within a few hours. Approximately 65–70 per cent of the cost involved in keeping a flock from day-old to the end of lay is attributable to its food; hence the importance of nutrition and the management of feeding systems in egg production.

## Diets

Chickens tend to eat to their energy requirements, especially egg-type stocks which show a closely related decrease in food intake with increasing dietary energy content. This means that other nutrients, like the amino-acids of protein, need to be included in the diets in proportion to energy. In practice, the main factor influencing the energy requirement of a flock is ambient temperature. Within the range from about 15° to 25°C the food intake declines at the rate of approximately 2 per cent for each 1°C rise in temperature and *vice versa*.

The relationship between energy content of the diet and food intake is not perfect and with *ad libitum* feeding, birds will marginally over-consume, especially if fed pelleted diets. Fatness caused by over-consumption is detrimental to egg production, so the food of growers and layers is usually provided in the form of a 'meal' or dry 'mash' which reduces consumption slightly and also extends feeding time markedly compared to pellets. This helps to occupy the birds and reduce the incidence of undesirable behaviour such as feather pecking. Meal is cheaper than pellets, but is more dusty and does not flow as well through feeding systems.

Typical diets for growing pullets and laying hens kept under intensive conditions are shown in Table 9.3. These diets are least-costed to provide the correct nutrient specifications using ingredients that are competitively priced on the UK market. Many egg producers do their own milling and mixing of diets and usually obtain least-costing services from the suppliers of the vitamin/mineral supplements. All food preparation should be undertaken in compliance with the relevant legislation and codes of practice on the control of Salmonella.

Looking at the ingredients listed in Table 9.3, the main sources of energy are the cereals wheat and barley. Much more maize would be included if it was available in the UK at comparatively low world prices. Maize products, maize germ and prairie meal, are included in the layer diets as useful sources of carotenoid pigments to enhance yolk colour. They are of particular value in contributing towards the colour base of yellow pigment which should be in the ratio of 2 or 3:1 of red pigment to produce yolks of acceptable orange colour. Most of the red pigment is provided as synthetic or extracted products

incorporated in the layer vitamin/mineral supplements.

The main source of protein in this example is soya bean meal. The usual source of animal protein is fish meal but this has been excluded on price. The protein sources must provide all 12 of the amino-acids essential to fowl (which they cannot synthesize). The two that are most likely to be limiting, methionine and lysine, are included in the nutrient specification to ensure that there are sufficient amounts of them present. This being the case, the chances are that all the other essential amino-acids are also present in adequate quantities.

Limestone (calcium carbonate) is included at a comparatively high level in the layer diets to boost calcium content to at least 40g/kg (4%) and produce good egg shells. The other mineral of particular importance in egg shell formation is phosphorus, which in practical diets is more likely to be in excess than deficient. An inclusion level of about 0.55 per cent total phosphorus should not be exceeded.

Metabolisable energy is the normal expression of energy used in the formulation of poultry diets. It refers to the proportion of total (gross) energy in a feedstuff that remains available for metabolism after allowing for losses in the faeces and urine.

## Practical feeding

The egg yolk remaining in the chick at hatching provides nutrients for about one day; after that food and water are essential.

The chick starter diet is fed from day-old to 6–8 weeks of age during which time each bird will consume some 1.0–1.5 kg. The change is then made to a grower diet of reduced nutrient density which is appropriate for the birds' requirements to point-of-lay at about 18 weeks of age. Consumption of this diet when fed *ad libitum* is approximately 5.0-5.5 kg per bird. The high-protein/high-energy layer diet meets the birds' exacting nutrient requirements for egg production and growth from point-of-lay to about 40 weeks – a period, moreover, when food intake is comparatively low. It is then appropriate to 'change down' to a diet of lower nutrient density, as total egg output (number x weight) has passed its peak and food intake has risen to a level slightly above the average for the full laying period of 115–118g per bird per day. A programme involving two layer diets, as presented in Table 9.3, is common. Some producers may use additional diets of even lower protein content for the final stages of lay, while others may simply feed one diet of high to medium nutrient density throughout the laying period. A further possibility is to feed a pre-layer diet of intermediate nutrient density between the grower and high-protein layer diets (from 16–17 weeks of age to the commencement of lay) to give a more gradual build-up in nutrient intake, particularly with regard to protein and calcium.

Under free-range conditions the main difference is that food intake increases markedly during cold weather to meet the high demand for energy. This may give rise to wastefully excessive intakes of protein, vitamins and minerals, unless the energy content of the diet is increased. However, this is less of a problem than it might seem as comparatively high levels of protein are required in the diets of free-range hens to produce the high proportion of large grade eggs needed to gain satisfactory premium

prices. A high content of linoleic acid (approximately 2.5%) also helps to maximise egg weight. The contribution of grazing to the free-range hens' nutritional requirement is small and tends to be discounted. Chickens are not herbivores and do not have the appropriate enzymes to digest grass. They derive a little nourishment in the form of carotenes and some B-complex vitamins from the grass and the range may yield a few seeds, worms and insects.

Laying hens are nearly always fed *ad libitum*. Any restricted feeding that is practised will be marginal, otherwise egg output is reduced disproportionately. Replacement pullets have often been rationed in the past to prevent them becoming too fat. However, overfatness is not a problem with present-day hybrid stocks fed meal and housed at a minimum temperature of about 20°C. Therefore, food restriction during rearing is unlikely to give any improvement in the subsequent laying period. In fact, unless carried out skilfully, it might have a negative effect on egg production which would offset the

**Table 9.3**   **Typical least-costed diets for baby chicks, growing pullets and laying hens.**

| Ingredients (kg/tonne) | Chick starter | Pullet grower | Layer: high-protein | Layer: medium protein |
|---|---|---|---|---|
| *Wheat, ground* | 500.0 | 350.0 | 450.0 | 375.0 |
| *Barley, ground* | 218.0 | 454.0 | 106.0 | 175.0 |
| *Maize germ meal* | - | - | 87.5 | 140.0 |
| *Prairie meal* | - | - | 50.0 | 37.5 |
| *Soya bean meal* | 215.0 | 85.0 | 180.0 | 150.0 |
| *Soya oil* | 5.0 | 3.0 | 12.5 | - |
| *Sunflower meal* | 25.0 | 75.0 | - | - |
| *Limestone* | 12.0 | 12.0 | 90.0 | 100.0 |
| *Dicalcium phosphate* | 12.0 | 8.0 | 9.0 | 8.0 |
| *Salt* | 0.5 | 0.5 | 1.0 | 1.0 |
| *Methionine* | - | - | 1.5 | 1.0 |
| *Vitamin/mineral supplement* | 12.5 | 12.5 | 12.5 | 12.5 |
| **Nutrients (g/kg)** | | | | |
| *Protein* | 184.0 | 147.0 | 182.0 | 164.0 |
| *Calcium* | 11.9 | 10.9 | 41.0 | 44.4 |
| *Phosphorus* | 6.3 | 5.6 | 5.4 | 5.2 |
| *Sodium chloride* | 3.7 | 3.8 | 3.8 | 3.8 |
| *Methionine* | 3.4 | 2.9 | 4.7 | 4.1 |
| *Lysine* | 9.3 | 6.3 | 8.3 | 7.5 |
| **Metabolisable energy (MJ/kg)** | **11.7** | **11.3** | **11.5** | **11.2** |

(**Source:** Computer Formulation Service, Ian Follows Feed Supplements)

saving in food cost. If practised, the timing and degree of rationing of pullets is critically important. Starting food restriction before 5–6 weeks of age will stunt early vital organ and skeletal growth and result in undersized hens which lay unacceptably small eggs. Starting feeding restriction after 15 weeks of age and checking the development of ovaries and oviducts is hardly conducive to maximising subsequent egg production. The time of any rationing should therefore be in the middle of the rearing period. This would also permit the exploitation of efficient compensatory growth, allowing birds to catch up and achieve breeders' target liveweights at point-of-lay.

It is good management practice to take samples of food from each delivery and store them for about six weeks. Checks can then be made on food quality if problems arise subsequently that could be nutritional in origin.

Further information on poultry nutrition can be obtained from several publications, such as the practical book by Leeson and Summers (1991) and the scientific symposium proceedings edited by Fisher and Boorman (1986).

**Feeding systems**

Food is normally delivered in bulk to storage bins outside the building, from which it is augered automatically to replenish supply hoppers inside. The bins should hold at least 7–10 days' food and should keep it dry to prevent the formation of aflatoxins.

The choice for deep-litter and other non-cage systems of housing is largely between chain and pan feeders. The former consist basically of a trough, usually supported on height-adjustable legs, although it can be suspended, with a moving flat chain in the bottom (see Figure 9.12). The troughing is typically arranged in two or three complete circuits and the chain drags food from the base of the supply hopper

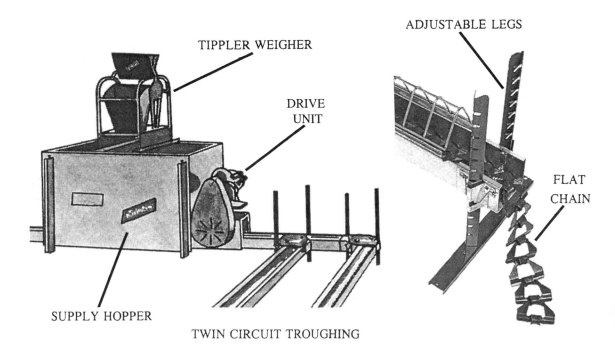

**Figure 9.12      Chain feeders**
*(Courtesy of Sterling Farm Equipment and Big Dutchman International)*

around the circuits. The frequency of running is controlled by time-clock to match the birds' food intake. The length of troughing provided for growing stock is typically 2.5cm per bird which, as there is access to both sides, gives each bird 5.0cm of feeding space. This allowance is about doubled for laying hens.

Pan feeders comprise rows of circular pans attached to a food supply pipe which is suspended from the ceiling and usually incorporates a centreless auger (see Figure 9.13). Height adjustment is by means of a manually or mechanically operated winch. Each line of pans requires a supply hopper, also suspended from the ceiling, and a motor at the far end to drive the auger. The last pan in the row has a sensor to detect the food level and permit automatic refilling. Pans are usually designed so that chicks can actually get through the grill into the pan base to feed in the early stages (see Figure 9.14 on page 218). The number of pans provided is based on an allowance of 2.0–2.5 cm of circular feeding space per bird for growing pullets and about double that for adult hens. Birds fit more compactly around a circle than along a line, which explains why the feeding space requirement with pans is about 50 per cent less than it is with troughing. Pan systems are now being

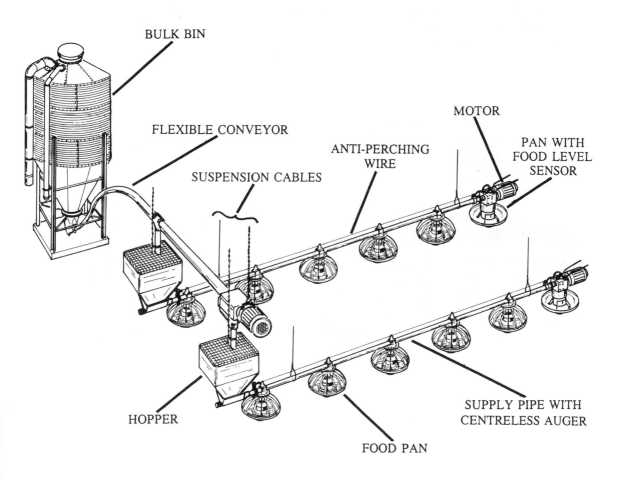

BULK BIN

FLEXIBLE CONVEYOR

MOTOR

ANTI-PERCHING WIRE

PAN WITH FOOD LEVEL SENSOR

SUSPENSION CABLES

HOPPER

FOOD PAN

SUPPLY PIPE WITH CENTRELESS AUGER

**Figure 9.13      Pan feeding system**
*(Courtesy of Roxell UK)*

more widely used even though they may cost up to 50 per cent more per bird to install than chain feeders. This is because pans have a number of operational advantages, particularly in terms of saving labour and preventing food wastage and soiling.

In small non-cage units, manually filled tube feeders are normally used. These are suspended individually from the ceiling and each incorporates a tubular supply hopper which the food flows down into the pan below.

**Figure 9.14**     **Pan feeder showing acessibility to baby chicks**
*(Courtesy of Roxell UK)*

In cage units birds are supplied with food from a trough which is accessible through the bars at the front of the cage. The minimum trough space necessary to comply with EU legislation for laying hens is 10cm per bird (MAFF 1994). Cage-reared pullets each require about 5–7 cm of food trough, but there is no regulation to this effect. The EU regulations for caged layers also stipulate that alternative ways of feeding must be available for use in the event of a mechanical breakdown. An increase in feeding space to 12cm per bird is necessary for laying hens of 2.2kg liveweight if all of them are to feed simultaneously. In practice, however, even if given the opportunity, all birds will rarely eat at once when provided with food *ad libitum*.

The traditional method of delivering food to caged birds, which is still used in the smaller units, is a manual push-along hopper filling the troughs by gravity (Figure 9.15a). This system is basically of poor design and wastes food. It is not able to maintain an even low level of food in the trough, at least with most meals. Therefore, mechanised systems with conveyors in the troughs are now commonly used which, as well as saving labour, provide a more positive uniformly low level of food distribution. Apart from flat chains in conventional troughs, other conveyors such as centreless augers and chain-link discs sunk into 'U'-shaped depressions in the bottom of troughs, are used to deliver food around each tier of cages (see Figure 9.15b,c and d). They run frequently, supplying food on a little and often basis and they redistribute uneaten food so that it is not left to go stale. Modern mechanised hopper/trough systems of much superior design to the traditional types are available and can also be operated on a little and often basis, but they are still liable to blockage problems with poorly processed batches of food.

**Prevention of food wastage**

Food constitutes by far the highest single cost factor in egg production, therefore the design and management of feeders must minimise wastage of food. Birds must be prevented as much as possible from 'playing' with or billing the food into heaps and spilling it over the edge of the feeder. They have a habit of pushing food sideways or dragging it towards them, perhaps an expression of natural foraging behaviour. On a daily basis the resultant wastage can seem innocuous, but it adds up over time to constitute a significant economic loss.

Prevention of food wastage is largely a matter of keeping the depth of food in the trough or pan at a low level. Ideally it should be no more than 1.5–2.0 cm, but this is impracticable in, for example, long runs of chain feeder where the aim should be nearer 3.0cm. Also, the feeders themselves need to be deep with adequate lips to contain the food. They may be fitted with special anti-waste devices such as grids, which is the case with pans but not with most forms

(a)

(b)

(c)

(d)

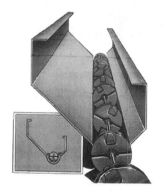

**Figure 9.15**      **Battery cage feeding systems**
    **(a) Hopper/trough** *(Courtesy of R J Patchett Ltd)*
    **(b) Flat chain** *(Courtesy of Valli International SRL)*
    **(c) Centreless anger** *(Courtesy of Roxell UK)*
    **(d) Chain-link discs** *(Courtesy of R J Patchett Ltd)*

of trough. With open, unguarded troughing it is best to apply the principle of feeding little and often, which is facilitated by mechanisation. Also, billing is restricted in troughs incorporating conveyors if food levels are kept below the top of the conveyor mechanism.

Positioning of the feeders is also important. Birds should not be able to scratch in the food and should be made to stretch to reach it – but not too far or they will develop neck sores by rubbing against the rim of the feeder. In batteries this requires careful design of the cage in relation to the food trough and in non-cage situations it entails having the feeders at the correct height, which in practice tends to be the height that positions the food at a level of about the birds' keel bone.

## Drinking

Water usage should be metered and recorded at the beginning of each working day. A change from the expected volume could be the first sign that something is wrong, for example, with the health of the birds or the formulation of their diet. A big departure from the daily norm should prompt an immediate search for the flood or blockage in the system! Modern recording and alarm systems can be installed to warn of a water supply failure, but checks also need to be made to ensure that water is actually available in every pen or cage. This involves making observations on the flow of water in all supply lines and keeping a look out for any places where food remains uneaten and/or eggs have not been laid as these might indicate a local failure in drinker operation. This takes the stockperson to every part of the unit and is conveniently combined with a check on the health and well-being of the stock.

As a general guide, birds housed at about 21˚C drink approximately twice as much by weight as they eat. The ratio widens, however, as the ambient temperature rises, due to a large increase in water consumption coupled with a decrease in food intake. The daily pattern in the consumption of water is closely related to that of food, showing peaks first thing after lights come on and particularly when 'cropping up' towards the end of the light period.

Most birds kept in cages in the UK are provided with nipple drinkers, that is, small valves screwed into plastic water piping which runs the length of

each tier of cages. Each pipe line is supplied from a small ball-valve-controlled header tank which delivers water at uniformly low pressure. These small tanks are in turn supplied from a main header tank which provides: 1) some water reserve; 2) water at a temperature raised towards that in the building, which helps to prevent condensation on the piping; and 3) a means of incorporating medicants and vaccines into the water. All header tanks should have securely fitting lids as it is essential that the water is clean for nipple drinkers to work successfully. Water with a high concentration of salts is unsuitable for use with nipple drinking systems as it tends to clog the valves.

Nipple drinkers are usually located at the rear of cages and should be positioned just above bird head height. At the rearing stage, one nipple is usually provided per 6–8 birds. During lay there is a legal requirement to provide at least two drinkers within reach of the birds in each cage (MAFF 1994). Nipples preferably have a trough or individual cups fitted underneath to catch the spillage, especially if sited over the accumulating droppings in a deep-pit unit.

The alternatives for caged birds are small valve-operated cup drinkers or continuous open troughs. The latter, in particular, have a useful role in hot climates where water consumption needs to be maximised. They are also needed where available water is unclean or contains a high level of salts. An open system is less hygienic and requires regular cleaning, which involves more labour than a self-cleaning nipple system. It is food in particular that contaminates the water, due to the habit of birds alternating feeding and drinking.

Birds kept in non-cage conditions are usually supplied with water from bowl or bell drinkers which are suspended individually from the ceiling and either distributed evenly over the available floor area or concentrated over a droppings pit. They are automatically operated on a weight basis. The drinking space requirement for growing pullets is approximately 0.75–1.0 cm per bird around the circumference of the bowl, (eg 6–8 drinkers of 38cm bowl diameter per 1000 birds). Laying hens require about 50 per cent more space. Ballast is usually placed within the bell to stabilise the drinker against knocks and reduce water spillage. Also to prevent spillage, the drinkers are suspended at the birds' back height and, after the first few days of brooding, the level of water is kept low at about 1.5cm. As with open troughs in cage units, the bowls need cleaning out regularly.

Nipple drinkers and small cups can be used in non-cage situations. In the tiered framework of percheries they can be fixed in position similar to cage installations. In open plan pens drinker lines are suspended from the ceiling, usually alternating with feeder lines, and adjusted for height (just above the birds' heads) by a winch, which saves labour. A pressure regulator is fitted at the beginning of each nipple supply line to provide appropriate low water pressure. Nipples tend to give less spillage of water than bell drinkers, which is of benefit in situations where litter is liable to become too damp. The conventional type of nipple, described earlier, is usually fitted with a comparatively large drip cup, but the more advanced multi-directional nipples are sufficiently drip free to dispense with the cups. This provides a very hygienic, completely closed system with no standing water.

Recommendations on the required number of drinking points for loose flocks vary greatly depending on the type and flow rate of drinker used. It is normal to provide nipple drinkers at the rate of one per 15–20 growing pullets or one per 10 adult layers.

## EGG 'HARVESTING'

It is of major economic importance that the eggs, once laid, are safely gathered in. Loss or damage of the final product is very costly. The task of preventing this is far more complicated and labour intensive with non-cage systems, especially free-range, than with batteries. This is one of the main reasons why, when comparing typical modern units, the labour requirement for the production of free-range eggs is some three or four times greater than that for battery eggs.

### The control of nesting

Firstly, it is important that the eggs are laid where they should be laid. Birds in conventional batteries have no option. They are obliged to lay their eggs on the sloping wire mesh floor within the confines of their cage. This does appear to cause some frustration to the birds, however, which probably explains why not all of them squat down when they lay, so that some shell damage occurs on impact with the floor. The eggs then roll down to collect in a 'cradle' at the front of the cages and as the birds are distributed uniformly throughout cage stacks, their eggs come together uniformly at the cage fronts. This absence of bunching into particular areas restricts damage from egg-to-egg impacts and means that eggs need only be collected once per day.

The situation is completely different, of course, with birds in loose flocks that are required to lay in nest boxes. To begin with the birds must be schooled to use the nests. There will always be at least 1–2 per cent 'floor eggs', but the proportion will be much higher without good schooling. This includes the following measures:

1. Rear pullets on litter with access to perches. Without prior experience of perching many birds, when the time comes, will not fly up onto the alighting rails or mesh platforms at the entrance to most types of nest

2. Move pullets to their laying quarters in good time, preferably by 16 weeks of age, so that they become familiar with the nesting accommodation before egg laying commences. Very difficult problems arise if pullets have started to lay before being housed. The nests should be kept closed during the pre-laying period, however, to prevent birds fouling them, particularly from using them as roosts

3. Keep free-range birds shut in the house until a week or two after they have come into lay. They are then let out, initially late in the day after most of the eggs have been laid. This helps them to recognise the house as the place where they lay their eggs

4. Give special encouragement to nesting in the early stages of lay. Birds are attracted by the presence of eggs, so for the first few days of lay they are often left uncollected. However, it is vital to pick up floor eggs regularly. With roll-away nests it is sometimes possible to prevent the eggs actually rolling away in the initial stages and, if the nest floors are mesh lined, adding a little straw or litter material will help to attract birds into them

5.  Block off temporarily corners of the floor that prove to be particularly popular initial nesting sites. Dark corners or other secluded areas may need permanently fencing off.

It is clearly important to make frequent visits to the house as the flock comes into lay. It may sometimes be necessary to deal with gregarious nesting behaviour, where many pullets crowd into a few favoured nest boxes, even to the extent sometimes of causing deaths from suffocation.

Although training the birds is very important, it does not, in itself, guarantee a satisfactorily high and sustained level of nest box use. The basic requirement for this is that the nests are at least as acceptable to the hens as egg laying sites, if not more so, than any other place in the house. A littered floor is the main counter-attraction, which is in keeping with the behaviour of birds in the wild to lay in shallow nests on the ground. However, the sites chosen by wild fowl are under cover and this probably explains why domestic hens show a preference for the seclusion of enclosed nest boxes, even if they are raised above the floor. Moreover, domestic hens do not demand the privacy of individual nests; they will lay as readily in large communal nest boxes designed for 50 or more birds (see Figure 9.16 a and b). Within the enclosure, big or small, hens do like to find some loose litter material which can be formed, by light scratching and/or pecking, into a shallow depression for laying, as in the wild. This is the case in conventional nest boxes, which are usually littered with woodshavings or some similar material.

Many producers, however, prefer to use the more modern roll-away nests which, because they have sloping floors, cannot retain litter and are therefore less attractive to the birds (see Figure 9.16c). The problem with conventional nest boxes is that eggs are left at risk from soiling, breakage and eating by a succession of birds coming in to lay. Also, the eggs are kept warm in this process which exacerbates the hygiene problem and is detrimental to albumen quality. The partial remedy is to collect the eggs 3–4 times a day and to adopt a weekly routine of replenishing nests with clean litter material. It is more satisfactory, however, if the eggs can roll down away from the birds, as in battery cages. As well as enabling less frequent egg collections, this also facilitates mechanisation of the collection process, which may be warranted in large units.

The choice of roll-away nests is essentially between those with floors of synthetic grass such as Astroturf, and those with wire or plastic mesh floors which allow droppings and other debris to fall through. Again, the preference of the birds conflicts with the requirement for clean hygienic egg production. Birds prefer the Astroturf; they seem to be attracted by the bristles and respond with nest building activities. However, materials like Astroturf become dirty and need periodic cleaning; they can harbour red mites and are gnawed by mice. The birds' preference among the types of mesh floor is probably a dished shape in comparatively soft plastic (see Figure 9.16d), but even with these refinements some birds may still opt to lay on a litter floor if given the opportunity.

The additional measures needed to prevent floor laying include the following:

1.  Provide a sufficient number of nests. Individual nests of about 30cmx30cm floor area are typically installed at the rate of one per five layers and a nesting space of about 100 cm$^2$ per bird is usually provided with large communal nest boxes

2.  Make all nests uniformly attractive so that birds do not prefer some nests and avoid others and thereby create an effective shortage. All nests must be of the same design, uniformly lit and ventilated. Prompt replacement of spent light bulbs or tubes is called for to maintain uniform light intensity at all times

3.  Make all nests readily accessible. Tiered banks of nests need to have sufficient headroom at the nest fronts between the runs of alighting rails or platforms (at least 30cm, preferably 40cm ) for the birds to fly onto them without difficulty. Where perches are installed between the banks of nests, the heights of the perches and adjoining nest rails should be offset so that the birds can fly upwards or downwards from one to the other instead of straight across

4.  Exclude birds from any secluded areas that might form attractive sites for egg laying, for example, from under the bottom tiers of nests.

(a)  (b)

(c)

(d)

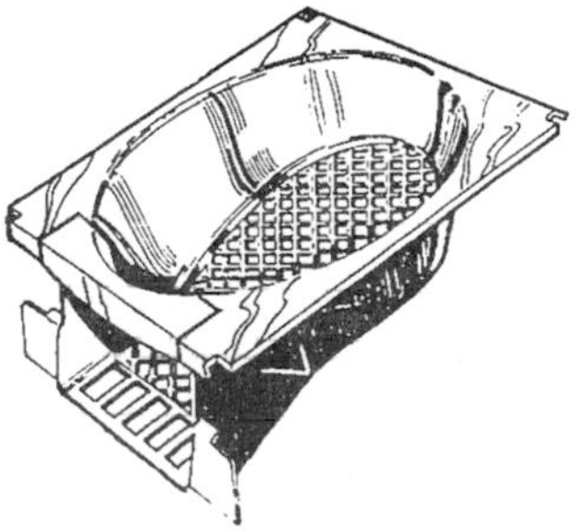

**Figure 9.16**    **Different types of next boxes**
   **(a)**    **Individual nests with wooden alighting rails**
       *(Courtesy of Willard Agricultural Ltd)*
   **(b)**    **Communal next box with wire mesh alighting platform**
       *(Courtesy of CES Potters Ltd)*
   **(c)**    **Individual roll-away nest with mesh floor - alternatively the floor could be
        lined with synthetic grass**
       *(Courtesy of CES Potters Ltd)*
   **(d)**    **Dish shaped plastic roll-away nest**
       *(Courtesy of Spiro GB Ltd)*

## Clean egg production

Dirty or cleaned eggs do not qualify as Class 'A' under the EU marketing standards. As a consequence their monetary value is reduced by some 60–75 per cent on average. Producing clean eggs includes preventing floor laying, as most floor eggs are dirty, and keeping the nests clean. On balance, the best combination of features for this is probably roll-away nests with mesh floors and a litterless house floor of raised slats or wire mesh. In the absence of litter the birds' feet are cleaner and less dirt is carried into the nests. However, this prevents the birds from scratching and dust bathing. If litter is provided, then it should be kept dry and friable which, as noted earlier, is easier said than done. The problem is alleviated to some extent in most deep-litter units by locating the nests over a droppings pit on the side remote from the littered area, so that the birds clean their feet as they walk across the raised platform to the nests. The alighting rails or mesh platforms in front of the nests also help in this foot cleaning process.

In free-range units the welfare case for littered floors is less strong since birds can scratch and dust bathe outside. The use of the range for this purpose can be augmented by having a verandah down each side of the house with a floor of sandy material. However, access to the outside does not help in the fight against dirty eggs. The dust bathing sites all around the house become quagmires in wet weather and measures need to be taken to minimise the amount of mud that reaches the nests. Positioning the nests well back into the house from the pop-holes increases the chances of the hens cleaning their feet *en route,* especially if this is over an uninterrupted platform of slats or wire mesh. The distance involved should be no more than about 4.5m, or floor eggs may become a problem. Outside it is of considerable benefit to have the pop-holes opening, via slatted ramps, onto a verandah with duck-boards leading out from the verandah exits over the worst of the muddy ground. Also, it clearly helps to prevent soiling by mud if most eggs are laid before the birds are let out at about 8.00am. This is achieved by providing supplementary light in the house from the very early hours of the morning.

In all situations, it helps to keep nests clean if they are closed off at night, something which also restricts the development of broodiness. The usual method of nest closure is to raise the hinged alighting rails or platforms, although automatic bird expulsion and nest closing systems are available.

The task of keeping eggs clean at the site of laying is much simpler in battery cages. Here the main potential soiling agent is dust which collects on the egg cradle at the front of the cages, especially if it holds a belt for mechanical egg collection. This dust will stick to newly laid moist eggs. Egg cradles need dusting down at regular intervals while egg belts are dusted automatically during the collection process by passing between rotating stiff-bristle brushes after delivering the eggs at the ends of the cage stacks. Faecal contamination of battery eggs is normally very low, but it can become a problem if the birds' droppings become particularly wet or sticky as a result of disease or faulty nutrition.

## Egg collection

The next task is to collect the eggs from where they have been laid with minimal further soiling or damage. Collection is either manual or mechanical, depending primarily on the size of the operation. As a very rough guide, it requires a minimum stocking capacity of about 10,000 layers for any individual house and some 40–50,000 layers on the site as a whole in order to justify the capital expenditure on **fully** mechanised egg collection. The system should incorporate link conveyors between houses that deliver eggs to a central point for automatic packing. These requirements are typically met on modern battery units, but not on free-range units, where eggs are usually collected by hand or semi-automatically. Free-range houses are of necessity comparatively small and widely spaced in order to provide at least one hectare of land per 1000 birds and allow for the fact that there is a limit to how far birds can be expected to range. Whichever method of collection is adopted, the eggs should be treated with care throughout the process, bearing in mind the fragile nature of their shells. The majority of downgrading is due to shell breakage. Hygiene is also important and staff should wash their hands before and after collecting eggs.

Manual egg collection is a time consuming job. An experienced collector, picking up two or three eggs in each hand, can work at the rate of some

1,500–2,000 eggs per hour. However, much depends on the dexterity of the collector and the proportion of poor quality eggs to be sorted out. Additional time has to be spent in non-cage units going round all the pens picking up the floor eggs, although this does provide another opportunity for inspection of the stock. In cage units this opportunity arises within the main collection process.

Once collected, eggs should be held in a well insulated store room, preferably refrigerated to maintain a temperature of 8-12°C. Apart from restricting bacterial growth, low temperatures reduce evaporative weight loss from the eggs and slow down the rate at which albumen goes 'watery'. The construction of stores should permit ease of cleaning, the exclusion of vermin and the smooth flow of egg trolleys in and out.

Mechanised collection speeds the movement of eggs from hen house to cool store. However, it does not involve merely switching on the conveyors and waiting at the end of the line for the packed eggs. Preparations for collection involve walking through the poultry houses checking that all the collection belts are in good working order and that there are no obstructions to the smooth flow of eggs. Soft-shelled and badly broken eggs are best removed from the belts to restrict soiling from broken egg contents. Also, it is during these preparations in battery units that a further opportunity is provided for the inspection of all stock. Once running, the operation of the system needs inspecting frequently.

In the most common system of mechanical collection, eggs are conveyed on belts from in front of the stacks of battery cages, or from the lines of roll-away nests in large non-cage units, to one end of the house. The belts used in non-cage situations need to be wider than those in batteries (about 20cm compared to 10cm wide) to accommodate the bunching of eggs that results from hens using some nests more than others. Belt speeds are adjustable and are varied according to the number of eggs and quality of their shells. With some comparatively small-scale operations, particularly in non-cage units, the collecting belts merely deliver eggs to tables at the ends of their runs for manual packing; that is semi-automatic collection. However, most eggs are transferred in some way from the belts to a cross conveyor for delivery to a separate building for packing automatically. The cross conveyor often links several houses to the packing room. It is usually rod-type and designed so that the eggs lodge securely between the rods. In this way egg-to-egg collisions are reduced and the conveyor can go up and down gradients (up to 20°). This, in conjunction with an ability to go round bends, means that conveyors can take complex routes between houses to the packing centre. The rod design also restricts egg soiling by presenting a minimal area of contact with each egg and allowing any debris or broken egg contents to fall through the gaps. Obvious cracked and dirty eggs are removed from the main flow at some convenient stage along the cross conveyor, usually at the far end just prior to packing.

### Egg grading

Eggs are graded on farm or at specialist packing stations for quality and size. The quality classes are outlined below.

*Class A eggs only may be sold fresh for human consumption. These eggs have a normal, clean, intact shell; an air cell not exceeding 6mm in depth; a clear, translucent, gelatinous egg white, free of foreign substances; and a stationary yolk which is visible under candling as a shadow only and which is free from foreign substances. There should be no perceptible development of an embryo and the egg should be free of all foreign odours. Washing of Class A eggs is not permitted.*

*Class B eggs have a similar specification to Class A, except the air cell should not exceed 9mm in depth; the shell does not have to be clean; and washing is permitted. Class B eggs are not sold as shell eggs for human consumption in the UK but are broken out, pasteurised and used to make egg products.*

*Class C eggs are those which may be visibly cracked and which do not satisfy the requirements of Classes A and B but are suitable for the manufacture of processed foodstuffs for human consumption to comply with Egg Products Regulations (SI 19193 No. 1520).*

*All eggs which do not fall into Classes A, B and C are classed as "Industrial Eggs" and may not be used for human consumption, either in the manufacture of foodstuffs or otherwise. Uses for "Industrial Eggs" may be in non-food products such as shampoo.*

The official weight grades for Class A eggs are as follows:

|  |  | *Weight per egg* |
|---|---|---|
| *Very large* | *XL* | *73g and over* |
| *Large* | *L* | *63-73 g* |
| *Medium* | *M* | *53-63 g* |
| *Small* | *S* | *under 53g* |

## FREE-RANGE MANAGEMENT

The land for a poultry range should be free-draining, preferably light and sandy. The ideal herbage is a good permanent pasture or a well established ley. The latter requires persistent varieties of grass, such as perennial ryegrass and red fescue. The sward should be kept short as crop impaction may result from birds eating long grass. Also, short grass provides less cover for predators, gives less encouragement to hens to lay outside and allows more sunlight to reach ground level to destroy parasitic worm eggs. Complementary grazing, for example with sheep, is one possible alternative to regular topping for keeping grass short during the spring and summer. Bare patches that form during winter can be broadcast with grass seed in the spring, while larger heavily used areas can be rotovated and re-seeded at convenient times in the break between flocks. Well established hedges and trees will give the birds some shade and shelter from winds, although they will provide predators with cover.

The area of range required for a given number of birds depends largely on the type of land and its management. The requirement of one hectare per 1000 birds needed to describe eggs as free-range within the EU (*Commission Regulation EEC No 1943/85* – see MAFF 1987) is minimal for maintaining the land in a satisfactory condition on a permanent basis. The main task is to prevent the land becoming heavily infested with parasitic worms, or 'fowl sick', in addition to preventing excessive wear and the development of a dust bowl or mud bath in the area immediately around the house. The traditional way of doing this is to accommodate the birds in small colonies of 50–100 in mobile houses which are moved over the range at about weekly intervals and on to a new range after a year or two.

This, however, is highly inconvenient for the provision of services and extremely labour intensive. A compromise is some form of comparatively inexpensive semi-permanent housing which can be dismantled and moved to a fresh range every one to three years. Polythene tunnel and straw clad houses with bird capacities in the thousands are sometimes used in this way. The use of semi-permanent housing has not become widespread, however.

Most free-range egg producers use permanently static houses to save labour and to facilitate servicing and the incorporation of modern automated equipment. Such units need some special provision, such as slatted platforms, hardcore, concrete or a covered verandah, to avoid excessively muddy conditions immediately around the pop-holes. Also, it is advisable to divide the range into paddocks for grazing in rotation to control the build-up of parasitic worms. Ideally, at least four paddocks should be established and the birds moved round to a fresh paddock every 5–6 weeks. It is far better for maintaining the land in a satisfactory state over the long-term if, for example, 1000 birds have rotational access to 4x0.25 hectare paddocks rather than continual access to a full hectare. It is also of benefit if the range can be left unstocked for at least six weeks in the break between flocks.

From the range management standpoint it is best to locate the house in the middle of the area, with paddocks all round it. If the house is positioned on one side for ease of servicing, then the paddocks can be arranged in a fan configuration spreading out from the house. However, this increases the maximum distance from the house to the perimeter fence – an important consideration when most birds are reluctant to travel more than about 100m over the range. A problem with some houses in the middle of paddock areas is that the roads to them have not been made to take heavy lorries. This may necessitate secondary transport across the range and double handling of point-of-lay pullets, old hens, eggs and food.

### Fencing
A perimeter fence is essential to contain the birds and to exclude foxes and other predators. The fence may also need to keep sheep, cattle and other livestock in place, either inside or outside the range. Wire netting of about 2m in height is often used, together with

some form of electric fencing. A single strand of electrified wire, about 30cm out from the netting and 30cm above the ground, is quite effective against foxes, whereas to withstand the attention of other farm livestock the wire netting needs supplementing with electrified plastic netting of at least 1m in height. This plastic netting is also suitable for dividing the range into paddocks within the perimeter fence. An alternative robust perimeter fence is comprised of seven or more strands of electrified wire, the bottom four spaced about 10cm apart with the top wires spaced more widely to bring the fence to a total height of at least 1m. With all forms of electric fencing there is the need to prevent shorting by keeping in check the growth of vegetation under and around the fences. Gateways through the perimeter fence should be effective barriers when closed and provide wide access when open to facilitate servicing.

### Ranging behaviour

One way of encouraging hens to range widely is to give a daily 'scratch' feed of whole grain over the more distant parts of the paddocks. This, however, attracts wild birds and animals which are liable to spread disease. Providing a 'scratch' feed of no more than about 5g per bird per day for a few weeks soon after the birds are first let out onto the range is perhaps an appropriate compromise.

Some birds are reluctant to leave the house at all, let alone range widely. They can survive inside where there is everything they need. It may be that they do not like to venture out into comparatively bright light, so some form of shade over the pop-holes may help. Another problem may be that there are insufficient pop-holes and/or that they are too small and patrolled by dominant hens. It is uncertain what constitutes an adequate number and size of pop-holes. The Farm Animal Welfare Council (1991) suggest that individual pop-holes should be no smaller than 30cm high and 75cm wide, but make no specific recommendation on their rate of provision. Other organisations recommend much larger sizes, while in past commercial practice a typical provision has been one pop-hole measuring 38cm high by 110cm wide per 300–400 birds. It is perhaps not surprising, however, if birds sometimes choose to stay inside, especially when the weather is wet and windy or hot

and sunny. Also, there are times when it is unwise to provide many large openings at floor level down each side of the house. For example, when an icy wind is blowing and creating a cold draught across the house, it cools the birds and it turns the surface of any litter into a damp 'cake' of hen excreta.

Some birds are reluctant to return to the house to be shut in at night, especially at the end of long summer evenings. If shut out, they are likely to become the victims of predators. Leaving the lights on in the house after dusk tends to entice the birds to go inside.

## DISEASE PREVENTION

Any disease that actually affects the birds is liable to wipe out the profit margin. Treatment is too late, therefore, so egg producers and their veterinary consultants are primarily concerned with the prevention of disease. This involves a wide-ranging programme of measures. The main ones are outlined below:

1. *Thoroughly clean and disinfect housing and equipment between batches of birds to break the build-up in disease organisms.*
   Operations include bagging off uneaten food and cleaning the bulk food bins; as well as draining the water tanks and drinker lines, flushing the system with a sanitant and then rinsing with clean water

2. *Close the unit as much as possible to vectors of disease.*
   Measures under this heading include using new materials for packing eggs and transporting them in trolleys dedicated to the poultry unit. On poultry farms, it is particularly important to keep wild birds out of the houses. Where poultry have access to range it is, of course, impossible to isolate them from wild birds, as well as many other animals and insects, which can carry infectious diseases and parasites

3. *Control rodents, insects and mites.*
   Mice are the main rodent problem, especially in deep-pit units. They thrive in the seclusion of the pits where there is warmth, a ready supply of food and ample nesting sites. Apart from spreading disease, mice eat the birds' food and cause costly damage by gnawing things, such as

insulation, electrical wiring, egg belts and nest linings. Producers often take out a contract with a company for rodent control. This should not lead to an abdication of responsibility, however, particularly in terms of keeping the place tidy and making the buildings as mouse-proof as possible by sealing every conceivable entry point, especially around the doors into pits. The Ministry's code of practice for the prevention of rodent infestation in poultry houses provides helpful advice (MAFF 1996).

Insect pests are not likely to be a problem in belt-clean battery units where the manure is removed frequently, but serious infestations can develop in deep-pit and deep-litter houses, especially those accommodating layers for the full production cycle. Flies can present a major problem; their populations can 'explode' to annoy workers and neighbours, to spread disease and mark surfaces, including egg shells. When this happens it becomes necessary to treat with a knock-down aerosol spray for a few days. To avoid this producers normally control the build-up of fly populations through the routine use of baits. Also, it is important to keep the manure as dry as possible and encourage its habitation by predatory *Carcinops* beetles. These are small, shiny black beetles, similar in shape to ladybirds, which feed on fly larvae. Further information on insects in poultry houses is provided in an advisory leaflet by MAFF (1980).

Red mites (*Dermanyssus gallinae*) tend to be the most troublesome of the ectoparasites of poultry. They spread disease organisms such as *Salmonella* as well as causing reduced performance and, in the severest infestations, even death. They are almost bound to infect layers on range and those kept intensively in non-cage units or in cages of wooden construction at some stage. They can be transmitted by wild birds, especially starlings and sparrows, and also by mammals. Although red mites live exclusively on the blood of birds, their control is further hampered by an ability to survive without food for over four months. Initial infestations often go unseen until large populations have become established. The parasite remains hidden in crevices when it is light and only comes out in

darkness to feed, when it goes the characteristic red colour. Early signs of trouble are red spots on eggs or birds pecking themselves around reddened vents. Treatment involves spraying the building and equipment with an insecticide and for effective control of the mite population this needs doing every 3–4 months with the birds *in situ*. However, the risk of harmful residues in eggs places restrictions on the type of insecticide that can be used. A compromise is to install special red mite 'traps' and only treat when mites are seen to be present

4. *Dispose of dead birds promptly and hygienically.*
It is normal to deal with dead birds at least once each day as part of the stock inspection routine. Plastic gloves should be worn when handling dead birds and the carcases should be placed in strong plastic bags. While doing this job any birds that are seen to be seriously ill or injured should be culled by dislocation of the neck. If the daily numbers of deaths and culls rise unexpectedly then samples of carcases should be retained for post-mortem examination. The most sanitary method of dead bird disposal is incineration, but efficient, smell-free incinerators are expensive to buy and run. The traditional on-farm alternative is a disposal pit, which is comparatively cheap and easy to operate, but gives problems unless located correctly to avoid water pollution and constructed properly to provide an air-tight container with a tightly sealed lid. Other possibilities include composting on site or collection of the carcases from the farm for maggot production. The containers used to hold carcases awaiting collection should be proof against leaks, pests, wild birds, and pets

5. *Medicate birds when necessary.*
Preventative medication does not feature widely as a disease control measure, but pullets reared on litter, or in any system where they have access to their droppings, are usually fed diets containing a coccidiostat. This is to protect them from coccidia (*Eimeria spp.*) which are microscopic protozoan parasites that infect the intestine, and spread as oocysts in the droppings. These oocysts are found almost universally wherever chickens are reared. The prime action of the coccidiostats used in pullet diets is to

suppress the growth of coccidia, not kill them, so that the birds are able to develop an immunity. The lack of immunity to coccidiosis in cage-reared pullets means that they are unsuitable for use in non-cage egg production systems. Laying birds should not be treated prophylactically with anticoccidial drugs because of the dangers from residues in the eggs. Short-term therapeutic treatment should be accompanied by withdrawal of the eggs from the human food market. As an alternative, vaccines against coccidiosis are becoming available.

Free-range birds have access to parasitic worms which may be spread initially by wild birds. Large infestations can build up resulting in 'fowl sick' land. The main problem is usually with roundworms (*Ascaridia*) which, once firmly established, can only be effectively controlled by routine treatment with an anthelmintic in the birds' food or drinking water. However, because of the risks of residues, it is necessary to withdraw eggs from sale for human consumption during and after the treatment for a period of about two weeks. Every effort should be made, therefore, to prevent the development of 'fowl sick' land and the need for this medication (see page 30-31).

6. *Fully vaccinate birds against all the major infectious diseases.*

Vaccinations feature prominently in disease prevention programmes. Those for replacement pullets are designed to develop immunity to last throughout the laying period as well as protect the birds as they grow (see Table 9.4). The specific details will vary depending on the local disease situation and should be worked out in consultation with a veterinary surgeon. Vaccinations are not normally done in the laying period, thus avoiding adverse reactions in terms of egg production and quality.

All the diseases in the programme shown in Table 9.4 are caused by viruses and, unless stated otherwise, the vaccines used against them are of the live type. Marek's disease is composed of a range of deadly avian cancerous conditions. Gumboro disease (infectious bursal disease) affects the birds' immune system and can result in high rates of mortality, typically at about 4-6 weeks of age. The infection also have a lasting immuno-suppressive effect. Newcastle disease (ND), infectious bronchitis (IB), infectious laryngotracheitis (ILT) and avian rhinotracheitis (ART) are all respiratory diseases which can kill in the acute form, but they also have chronic debiliting effects and predispose to secondary bacterial infection. In addition to the respiratory tract, the viruses also infect the hen's oviduct and impairs the formation of albumen and shell. This results in eggs with 'watery' whites and/or thin, deformed, depigmented shells. IB is the most problematical of these four in practice, especially the variant forms which require special attention within the vaccination programme. ND (fowl pest) is largely a scourge of the past, and is now a rare occurrence in the UK. It is a notifiable disease subject to the *Diseases of Poultry Order 1994*. Epidemic tremor (infectious

**Table 9.4      An example of a vaccination programmes for replacement pullets.**

| Age of birds | Vaccination |
|---|---|
| Day-old | Marek's disease - individual injection at the hatchery |
| 2 weeks | Gumboro disease - in the drinking water |
| 3½ weeks | Gumboro disease - in the drinking water |
| 5 weeks | Newcastle disease and infectious bronchitis combined - in the drinking water |
| 7 weeks | Infectious laryngotracheitis - by spray |
| 9 weeks | Newcastle disease and infectious bronchitis combined - in the drinking water |
| 11 weeks | Avian rhinotracheitis - by spray |
| 13 weeks | Epidemic tremor - in the drinking water |
| 16-18 weeks* | Newcastle disease, infectious bronchitis and egg drop syndrome combined - individual injection of oil-based dead vaccine |

\*   At the time of transfer from rearing to laying quarters.

avian encephalomyelitis) is a nervous complaint that is fatal to young chicks. Its effect on adult hens is to cause a temporary (2-3 weeks) drop in egg production of 5-10 per cent. The 'egg drop syndrome' (EDS) virus, as the name suggests, also causes reduced egg production. In addition it has an adverse effect on egg shell quality, causing a high proportion of soft-shelled eggs.

Increasingly, farmers are adding to these already complex programmes by vaccinating pullets against Salmonella (specifically *Salmonella enteridis)* to guard against the occurrence of this food poisoning bacterium in eggs. However a standard vaccination procedure has yet to be developed. Currently, prior to the licensing of live vaccines for use in the UK, dead vaccine is being administered. This involves at least two individual injections which is an expensive, time consuming process. Moreover, the birds are stressed by the additional handling and any adverse reactions to the vaccine.

Further details on the diseases of poultry can be found in a number of text books, for example, Jordan and Pattison (1996).

Manufacturers provide instructions on vaccine storage and administration. The technique of mass administration of quick acting live vaccines as a spray or via the drinking water simplifies the task of treating large flocks. The ILT vaccine is usually given as a spray to large flocks, but the eye drop method is the safer effective choice if time and labour are available. Ideally, there should be at least two weeks between vaccinations to avoid the immune response from one interfering with that of another, but this is not always possible. Another complication on timing is with the first vaccination in cases where initial immunity is acquired through the egg protection. For example, baby chicks have maternal antibody protection against Gumboro disease; this begins to wane at about 2 weeks of age. A vaccination at this stage may not be fully effective, but leaving it a week or two longer may be too late. Hence the precautionary double treatment in the example given. With regard to ND and IB, immunity is built up in stages during rearing and then it is necessary to inject each bird at point-of-lay, usually in the leg, with a combined oil-based dead vaccine to provide a strong immunity against both diseases that will then last for the entire laying period.

The results of the vaccinations should be monitored regularly and the programme amended if necessary. This involves taking blood samples, usually from a prominent vein on the underside of the wing, and sending them to a veterinary laboratory for tests on antibody levels. It is particularly important for producers to routinely check the immunity status against ND and IB, especially at point-of-lay and at intervals through the laying period.

**Beak trimming**

Beak trimming, being a mutilation, is a controversial issue. However, it is sometimes the only effective way of preventing deaths from cannibalism and can, therefore, make a positive contribution to bird welfare. No producer wants the trouble and expense of beak trimming unnecessarily, but if cannibalism does arise, which can be unpredictable, the adverse consequences both economically and in terms of welfare are severe.

In the UK, beak trimming usually involves removal of no more than one-third of the beak at its tip. This blunts the beak and greatly reduces its effectiveness as a gripping device. Beak-trimmed birds are therefore far less able to puncture skin, pull feathers, tear flesh and break egg shells. If carried out routinely, the best time for beak trimming appears to be when the birds are 5–10 days old. Done skilfully at this early age it causes little apparent distress, the birds actively feeding and drinking within seconds of the operation. The normal technique is to gently push the beak into an appropriately sized hole in a metal plate and bring a red hot blade down behind the plate to cauterise the tip. Any beak trimming should be done by or under the supervision of a skilled operator and must be carried out as prescribed in the *Veterinary Surgery (Exemptions) Order 1962.*

The main alternative time for routine beak trimming is at the hatchery prior to delivery of the chicks, which may be more convenient. However, it may be unwise, bearing in mind all the other stresses on the day-old chick, to trim its beak before it has had the opportunity to find and take food and water. On the other hand, if the operation is delayed until the beaks are well developed, it is much more difficult to perform, is clearly more stressful to the birds and may result in chronic pain. This calls into question the wisdom of recommending in the bird welfare code (MAFF 1987) that beak trimming should be carried out only as a last resort, which implies waiting till the birds are actually cannibalising one another. Moreover, once they start it is very difficult to stop them. Producers need to assess the risk of cannibalism developing and if this is moderate to high, then the safest thing to do is to beak trim the birds at about one week of age. This means that producers should be aware of the factors affecting the incidence of cannibalism and the extent to which they can be controlled in their circumstances.

The following is a summary of the many inter-related factors that predispose birds to cannibalism:

1. Aggressiveness – although there is not much to choose between the main brown hybrid laying stocks. White birds tend to be more prone to pecking than brown, while both are markedly aggressive in comparison to docile meat-type birds
2. High light intensity – this is a major factor, especially relatively bright shafts of sunlight
3. High ambient temperature and /or low relative humidity (below about 50%) - the latter may cause skin irritation
4. Crowding – this relates to (3) and causes pressure on living space, including that for feeding, drinking, perching and nesting as well as floor space
5. Large groups/colony size - has a bigger effect in practice than high stocking density
6. Unstable social groups - this relates to (5) and means also that strange birds should not be introduced into established groups
7. Onset of lay - the bulging vent region associated with initial egg laying is a target for pecking. A bird with prolapse is particularly at risk
8. Malnutrition - this may cause a craving for deficient nutrients, for example, sodium chloride, crude fibre or essential amino-acids
9. Poor feathering - rough, loose feathering related to (3) and (8) and inadequate feather cover, as when changing from down to feathers, or moulting
10. Skin damage - injury from badly designed or constructed accommodation or small wounds caused by parasites
11. Boredom - particularly when feathers are the only alternative thing for effective pecking apart from food, as in cages, and/or when the time spent feeding is short, as with pellets compared to meal
12. Lack of natural controls - in the wild, fighting among chicks is suppressed by the broody hen and among a group of hens, by the dominant male.

In practice, high light intensities and large, unstable social groups are the main causative factors. When they are in combination the development of cannibalism is practically guaranteed at some stage. This explains why most free-range birds are beak trimmed. Caged layers, on the other hand, because they are kept in very small stable groups, do not require beak trimming provided that light intensities can be maintained at uniformly moderate levels of between 5 and 10 lux at the cage front. It is difficult to maintain minimal light intensities of 5–10 lux in the colonies of non-caged intensively housed layers without experiencing pecking problems, particularly in the early stages of lay. Keeping light intensities fairly low until egg production is well established can help. There should be no problems of cannibalism with growing pullets as their pecking activity can be controlled by subdued lighting.

**THE WAY AHEAD**

The main issue concerning the future of egg production systems is the place of battery cages. Many people do not like what they see when looking at hens in battery cages and some regard them as ethically unacceptable. However, the banning of battery cages would have serious economic consequences and, moreover, would be of doubtful advantage to bird welfare. All the alternative systems have their own problems.

It is most unlikely that a more economic system of producing high quality eggs than batteries will be developed in the foreseeable future. This means that eggs produced under alternative systems will continue to be more expensive. Currently in the UK, despite concerns over bird welfare, the cheaper battery eggs only command about 80 per cent of the market. If cages were banned in the UK or throughout Europe, this market would be supplied from caged flocks kept elsewhere in the world. The welfare objective of the ban would have been defeated and the bulk of the UK egg industry destroyed.

One reason battery eggs are cheaper to produce is that egg production per hen housed is comparatively high. This is the combined effect of less mortality and greater egg production per survivor. The former is clearly of advantage in terms of welfare. It can be argued that this is also true of the latter, as lower egg production per bird in non-cage systems results primarily from morbidity associated with the same factors that cause the higher mortality, ie social order stress, parasitic infestations, infectious disease challenge and physical injury. Thus, as stress levels rise egg production per bird falls. High egg production per individual, as opposed to the flock, is probably one of the most sensitive indications we have that a hen has overcome the stresses to which it is subjected – that it is not distressed. Egg production is a reproductive process and hens will strive to reproduce whatever the conditions, but they will be less successful in this endeavour, not more so, when subjected to undue stress. As batteries give the highest egg production per bird of any practical system yet devised, it seems inappropriate to producers that they should be banned on welfare grounds.

The biological mechanism through which stress influences egg production is not fully understood. It should perhaps receive more attention from research workers in the future. Acute stress, such as a frightening experience or a sudden marked change in ambient temperature, will result over the next day or two in some hens not laying. They appear to 'switch off' the egg production process. The most plausible explanation is that the stress-induced release of adrenalin causes inhibition of ovulation or immobility of the oviduct. The latter would leave the ovulated yolk 'stranded' in the body cavity from where it would be reabsorbed (the phenomenon of internal laying) instead of being engulfed by the infundibulum and conveyed through the oviduct (see Figure 9.10). With chronic stress, the long-term release of adrenalin and attendant shunting of the blood into the main muscles away from the viscera, including the reproductive tract, may have an adverse effect on the overall efficiency of egg formation. The reproductive tract of the hen normally receives over 20 per cent of the cardiac output and any interference with this blood flow will not be conducive to maximum egg production.

Research into alternative systems to cages for housing laying hens (Savory and Hughes 1993) will undoubtedly continue. The conclusion to be reached in the light of present knowledge still seems to be the same as that reached by members of the Brambell Committee back in 1965. They concluded, somewhat reluctantly, that 'a modified battery system may be as good as or better than loose housing' where in their judgement the inherent risks to the bird 'are as great as the deprivations of the battery'. In the time since the report of the Brambell Committee, several modifications have been made to conventional cages to improve bird welfare and there has been research into more radical changes to enrich the cage environment.

One available modification to the conventional cage is to fit a narrow abrasive strip to the baffle plate below the food trough against which the birds work their feet when feeding. This shortens and blunts the birds' claws and gives less risk of trapped or broken claws and infected toes. Under natural conditions, ground scratching shortens the claws, but they remain sharp. Research into further improvements will no doubt continue.

Modifications to cages which permit more freedom of behavioural expression are still in the experimental stage. However, a perch can be installed as an optional extra in some commercial cages (about 15–18 cm from the back of the cage at a height of 5.5–7.0 cm above the floor – see Figure 9.5). This provides relief from the sloping floor and reduces foot damage. It might also increase leg bone strength. Unfortunately, some birds lay when on the perch, which increases the proportion of cracked and dirty eggs. A form of perch that can only be used by the birds for roosting at night is perhaps the answer. Work is being done on more complex modifications, including facilities for nesting and dust bathing (Sherwin 1994). These enriched cages, especially if designed to retain the major welfare advantage of socially stable groups of only four or five birds, will cost substantially more to install than conventional cages. Also, more eggs are likely to be downgraded due to breakage and soiling. Unless these extra costs are offset by greater egg production, a premium price for the eggs will be required to make the modified cages economically viable. In order to obtain a premium the welfare improvements must register with consumers. A problem with this is that, in their view, a modified cage will always remain a cage.

## REFERENCES AND FURTHER READING

**Appleby M C, Hughes B O and Elson H A** 1992 *Poultry Production Systems: Behaviour, Management and Welfare.* CAB International: Wallingford, Oxford, UK

**Austic R E and Nesheim M C** 1990 *Poultry Production.* 13th edition. Lea & Febiger: Philadelphia, USA

**Brambell F W R** Chairman 1965 *Report of the Technical Committee to Enquire into the Welfare of Animals kept under Intensive Livestock Husbandry Systems.* Cmnd 2836. HMSO: London, UK

**Coles R** 1960 *Development of the Poultry Industry in England and Wales 1945-1959.* Poultry World: London, UK

**Farm Animal Welfare Council** 1991 *Report on the Welfare of Laying Hens in Colony Systems.* FAWC: Tolworth, UK

**Fisher C and Boorman K N** (eds) 1986 *Nutrient Requirements of Poultry and Nutritional Research.* Butterworths: London, UK

**Freeman B M** 1984 Transportation of poultry. *World's Poultry Science Journal 40:* 19-30

**Gregory N G, Wilkins L J, Elepheruma S D, Ballantyne A J and Overfield N D** 1990 Broken bones in domestic fowls : effect of husbandry systems and stunning method in end-of-lay hens. *British Poultry Science 31:* 59-69

**Heil G and Hartmann W** 1994 Combined summaries of European random sample egg production tests completed in 1991 and 1992. *World's Poultry Science Journal 50:* 187-189

**Jordan F T W and Pattison M** (eds) 1996 *Poultry Diseases.* 4th edition. Baillière Tindall: London, UK

**Leeson S and Summers J D** 1991 *Commercial Poultry Nutrition.* University Books: Guelph, Ontario, Canada

**Lewis P D, Perry G C, Morris T R and Midgley M M** 1992. Intermittent lighting regimes and mortality rates in laying hens. *World's Poultry Science Journal 48:* 113-120

**Ministry of Agriculture Fisheries and Food** 1980 *Insects in Poultry Houses.* Leaflet 537. HMSO: London, UK

**Ministry of Agriculture Fisheries and Food** 1987 *Codes of Recommendations for the Welfare of Livestock: Domestic Fowls (amended 1988, reprinted 1992).* Leaflet 703. HMSO: London, UK

**Ministry of Agriculture Fisheries and Food** 1994 *The Welfare of Livestock Regulations 1994* (Statutory Instruments 1994 No. 2126, Animals – Prevention of Cruelty). HMSO: London, UK

**Ministry of Agriculture Fisheries and Food** 1996 *Code of Practice for The Prevention of Rodent Infestations in Poultry Flocks.* HMSO: London, UK

**Ministry of Agriculture Fisheries and Food** 1998 *Codes of Good Agricultural Practice for the Protection of Water.* PB 0587. HMSO: London, UK

**North M O and Bell D D** 1990 *Commercial Chicken Production Manual.* 4th edition. Van Nostrand Reinhold: New York, USA

**Rose S P** 1997 *Principles of Poultry Science.* CAB International: Wallingford, Oxford, UK

**Roush W B** 1986 A decision analysis approach to the determination of population density in laying cages. *World's Poultry Science Journal 42:* 26–31

**Savory C J and Hughes B O** (eds) 1993 *Proceedings of the Fourth European Symposium on Poultry Welfare.* Universities Federation for Animal Welfare: Potters Bar, UK

**Sherwin C M** Ed 1994 *Modified Cages for Laying Hens.* Universities Federation for Animal Welfare: Potters Bar, UK

# 10 Broiler chickens
## D W B Sainsbury

## THE UK INDUSTRY

### Trends within the industry

Broiler production began in the USA and only came to the UK in the early 1950s. It has grown rapidly and spread throughout Europe. In the last three decades, the methods of managing, housing and feeding broilers have not altered a great deal. The number of broilers reared annually by UK producers continues to rise steadily and now stands at approximately 600–700 million; this equates to about 10 birds/capita/year.

About 10 years ago some farmers attempted to rear broilers in tiered cages. They rejected this method because the capital cost was high and there was damage to the birds, especially of the breast muscles, leading to down-grading. In some Eastern European countries and the former Soviet Union caged broiler production is common.

Very few farmers grow broiler chickens under free-range conditions. Such production has increased during the last few years in some countries. At the beginning of the 1990s, 15 per cent of broiler production in France was free-range. In the UK only 1–2 per cent of throughput was free-range, and most of this was in Northern Ireland (Elson 1993). Producers grow these birds more slowly so that they are aged between 60 and 70 days when marketed and have a more mature flavour. In the UK the greatest demand for free-range broilers has been in the south east of England; consequently many large supermarkets in the area responded to the demand (Elson 1993).

### Structure of the industry

Broiler flocks are usually large, typically about 10–30 thousand birds. Many sites consist of several of these large sheds. Approximately 60–70 per cent of broiler production is under the complete control of the processing companies. Most of the remaining producers are individual growers supplying these same processors under contract. In intensive broiler systems the emphasis is on rapid growth rate and efficient food conversion, because profit margins are generally low. Intensive producers can turn over five or six crops of broilers a year, each living for an average 42 days.

The average broiler weighs 1.5–2.3kg liveweight but can range from about 1.3kg up to about 4.5kg liveweight. The heavier types are often called roasters or capons, although artificial caponization either by surgery or by hormone administration has long been illegal in the UK (ie *The Welfare of Livestock (Prohibited Operations) Regulations 1982* and the *EEC Directive 81/602* respectively).

## BREEDING AND GENETICS

Broiler chickens are carefully selected hybrids usually derived from two basic breeds, White Rock and Cornish Game. Modern broiler strains weigh up to 2.3kg and take only some 45 days, with a food conversion efficiency of 1.9–2.1. The docility and low locomotion of modern broiler strains does not make them suitable for free-range systems.

### Broiler breeders

The selection of broilers for increased growth rate has resulted in an increase in appetite. Breeders need to control the increased food intake of broiler parent stock because otherwise it results in the birds becoming obese and reducing reproductive performance. Food restriction is initiated when birds are 1–3 weeks old. It results in adult birds that are approximately 45–50 per cent lighter than those on *ad lib* diets. Food restriction can improve fertility, hatchability and egg quality, and has important health benefits. Birds on such rations are more resistant to disease, they have fewer joint, bone and foot problems, and have much lower mortalities.

The two most common restriction programmes are to feed every other day, or feed a limited amount every day. Breeders normally feed males every other day. Food withdrawal is a potent physiological stressor in fowl. Broiler breeders are chronically hungry, the males are more aggressive and the females are more fearful and have a higher rate of stereotypic pecking. They also over drink, so farmers need to restrict access to water.

Breeders rear males and females separately, but mix them several weeks before sexual maturation. However, it is necessary to prevent the males and females consuming each others' rations.

A recent report (Farm Animal Welfare Council 1998) has highlighted some of the problems associated with modern broiler breeder production.

## MANAGEMENT

Producers frequently rear cockerels and pullets together 'as hatched'. However, it is now very popular to sex the chicks at one day old and grow the pullets and cockerels separately ('sexed growing' or 'separate sex growing'). Such a system has several advantages. The growth rate and feed conversion efficiency of the female falls off at lower weights than in males of the same strain, so ideally producers should market pullets earlier. Farmers who divide their house into two sections, holding equal numbers of either males or females, can remove the pullets early at the lighter marketable weight. They can then give the extra space to the cockerels. This system ensures more uniform batches for processing, with less damage at the factory. Keeping the sexes separate also reduces feed costs, as it allows producers to feed males and females the diet best suited to their requirements.

Poultry keepers can assess a day-old chick's sex in three ways:

1. Vent-sexing, which requires great skill and is slow and rather stressful to the birds

2. Feather-sexing, in those crosses in which female day-olds have longer wing feathers than the males

3. Colour-mating, in those crosses in which the males have white or yellow-coloured down while the females are a buff or red colour.
   Breeders usually feather-sex broiler chickens.

## ENVIRONMENTAL AND HOUSING

The preface to *The Codes of Recommendations for the Welfare of Livestock: Domestic Fowls* (MAFF 1987) requires birds to be able to exercise their basic physiological needs. In general, broilers can do this, providing the management is satisfactory and staff maintain the environment and the litter in good condition. The litter floor permits normal behaviours, such as being able to exercise, dustbathe and choose warmer areas in the house for their resting periods. However, other factors, such as skeletal problems, very dim lighting and limited space as birds grow larger, may restrict normal behaviour. Farmers adjust lighting to combat stress, undue excitement or aggression. It is important to avoid excessive sudden noise to prevent panic, which can lead to a pile-up and some birds suffocating. The attendant must learn to move slowly along the stock to prevent any panic moves by the birds. Uniform and quiet conditions are essential and it is important to make all changes in the environment as gradually as possible.

### Temperature
*Brooding*
Rearing broilers requires artificial heat supplied by a brooder. Producers often confine the source of warmth to an area of the house, providing about 35°C at one day old and reducing this by 3°C a week (see Table 10.1). The brooding area should be as large as possible, but preferably up to a third of the total house. This allows a wide distribution of birds, which helps to improve growth and reduce the likelihood of disease.

To ensure full use of the house, the ambient temperature is as important as the brooder temperature. For best all round performance ambient temperature should be about 25–30°C. Below and above this range weight gains and food conversion efficiencies fall. Birds perform best when the farmer reduces the house ambient temperature from 30°C during the first week to 27˚C in the second, and 24˚C in the third week, and then to 21˚C from the fourth week onwards. Performance is worst when brooder temperature is correct but the house temperature is below 20˚C; the chicks are reluctant to venture away from the brooder heat to feed and drink. However, too high a temperature restricts appetite and retards activity and growth.

**Table 10.1        Recommended brooder temperatures.**

| Age (days) | Temperature (˚C) |
|------------|------------------|
| 1-7        | 32-35            |
| 8-14       | 29-32            |
| 15-21      | 26-29            |
| 22-28      | 23-26            |
| 29-35      | 21-23            |
| 36 onwards | 21               |

To ensure a good distribution of chicks within the brooding area, the temperature must be uniform and there should be no draughts at floor level. Overhead radiators give the most satisfactory results because their fine thermostat control and adjustable height offers flexibility (Figure 10.1). They also serve the dual purpose of brooding and space heating.

Alternatively producers can use hot air blowers; they are simple, cost little to run and are efficient at heating. The initial temperature of the whole house should be 31˚C, which is a compromise between the ideal brooder and house temperature. Stockpersons should lower the temperature by about 0.5˚C/day until it reaches about 21˚C. It is important to make all changes steadily and regularly to avoid stress to the birds. Blowing air great distances from end to end of the house creates considerable draughts and uneven air temperatures. With such a system it is necessary to lift the temperature several degrees to compensate for these problems. However, it is both costly and unsatisfactory for the productivity and health of the chicks. Ducting the hot air along the length or width of the house, or integrating it with the ventilation system, overcomes these problems. The latter system mixes the heat with the incoming fresh air, perhaps through a central intake duct under the ridge. The very dry air conditions that hot air systems produce

are not entirely favourable to the birds' health and well-being.

With a whole house heating system the chicks should be evenly distributed and not concentrated in one area. With brooders it is important to monitor the distribution of chicks under each heater (see Figure 10.1).

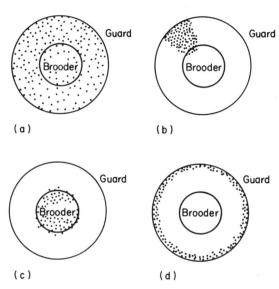

**Figure 10.1        Brooder conditions and chick behaviour (a) Just right, (b) too draughty, (c) too cold, (d) too hot.**

### Post-brooding

The brooding period ends at about three weeks of age, after which the birds need encouraging to use the whole of the house. It is still necessary to heat the building, other than in warm weather, to ensure favourable environmental conditions. Farmers can use the brooders to heat the whole house. Raising radiant heaters towards the ceiling of the house gives a wider spread of heat. Generally, ambient house temperature for broilers should be a minimum of 21˚C.

### Flooring and litter

It is vital that the rearing environment has a hygienic floor, which is also suitable for the birds both physically and behaviourally. Poultry keepers have tried many types of flooring and litter but the best is still about 10–15cm of litter on a well-built, damp-proofed concrete floor. The best litter is soft wood shavings, but a close second is chopped straw. Producers provide new clean litter for each batch; if they retain it mortality is higher and there is an excessive production of ammonia. Good litter management is vital, as it is very likely to become wet and caked, especially around the drinkers. Such litter is not only bad for the health of the birds, but may also bruise the flesh and lead to down-grading of carcasses. It is often necessary, therefore, to turn the litter physically, a time-consuming task if done manually. A small electrically powered cultivator, which is quiet and fumeless, is more satisfactory. Often when the litter has become very poor in its condition it is necessary to add more litter – either soft wood shavings or chopped straw.

Some producers have tried to use elevated perches and various types of raised perforated floors, but none are in widespread use.

## Light

Normally producers give broilers 23 hours of light and one hour of darkness in each 24 hours. This period of dark is necessary to train the birds to darkness. If this is not done and the birds suddenly find themselves without light, they tend to panic, crowd into corners and suffocate. Alternatively, farmers use a programme of intermittent lighting, which can improve growth and food conversion efficiency. A suitable example is as follows:

| | |
|---|---|
| 0-3 weeks | continuous lighting (with 1 hour off in 24 hours) |
| 3-5 weeks | cycle 3 hours on - 1 hour off |
| 5-7 weeks | cycle 2 hours on - 2 hours off |
| 7 weeks onwards | cycle 1 hour on - 3 hours off. |

With such a programme, it is vital to provide plenty of feeder and drinker space. This is because when the lights come on there will be a strong demand by all birds simultaneously for food and water.

Initially, light intensity must be high to encourage the young chicks to seek food and water, but it is possible to reduce this after about 5–6 days. It is important that the intensity is uniform to encourage even use of the floor. Suitable intensities for broilers are as follows:

| | |
|---|---|
| 0-5 days | up to 30 lux |
| 6-10 days | 10-15 lux |
| 11-30 days | reducing gradually to 5 lux |
| 30 days onwards | reducing further, if necessary, to 2-3 lux. |

Farmers may use either fluorescent tubes, tungsten bulbs or long-life bulbs. Ideally all systems should have a dimming mechanism, to allow variations in light intensity. Producers vary the light intensity to help control the behaviour of birds. For example, it is possible to quieten overactive birds by lowering the intensity. Alternatively if they are ill and inclined to huddle together, increasing light intensity (and warmth) is beneficial.

## Ventilation

Good ventilation is vital to always provide the birds with a constant and uniform supply of fresh air. It will also extract from the house the products of respiration and the moisture and gases arising from the litter and droppings, including airborne droplets containing pathogenic microorganisms. High ventilation rates are necessary to keep the birds cool in hot summer conditions. The system, however, must be capable of reducing these rates in cold weather, to prevent the birds being chilled, while still expelling the unwanted gases and moisture.

Producers can keep broilers in open, free ventilated natural convection housing (Figure 10.2) or, as is more usual, in controlled environment housing (Figure 10.3).

### Naturally ventilated heating

Controllable inlets and outlets vary the amount of ventilation and air movement in such houses (see Figure 10.2). The roof should be well insulated, with at least 15cm of glass fibre, 10cm of expanded lightweight plastic, or the equivalent. The outside should be clad in light-coloured heat reflective materials.

Naturally ventilated houses should be narrower than 15m with a ridge height greater than 4m. It should have a generous opening of up to 0.6m at the ridge, suitably capped to prevent rain entry. A sharp-angled pitch to the roof aids ventilation. An overhang on the roof, of up to 1m beyond the eaves, gives protection from the hot midday sun.

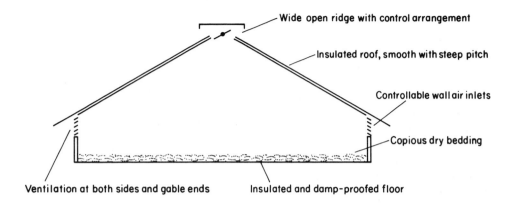

Wide open ridge with control arrangement

Insulated roof, smooth with steep pitch

Controllable wall air inlets

Copious dry bedding

Ventilation at both sides and gable ends

Insulated and damp-proofed floor

**Figure 10.2     Cross-section of freely ventilated natural convection house suitable for broilers.**

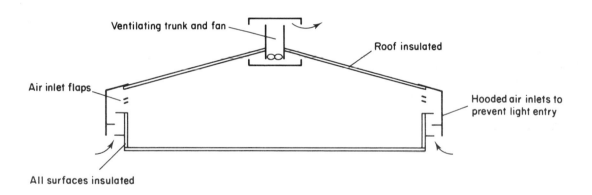

Ventilating trunk and fan

Roof insulated

Air inlet flaps

Hooded air inlets to prevent light entry

All surfaces insulated

**Figure 10.3     Cross-section of typical conventionally ventilated, controlled environment house.**

Inlet ventilators should extend along the whole length of the house on both sides. The farmer should be able to control these automatically by thermostats emergency fail-safe release mechanism for high temperatures.

In very hot weather, high air speeds are of great benefit as they increase the heat lost by convection. and motors. However, they should have an It should be possible to open much, if not all, of the sides of the house without restriction, especially in very warm areas. Farmers should site naturally ventilated buildings so that it allows them to make maximum use of winds flowing through the house when necessary.

Naturally ventilated houses are gaining in popularity. However, it is important to remember that they require a higher standard of management care to cope with changing weather conditions. It is impossible to compensate for all changes automatically. Furthermore there are risks in very hot weather and it is advisable to have some circulating fans available to supplement the air movement in hot, humid, windless conditions.

## Controlled environment housing

Most broiler producers house their birds in this system (Figure 10.3). The principal requirements for controlled environment housing are as follows:

1. A high standard of thermal insulation
2. Mechanical ventilation to provide a maximum of 7m³/h/kg bodyweight
3. Positive pressure systems to allow good control of air movement
4. Ceiling or other circulating fans to help air circulation
5. Reliable electricity supplies and stand-by generators
6. Substantially built structures to reduce the diurnal variation in temperature, being slower to warm up by day or cool down at night
7. Increased ventilation rate at night during warm periods to cool the entire structure and so delay the heating up of the house the following day
8. Full control of artificial lighting both for duration and intensity
9. Smooth surfaces to enable effective cleaning and disinfection between batches of birds.

## Environmental cooling systems

During very hot weather, special measures may be necessary to avoid birds overheating. Measures currently in use include:

1. Introducing free standing reciprocating circulation fans, usually mounted on the floor
2. Trickling water over the roof using a perforated plastic pipe
3. Painting the external surfaces with reflective paint
4. Providing a sufficient overhang at the eaves to keep the sun off the walls
5. Using construction materials that are not only good insulators but also have a good heat capacity.

## Thermal insulation

Apart from saving food and fuel, maximizing production, eliminating condensation and helping air-flow, good insulation generally stabilizes environmental conditions. However, in spite of the larger number of houses with insulation, designers and builders still make some serious errors. These strongly affect the success of the building, and the health and well-being of the animals. A well built house will last longer. The main error in shed construction is in the sealing of the insulation and of the cavity, if there is one, against damp penetration. The damp does not normally come from the outside as the external cladding is usually adequate. Vapour penetrates from within the internal air space of the house, which will be warm and laden with moisture from the birds. If the sealing is ineffective, the moisture constantly penetrates the fabric and condenses somewhere within the cavity. Many more traditional methods of insulation have a cavity between the inner and outer claddings. This not only causes massive deterioration in almost any material, but also reduces its insulating qualities. Moreover, it is best to avoid this arrangement as it is an ideal habitat for vermin and insects. More recent methods of insulation, in which materials such as polyurethane are bonded to the inner and outer claddings, avoid the gap altogether. They are prefabricated in the factory and are easy to erect. It is important to ensure that the joints between the boards are moisture sealed.

In temperate Western European climates, the principal problem is to conserve heat. Insulation can also keep temperature down during warmer weather. Good thermal insulation is becoming even more important with the current enthusiasm for natural ventilation.

It is essential to thermally insulate all round the building and ensure that there are no weak points; these otherwise create damp areas of great harm to the birds.

## Stocking density

Stocking density is critical to the health, welfare and production of broilers. Too few birds in a house may make environmental control, especially of temperature, difficult to maintain. Too high a stocking density will impede growth, impair access to food and water and increase the risk of disease. Also, the litter is almost certain to become wet and/or caked. Stocking density is an important factor to optimize the profit margin. Plumage condition is worse and carcass quality and weight is poorer at high densities (Fris Jensen 1993).

In open non-controlled environment houses the maximum stocking density should be 9-12 birds/m² (14-18kg/m²). In closed controlled environment houses it is possible to increase the density up to a maximum of 18–21 birds/m² (27–32kg/m²). *The Codes of Recommendations for the Welfare of* *Livestock: Domestic Fowls* (MAFF 1987) advises that the maximum stocking density for broilers should be 34kg/m².

**Table 10.2    Recommended nutrient levels for broilers.**

| Nutrient | Level in rations (%) | | | |
|---|---|---|---|---|
| | Starter | Grower | Finisher 1 | Finisher 2 |
| *Protein* | 22 | 21 | 19 | 17.5 |
| *Lysine* | 1.26 | 1.20 | 1.06 | 0.90 |
| *Methionine* | 0.58 | 0.54 | 0.48 | 0.42 |
| *Methionine plus cystine* | 0.96 | 0.92 | 0.84 | 0.75 |
| *Tryptophane* | 0.22 | 0.21 | 0.20 | 0.19 |
| *Calcium* | 0.90 | 0.90 | 0.90 | 0.90 |
| *Available phosphorus* | 0.45 | 0.45 | 0.40 | 0.40 |
| *Salt* | 0.32 | 0.32 | 0.35 | 0.35 |
| *Sodium* | 0.17 | 0.17 | 0.17 | 0.17 |
| *Chloride* | 0.16 | 0.15 | 0.17 | 0.17 |
| *Potassium* | 0.40 | 0.40 | 0.40 | 0.40 |
| *Linoleic acid* | 1.50 | 1.30 | 1.20 | 1.10 |
| *Metabolizable energy (Kcal/kg)* | 3060 | 3100 | 3125 | 3125 |

**Table 10.3    Recommended trace mineral mix for broilers.**

| Element | Parts per million |
|---|---|
| *Manganese* | 75 |
| *Zinc* | 60 |
| *Iron* | 20 |
| *Copper* | 10 |
| *Iodine* | 1.25 |
| *Selenium* | 0.15 |

**Table 10.4    Recommended vitamin levels for broilers (per tonne of feed).**

| Vitamin | Unit | Level |
|---|---|---|
| *Vitamin A* | million IU | 13 |
| *Vitamin D3* | million IU | 4 |
| *Vitamin E* | thousand IU | 20 |
| *Vitamin K3* | grams | 2 |
| *Vitamin B1* | grams | 1 |
| *Vitamin B2* | grams | 8 |
| *Vitamin B6* | grams | 2 |
| *Vitamin B12* | milligrams | 10 |
| *Biotin* | milligrams | 150 |
| *Choline* | grams | 250 |
| *Folic acid* | grams | 1 |
| *Nicotinic acid* | grams | 30 |
| *Pantothenic acid* | grams | 10 |

IU = International Unit

## FREE-RANGE PRODUCTION

The methods of housing free-range broilers are similar to the usual naturally ventilated broiler chicken house. In addition, birds have access to pasture through several popholes. Producers brood the birds inside, without allowing them outside, usually for the first three weeks. By law (*EEC 1906/90*) free-range broilers must have continuous daytime access to pasture for at least half of their lifetime. Growth of the birds is slower due to the ration having a lower energy value and protein content, or they are of a slower growing strain.

## EMERGENCIES

Intensively housed poultry in controlled environment houses are dependent on a reliable supply of electricity, heating equipment and other machinery. This is necessary to supply food and water and to keep the ventilation, heating and lighting functioning at all times. It is essential to install fool-proof arrangements to warn the stockperson of any equipment failure. Such a system should provide alternate means of keeping vital services going. On a large site it is usual to have a stand-by electrical generator that takes over automatically if the mains electricity fails. If this is not practical then farmers must provide alternate natural ventilation and battery operated lighting. Heating is usually by gas from a storage tank on site; this is a reliable source of fuel unless the weather interrupts the supply vehicles. Keeping the tank full when bad weather threatens avoids this problem.

Major priority needs giving to fire precautions, as the risk is quite high. Most fires arise from electrical faults with short circuiting creating the greatest risk. *The Codes of Recommendations for the Welfare of Livestock: Domestic Fowls* (MAFF 1987) detail the welfare risks from the failure of electricity and the breakdown of equipment and emphasize the dangers of overcrowding.

## NUTRITION AND FEEDING

Producers always feed broilers *ad libitum* from one day old to finishing. The starter ration is usually in the form of a crumb, and contains 22–23 per cent crude protein; it is suitable for birds up to 3–4 weeks of age, but during the later part farmers feed the diet

as a small pellet. Broilers then switch to a grower pellet, containing about 21 per cent protein, followed by finisher pellets of 19 per cent protein. If producers take birds beyond 50 days they will feed roaster or capon (finisher 2) diets that contain 1–2 per cent less protein than normal finisher rations. Table 10.2 gives the essential nutrient level of these rations, and Table 10.3 and 10.4 gives the mineral and vitamin requirements. White fishmeal and various types of soya preparation are the main protein rich ingredients in these diets. Wheat, maize, barley and milo form the basis of the cereal fraction. Computers are often used to calculate the exact composition of a ration; it is possible to programme for a ration that is a safe compromise between the need to economize on cost and the need for optimal biological efficiency.

### Feed additives
Producers feed all broilers a coccidiostat at a low level for most of the growing period. Most also normally use a growth promoter, these are antibiotics that have no therapeutic application or activity either in man or animals (eg virginiamycin).

### Feeding systems
Producers feed day-old chicks the crumb ration on feeder pans, egg trays or chick-box lids. They provide these at a rate of one for each batch of 100 chicks up to about seven days of age. It is important to accustom the birds to the trough pans or other automatic feeder systems as soon as possible. This is because they will rely on these exclusively after the first few days. Birds require 2.5cm of trough space each or 15–20 38cm diameter pans per 1000 birds.

**Table 10.5**  **Water consumption in broilers.**

| Age (days) | Consumption/1000 chicks/day (litres) |
|:---:|:---:|
| 7 | 35 |
| 14 | 57 |
| 21 | 76 |
| 28 | 99 |
| 35 | 129 |
| 42 | 160 |
| 49 | 186 |
| 56 | 208 |

**Water**

Water makes up 60–70 per cent of the weight of the chicken. It is important to avoid dehydration (see Table 10.5 for water consumption figures at different ages). Traditionally producers provide day-old chicks with 20 drinkers per 100 birds, made up of six permanent, 40cm diameter, automatically filled, hanging drinkers. They position these in the brooding area, with 14 chick founts. It is important to place the water close to the chicks. The stockperson must empty, clean and refill the chick founts daily with fresh uncontaminated water. On the third day it is possible to move the founts closer to the automatic drinkers and then remove them altogether at five days.

A more recent and popular arrangement is to use plastic apple trays, which the stockperson fills by hand. The chicks use these for the first few days until they can use the automatic founts. Also, broiler growers are increasingly changing from circular fount type drinkers to lines of nipple drinkers suspended from the roof. These have several advantages: they are hygienic, easy to maintain, and simple to clean. In addition, there is far less wastage and spillage on the litter, which helps considerably to maintain litter condition.

Throughout the growing period the stockperson should position the drinkers at the height of the bird's back. They will gradually need to raise this to avoid excessive spillage. It is important to adjust carefully the depth of the water in the drinker, for the same reason.

## HEALTH AND DISEASE

Over the last 40 years mortality rate in broilers in the UK has steadily fallen from an average of well over 10 per cent to about four per cent. However, averages conceal the fact that at one time there were frequent occasions when mortalities rose to 40 per cent or more. However, this would be very unlikely today; indeed many good broiler growers average mortalities of little over two per cent. This is attributable to:

- careful environmental control; namely, well constructed and thermally insulated buildings with good heating and ventilation
- the practise of an 'all-in, all-out' policy combined with thorough cleaning and disinfection.

- the strategic use of medicines.

The major problems affecting the health of modern broilers are ascites and skeletal disease (Thorp and Maxwell 1993). Both are the consequence of interactions between genetic and environmental factors.

**Hygiene and disease prevention**

Probably the single most important factor in keeping poultry healthy is maintaining good hygiene. Healthy breeding stock and hygienic hatchery management ensure that newly hatched broiler chicks are free of disease. If hygiene is also good in the broiler house, it will ensure an excellent chance of rearing healthy birds.

For good hygiene, it is important to maintain standards at every process in the growing operation. The best way to prevent disease coming on to the site through visitors and vehicles is to have a closed unit. This involves restricting visitors to a minimum and ensuring that those that do come, use wheel dips and foot dips, and wear protective clothing.

It is, however, impossible to eliminate airborne infections. Infective particles can travel many miles in the wind from diseased sites. In practice this a real risk because broiler-growing tends to be concentrated in certain areas. In dealing with any disease condition in broilers it is vital to recognize the trouble at once, diagnose rapidly and treat speedily; this will reduce any losses.

Another way to maintain hygiene is to operate an 'all-in, all-out' system. This process entails the complete stocking of a site or unit with day-old chicks within a very short period. Subsequently, at the end of the growing or production period it is vital to depopulate the site completely. It has enormous advantages:

1. The clearing of all living animals off the site can be the biggest factor in eliminating most, if not all, organisms capable of causing disease. The more frequently this depopulation, the more likely it is to prevent any build-up of disease

2. Once the farmer has depopulated the site, it is possible to rigorously disinfect and fumigate everything. This will help towards effectively eliminating bacterial, viral, fungal and parasitic infection

3. If all birds on a site are within a close age range this makes for a more uniform state of immunity to disease. Mixed ages lead to an immunological confusion, so that the effects of vaccines or treatment are less satisfactory. It also makes it much more difficult to administer disease prevention programmes correctly

4. There are sound and practical husbandry advantages. Filling a site at one time means there will be no further unavoidable disturbance until depopulation. If units have birds coming and going at irregular intervals, the disturbances can lead to health and behavioural upsets

5. It may be considered an advantage if the operator takes a break in the exacting task of management. Breaks between batches allows farmers to maintain equipment properly. This leads to far fewer breakdowns

6. With the great increase in intensification, the fly population rises in the manure and other organic matter that is continuously accumulating; this causes serious problems. Large sites often have poor arrangements for muck disposal. Smells can cause highly objectionable conditions for nearby residents. An 'all-in, all-out' system can limit the feasible size of a unit, one advantage of which is it can make it simpler to eliminate insects and rodents.

The merits of depopulation are inversely proportional to the age of the birds. It is less important with adults because, by the time a bird is mature, it will probably have a satisfactory immunity to most diseases.

## Hygiene programme for broiler sites

1. Remove all birds from the site
2. Clean out the litter from all houses and remove it totally from the site
3. Clean all the dust and dirt from the building, paying special attention to less obvious places, eg air inlets and fan ventilator boxes
4. Place all removable equipment, eg brooders, drinkers and feeders, on a clean area around the building for disinfection
5. Wash down all interior surfaces of the house with a detergent disinfectant, preferably using a pressure washer
6. Apply a specific, broad spectrum disinfectant to soak all the lower parts of the house to destroy viruses, bacteria, mycoplasmas, fungi and parasites
7. If there are litter beetles and other insects in the house, use an insecticide. Insects act as carriers for all poultry infections
8. Similarly clean and disinfect all the equipment
9. When the house interior is dry, set up the equipment and put in the litter. Then close the house and fumigate with formaldehyde gas after the house has been warmed to 20°C, or spray with a disinfectant specifically produced for this purpose
10. After 24 hours, open the house and ventilate to remove the remaining gas. The disinfected building is then ready for the chicks.

## Medication

Apart from coccidiostats and growth promoters, producers use medicines sparingly and most flocks receive no treatment. Good nutrition is another important factor in maintaining good health.

It is very important to recognize ill-health rapidly and act immediately to diagnose and treat the ailment, and so limit the risk of contagion. This is especially important in modern broiler units because there are usually many thousands of birds in close contact. Sometimes the most appropriate procedure after diagnosis is to send the birds as quickly as possible for slaughter before there may be a greater degree of disease. As with all emergency slaughtering, such action requires very careful professional - largely veterinary - judgement. The decision must be based on a balance between the economic and welfare aspects.

## Vaccination and preventive medication

It is most common to vaccinate broilers at day-old, frequently before they leave the hatchery, and again once or twice more during their lifetime. The precise programme of vaccination depends on the specific disease challenges in a particular area. The vaccines are usually live and farmers administer them either in the drinking water or by a coarse spray. Inevitably there is some stress on the birds from these activities. It is also common practise to give a short period of medication for the first few days of the chicks' life.

This helps to reduce or eliminate potential pathogenic organisms passing from the breeders via the hatchery. Broiler growers use probiotics at such periods to reduce stress and help the birds overcome infections.

Farmers use vaccines sparingly for broilers in the UK. Almost all birds receive live infectious bronchitis vaccine at one day old in the hatchery, usually as a large particle spray. Some producers may vaccinate against infectious bursal (Gumboro) disease and avian rhino-tracheitis. As the growth period is so short, broilers obtain most of their immunological protection passively from their dam. It is therefore important to ensure that mothers transfer good immunity to their chicks.

**Management related problems**

This is the principal area of concern regarding the welfare of broiler chickens. Considering the numbers of animals it affects, it is one of the largest welfare problem areas in farming.

*Ascites*

Ascites occurs in young fast-growing broilers and is a common cause of death or carcass condemnation. Ascites is usually manifest as an accumulation of serum-like fluid in the abdomen. It occurs more at high altitudes and during winter. Producers can lower the incidence and severity of ascites by improving ventilation and/or reducing growth rate.

*Skeletal disorders*

Selection pressure for production in modern poultry strains has placed increasing demands on skeletal integrity (Thorp and Maxwell, 1993). The two main categories of skeletal disorders (also called leg weakness) affecting broilers are: bone growth disorders, which affects young birds, and arthritis, which is prevalent among broiler breeders. The incidence of leg weakness sufficient to impair movement moderately to severely reaches approximately 25 per cent in heavy strains by the time of slaughter (42 days). However, it is below five per cent in strains less intensively selected for rapid growth and which reach slaughter weights at 70-90 days.

In birds over about 35 days the growth of the structure supporting the bird – bones, tendons and ligaments – often does not keep pace with the growth of muscle and fat. To prevent this occurring, it is necessary to slow the growth rate of the birds by careful attention to their nutrition. This involves reducing the energy value of the feed and adjusting upwards, where necessary, the quality and quantity of proteins, vitamins and minerals. Lighting programmes are effective in reducing skeletal disorders and metabolic disease (Fris Jensen 1993). Other means such as lowering stocking density and/or reducing ambient house temperatures may also help the problem. All these measures will also reduce the incidence of heart attacks, which have the same root cause. In the longer term it should be possible for the geneticists to ensure a balance between development of bones, tendons, ligaments, muscles and fat. This is a major area of priority in the welfare of broiler chicken.

*Hock-burn and breast blisters*

Bad litter condition causes hock-burn. In this, caked litter accumulates the wet droppings on the surface, causing inflammation of the skin over the hock. This may progress to ulceration followed by scabs over the ulcers. Hock-burn is extremely painful for the bird and is a serious welfare problem.

Caked litter can also affect the breast of the bird and lead to breast blisters, which may eventually become infected and will lead to abscess formation. Leg weakness accelerates its incidence, because such birds spend more of their time sitting on the damp litter.

*Feather pecking*

Unlike laying hen units, feather pecking and cannibalism are rare on broiler units. This is because the birds are more docile, they are harvested at an early age, they have access to litter and the light intensity is generally very low.

**HANDLING AND TRANSPORT**
**Stockmanship**

*The Welfare of Livestock (Intensive Units) Regulations 1978* requires the stockperson to inspect their livestock and equipment at least once daily. In practice they will usually inspect both the stock and the equipment several times every day.

A major requirement in all animal accommodation is to provide a first class service room for the stockperson. This is necessary for all personnel to maintain good standards of hygiene, to store equipment and medicines safely, have access to hot water and maintain records in an efficient way.

## Catching

Producers usually arrange to depopulate broiler sheds at night for slaughter the following day. Dimming the lights helps to keep the bird calm while the catchers pick up three or four birds in each hand. They then place them in containers brought into the house on a fork lift truck, which afterwards lifts them on to the lorry outside. There are also some mechanical methods of gathering birds up from the floor. They involve plastic or rubber covered arms that slowly revolve and draw the birds towards an operator, who then places them in a crate. Few producers at present use them, though further developments are taking place. In the future they may become more widespread.

## Transport

Chickens are very susceptible to stress during transportation; it is vital therefore to take great care to limit stress at this time. Generally, broilers do not travel more than about 50 miles to the processing plant. Even so, severe losses can take place if producers, stockpersons, hauliers and slaughter staff do not handle the whole procedure with understanding and skill (see Appendix).

## THE WAY AHEAD

Despite a Government review on the welfare of broiler chickens (Farm Animal Welfare Council 1992) significant changes do not appear to be imminent.

There is an increasing enthusiasm for the natural ventilation of broiler houses and for techniques that reduce the amount of vaccination or medication a bird receives. There is a real possibility that, within the next few years, improvements to methods of management and hygiene will enable farmers to rear broilers without any routine medication. Some impetus for this trend has come from the official action, stimulated by consumer awareness, to ensure that there are no medicine residues in the meat. However, growers are also aware that they can reduce costs by using fewer additives and placing

more reliance on high standards of management and hygiene. Using more simple systems of environmental control could also help in this effort. There is also a trend towards smaller units with the birds sub-divided into smaller groups. This makes practical use of the evidence that shows that the best results biologically come from farms of a modest size under the management of one good stockperson. A further bonus of small livestock farms is that it is easier to use the litter as manure on local land. This also removes the nuisance factor to the local human population.

## REFERENCES AND FURTHER READING

**Elson A** 1993 Housing systems for broilers. In: Savory C J and Hughes B O (eds) *The Fourth European Symposium on Poultry Welfare,* pp 177-184. Universities Federation for Animal Welfare: Hertfordshire

**Farm Animal Welfare Council** 1992 *Report on the Welfare of Broiler Chickens.* PB 0910. MAFF: London, UK

**Farm Animal Welfare Council** 1998 *Report on the Welfare of Broiler Breeders.* PB 3907. MAFF: London, UK

**Fris Jensen J** 1993 Stocking density, lighting programmes and food intake. In: Savory C J and Hughes B O (eds) *The Fourth European Symposium on Poultry Welfare,* pp 185-194. Universities Federation for Animal Welfare: Hertfordshire

**Mench J A** 1993 Problems associated with broiler breeder management. In: Savory C J and Hughes B O (eds) *The Fourth European Symposium on Poultry Welfare,* pp 195-207. Universities Federation for Animal Welfare: Hertfordshire

**Ministry of Agriculture, Fisheries and Food** 1987 *Codes of Recommendations for the Welfare of Livestock: Domestic Fowls (reprinted 1992).* PB0076. MAFF: London, UK

**Thorp B H and Maxwell M H** 1993 Health problems in broiler production. In: Savory C J and Hughes B O (eds) *The Fourth European Symposium on Poultry Welfare,* pp 208-218. Universities Federation for Animal Welfare: Hertfordshire

**Appendix:**
*Codes of Practice for the Catching and Transport of Live Chicken* (as formulated by a major broiler company, and reproduced with their permission).

1. *Catching*
   (a) Live chicken must be handled with extreme care at all times to ensure freedom from stress, injury and unnecessary suffering.
   (b) Birds must never be roughly handled by driving, kicking or throwing and quiet movements are essential to prevent smothering, bruising or more serious damage.
   (c) The top drawer of the module must be loaded first and filling should work downwards to avoid injury on closing the drawer.
   (d) Birds must be lifted into the drawers carefully and never be thrown into them.
   (e) The modules must be loaded on to the transporter with great care avoiding violent manoeuvres, eg heavy landings. All loaded module exit doors must be securely locked, standing of loaded modules must be kept to a minimum.
   (f) It is essential to catch birds in one hand at a time and never in both hands at once.
   (g) All staff involved with the catching and transport should be instructed in the relevant Animal Welfare Legislation.

2. *Transportation*
   (a) The vehicle and associated equipment must be inspected before every journey to make certain that all parts are mechanically sound and in good working order.
   (b) Scheduling of loaded transporters into the plant should be such that transport and waiting times are kept to a minimum. All loads must be processed in strict delivery rotation.
   (c) It must be established that the flock is in satisfactory health and is fit to travel.
   (d) The weather conditions will determine the loading densities and a recommended schedule for modular crates is as follows:

| Live weight (lbs) | Normal temperatures | Abnormally hot conditions |
|---|---|---|
| | (Number of birds per standard modular crate) | |
| 4.1 and below | 30 | 27 |
| 4.2-4.5 | 27 | 24 |
| 4.6-5.0 | 24 | 21 |
| 5.1 and above | 21 | 18 |

   (e) In the event of temperatures near or below freezing or when it is cold and there is heavy rainfall and there is a journey in excess of about 20 miles, then the modules must be protected with appropriate sheeting.
   (f) On arrival at the factory the vehicle must be unsheeted immediately weighed and parked in the special holding bay.
   (g) The Foreman in charge and his staff should be aware of welfare regulations and must be informed of the procedure to take whenever a breakdown occurs in the transport or there is some unavoidable hold-up.

3. *Unloading*
   (a) All crates must be opened carefully and the birds removed with great care to avoid pain or injury.
   (b) Any birds that break loose must be collected immediately.
   (c) Any birds unfit or unsuitable for slaughter may be removed for slaughter elsewhere.
   (d) All birds found to be dead on arrival must be placed in the receptacle provided.
   (e) If there is a plant breakdown so loading cannot continue, the Official Veterinary Surgeon or Senior Poultry Meat Inspector must be informed immediately and appropriate action taken after due consultation with the Factory Management and Agricultural Division (of the company).

# 11 Turkeys
## S A Lister

## THE UK INDUSTRY

The modern domesticated turkey originated from wild birds in North and Central America. Explorers mistook them for a type of peacock and called them 'Toka', the Tamil name for peacock, eventually anglicised to turkey. Settlers took the turkey back to Europe in the 16th century. Later, in the 19th century, the European domesticated turkey was reintroduced into the USA and crossed with wild turkeys. Further breeding has led to the modern hybrid, although compared with other poultry and farm animals, the turkey has only been domesticated for a short period. As a result, it may be more sensitive to environmental factors and managemental stress, requiring consistently higher levels of stock management skills.

Initially, birds were fattened mainly in Cambridgeshire and Norfolk, areas traditionally associated with cereal growing. More intensive housing of fattening birds really only began in the 1960's. Only a few primary breeding companies now supply most of the turkeys reared worldwide – British United Turkeys (BUT), Nicolas, and Hybrid Turkeys. These companies maintain and develop genetic stock through pure line research and supply parent stock as poults or hatching eggs to breeder multipliers. Smaller breeders still service more specialised markets. The overall trend in the industry is to breed broad breasted strains which produce high breast meat yield for the added value further processed market. This breeding strategy is achieved through separate male and female selection programmes, maintaining a balance between growth rate characteristics for the male line, and fecundity for the female.

Turkey meat consumption in the UK has risen over the last decade. The UK remains a net exporter of turkey meat although imports are increasing. Traditional turkey production was to serve the seasonal Christmas and Easter market, but now over 90% of output is non-seasonal, year round production with volume sales of value added products and portions greatly outstripping whole bird sales.

## REPRODUCTIVE PHYSIOLOGY

Modern breeds are brought into lay between 32 and 34 weeks of age with a laying cycle of 20 to 24 weeks. The age of onset of lay is controlled by manipulation of day length. Hens are reared on 14 hours light per day at a minimum of 60 lux. At 18 weeks, they are conditioned to a shorter day length of 7 hours light, again at 60 lux. Hens will not then come into lay until receiving more than 10 hours of light each day. During this conditioning phase, it is essential that birds are in lightproofed accommodation, especially in bright summer months. To stimulate the onset of uniform egg production, hens are then given a one step jump to 14 hours light at a minimum of 100 lux. This is maintained throughout the rest of lay. Since most breeding hens are kept in pole barns, natural light must be supplemented by bright uniform artificial light to maintain production.

The rate of egg production is dependent on the strain, weight of hen at point of lay and consistency of bodyweight throughout the lay.

Stags do not require a shortening day length to become reproductively active and can be maintained on 14 hours light at a minimum of 50 lux throughout rear. There is work to suggest that light restriction and conditioning late in the rearing period can help to synchronise peak semen output with the onset of lay in the hens. Day length may be reduced to 10 hours or even 6 hours light between 14 and 25 weeks and increased to 14 hours light at 25 weeks. Fertility in well managed flocks is maintained between 93 and 97%. Examples of typical reproductive performance are shown in Table 11.1.

**Table 11.1    Reproductive data (data for 24 week laying cycle).**

| | |
|---|---|
| Peak hen housed production (at 35 weeks of age) | 81% |
| Settable eggs per hen to 24 weeks | 115 |
| Poults per hen to 24 weeks | 94 |

## NUTRITION
### Breeders

During rearing, the feeding of breeders is aimed at increasing bodyweight in a controlled manner. There is far less need to restrict female bodyweight in turkeys than in broiler breeders. As a result, hens are fed virtually *ad libitum* on a fairly low energy ration to achieve target bodyweights for age (Table 11.2). Feeding progresses from a starter through various grower and rearer rations (Table 11.3). Stags are restricted to achieve target body weight (Table 11.2) for selection at 16 weeks of age. Feeding may then be altered either to increase or decrease weight gain to achieve target weight at onset of production of around 25Kg.

During lay, hen diets must provide adequate energy, amino-acid and calcium intake to ensure good peak production. During periods of high ambient temperature, nutrient density may need to be increased to maintain production (Table 11.4).

*Commercial stock*

The UK market requires a large weight range, from small, medium and large whole birds to 'tonnage' turkey producing large stags between 20 and 24 weeks of age, of high yield for further processing. Typical achievable growth rates are shown in Table 11.5. Growth rate is obviously related to nutrient intake and the typical nutrient requirements are shown in Table 11.6. Better early live weight is gained by feeding crumbs for the first 2–4 weeks of life, moving onto an increasing size of pellet of 3.5 to 4.5mm diameter. As with all stock, food presentation is important in relation to intake. Pellet quality and the avoidance of dust is essential. Adequate feeder space is required to maintain even flock growth. There should be 3cm trough space per bird or 1 tube feeder/pan per 40 birds. Water intake can be a severe limiting factor for growth, with feed intake being closely correlated to water consumption, birds drinking about twice as much water by weight as food. Unrestricted and constant availability of clean drinking water is essential, with at least one bell type drinker per 100 birds.

**Table 11.2                Target bodyweights for breeding stock (based on BUT8 strain).**

| Age (weeks) | Energy in feed (MJ ME/Kg) | Female bodyweight (Kg) | Male bodyweight (Kg) |
|---|---|---|---|
| 4 | 11.8 | 0.71 | 1.18 |
| 8 | 12.0 | 2.22 | 4.36 |
| 12 | 12.1 | 4.11 | 8.96 |
| 16 | 12.1 | 5.97 | 14.15 |
| 20 | 12.1 | 7.61 | 19.17 |
| 25 | 12.1 | 9.23 | 23.97 |
| 29 | 12.1 | 9.80 | 25.69 |
| End of lay | | 9.51 | 29.99 |

**Table 11.3**  **Nutrient targets for rearing parent stock.**

| Ration | Starter | Grower 1 | Grower 2 | Rearer |
|---|---|---|---|---|
| Age (weeks) | 0 to 4 | 4 to 8 | 8 to 12 | 12 to 16 |
| Metabolisable energy (MJ/Kg) | 11.8 | 12.0 | 12.1 | 12.1 |
| Crude Protein (%) | 26 to 28.5 | 23 to 25 | 18 to 20.5 | 18 to 18.5 |
| Lysine (%) | 1.57 | 1.21 | 1.00 | 0.86 |
| Methionine (%) | 0.60 | 0.48 | 0.41 | 0.38 |
| Calcium (%) | 1.30 to 1.35 | 1.20 to 1.25 | 1.10 to 1.15 | 1.05 to 1.10 |
| Available Phosphorus (%) | 0.75 | 0.70 | 0.65 | 0.55 |

**Table 11.4**  **Nutrient targets for breeding stock.**

| Diet | Pre-breeder | Breeder 1 | Breeder 2 |
|---|---|---|---|
| Males | from 16 weeks | - | - |
| Females | 14 to 29 weeks | from 29 weeks in Winter | from 29 weeks in Summer |
| Metabolisable energy (MJ/Kg) | 11.7 to 12.1 | 11.5 to 12.6 | 11.7 to 12.6 |
| Crude Protein (%) | 12.5 to 13.0 | 14.6 to 16.0 | 16.9 to 18.1 |
| Lysine (%) | 0.56 to 0.58 | 0.68 to 0.74 | 0.78 to 0.84 |
| Methionine (%) | 0.22 to 0.23 | 0.34 to 0.37 | 0.38 to 0.40 |
| Calcium (%) | 0.87 to 0.90 | 2.34 to 2.55 | 2.68 to 2.87 |
| Available Phosphorus (%) | 0.35 to 0.36 | 0.44 to 0.48 | 0.46 to 0.49 |

**Table 11.5**  **Growth rates for commercial stock (based on BUT T8 strain).**

| Age (weeks) | Stags (Kg) | Hens (Kg) |
|---|---|---|
| 4 | 1.05 | 0.91 |
| 8 | 3.81 | 3.07 |
| 12 | 7.65 | 5.67 |
| 16 | 11.52 | 7.95 |
| 20 | 15.18 | 9.39 |
| 24 | 18.65 | – |

**Table 11.6**     **Nutrient requirements for growing turkeys.**

| Age (weeks) | 0 to 4 | 4 to 8 | 8 to 12 | 12 to 16 | 16 to 20 | 20 to 24 |
|---|---|---|---|---|---|---|
| Metabolisable energy (MJ/Kg) | 11.80 | 11.90 | 12.00 | 12.10 | 12.20 | 12.40 |
| Crude Protein (%) | 28.00 | 26.00 | 22.00 | 19.00 | 16.50 | 14.00 |
| Lysine (%) | 1.60 | 1.50 | 1.30 | 1.00 | 0.80 | 0.65 |
| Calcium (%) | 1.20 | 1.00 | 0.85 | 0.75 | 0.65 | 0.55 |
| Available Phosphorus (%) | 0.60 | 0.50 | 0.42 | 0.38 | 0.32 | 0.28 |

### Environment/housing

*Commercial birds*

Most turkeys in the UK are housed on litter. Generally, climatic conditions are not suitable for outdoor commercial rearing, although some smaller scale free range production is practised. Such systems have been used extensively in the USA. Birds may be grown on a 'day old to death' system, being reared and grown on in the same house, or on a 'brood and move' basis, birds reared in specialist brooding houses to 6 weeks of age being moved to pole barn or controlled environment fattening houses. Concrete floors are favoured but earth floors are still common. Litter type is usually initially shavings or paper. Turkeys do not scratch to the same extent as chickens and, therefore, require only a thin layer of litter when environmental control is good. Further littering down with shavings or, in the case of pole barns, often straw, is required to keep conditions underfoot dry, especially around feeders and drinkers. Most birds are fattened in controlled environment housing with highly developed, fan assisted systems for controlling heat, ventilation and humidity. However, considerable numbers of birds are still grown (especially for the Christmas trade) in naturally ventilated pole barns, having been reared in control environment brooding houses. Pole barn accommodation varies considerably but essentially consists of a roof supported by poles about 3 metres apart. The open sides should be covered with netting to prevent wild bird or vermin access. In winter months, a degree of climatic control in the house may be achieved by plastic or fabric 'curtains'.

The *Codes of Recommendations for Welfare of Livestock: Turkeys (MAFF 1987)* recommends a maximum stocking density of 38.5Kg/m in controlled environment housing and 24.4Kg/m in naturally ventilated accommodation. However, present day ventilation and climate control in controlled environment housing is well advanced and allows effective removal of heat, carbon dioxide, dust and humidity at higher stocking densities. Recognising this, the recent Farm Animal Welfare Council's *Report on the Welfare of Turkeys* (FAWC 1995) accepts that producers should be allowed to stock at densities in excess of 38.5Kg/m where control of house environment is sufficient to avoid heat stress.

Feeders and drinkers are usually suspended, feed pans being tube or auger filled and drinkers of the bell type. Both should be gradually raised so as to be always just above the level of the growing birds back. Drinkers are initially well filled to ensure adequate intake but later the water level is lowered to prevent spillage.

Housing should be constructed of materials allowing effective cleansing and disinfection at depopulation. This should include the area outside the sheds which should be concrete or at least well drained. The removed litter is usually applied to the land, although increasing amounts are being utilised as fuel in power plants.

Effective vermin control at turnaround and when stock is present is essential to reduce the likelihood of fowl cholera *(Pasteurellosis)*.

*Breeding stock*

Breeders tend to be kept in pole barn type accommodation with pens either side of a central walkway. Straw litter is used. Turkey hens require an adequate nest at the onset of lay. These are usually wooden in construction with solid sides and base, but many have an open top. There should be at least one nest for every four hens, being large enough for the bird to turn around ie measuring approximately 45cm wide by 60cm deep and 60cm high. Nests are fitted with a trap front to prevent more than one hen entering at any one time. Nest box litter, usually straw, should be kept dry and be replenished frequently to maintain an adequate depth.

## Management

*Young stock*

The brooding period is one of the most critical in the turkey's life. Day old poults must be placed in accommodation that has been thoroughly cleansed and disinfected with an approved disinfectant. Fresh litter, free of dust, chemicals and, importantly, fungal spores must be used.

Day old birds must be well presented with feed and fresh, clean water to ensure the best possible start. Poults should be placed within surrounds of wire or cardboard, usually 300 to 350 poults in a 4.5 metre diameter surround in a preheated house. Each surround should provide spot heat from a single brooder at around 38-40°C, graduating out to a house temperature of about 25°C. This graduation of temperatures allows poults to seek their own comfort zone. The behaviour of the birds in their distribution and noise indicates their state of comfort – excessive chirping may suggest that the birds are not comfortable. Poults should move freely and not huddle.

Feed should be supplied on trays (3 per 100 poults) to allow poults to walk over the food and find it easily. Feed and water should not be added too soon before placement or directly under the spot brooder as feed quality deteriorates at high temperatures and high water temperature can restrict intake. Water is usually supplied as two or three bell drinkers and four temporary font type drinkers, or on disposable apple trays.

*Older commercial birds*

The surrounds are usually removed at 7-10 days of age, and birds are then given access to the whole house with one drinker per 100 birds and one tube or pan feeder per 40 birds. Chain feeders are not suitable for turkeys. House temperature is then around 25°C and is dropped by about 2°C per week to 18-20°C at around 5 weeks of age. This may be dropped lower for growing stags to promote better feather growing, increased activity and a reduction in leg problems. Birds going into pole barns should be 'hardened off' prior to moving out of the brooder house, being off heat by 5 weeks of age at the latest to avoid huddling.

Birds fattened in controlled environment houses are not routinely beak trimmed, careful environmental control reducing aggression and the need to trim. Birds going to pole barns are more often beak trimmed, either at day old or up to 3 weeks of age. The recent *FAWC Report on the Welfare of Turkeys* favours cold cutting of the upper beak up to 21 days of age.

*Breeding stock*

Stock destined for breeding tend to be broadly brooded in the same way as commercial stock. The nutrition, lighting and housing requirements needed to stimulate egg production have been discussed in previous sections.

Trap nest fronts should be tied up for the first few days of lay to allow free access and reduce the incidence of floor eggs. During early lay all hens sitting in corners should be caught and placed in nests again to reduce floor eggs. Broodiness can be a problem as hens reach peak production. These broody birds spend more time on the nest. All birds should be pushed out of the nest at each egg collection. All dark areas in the pen should be eliminated. Rotating the flock through laying pens can break the habit but this should not be done prior to peak production. Individuals identified as broodies should be removed to a well lit broody pen and observed for recommencing of normal laying behaviour when she can be returned to the laying pen.

Stags are penned in separate accommodation at lower light intensity and are given specific rations to control weight gain.

Breeding in commercial turkeys is by artificial insemination, stags being 'milked' (stimulated to produce semen) up to three times a week. Insemination of hens is done about 2 weeks after the stimulating effect of increased day length and commences when around 60–90% 'squat' (ie adopt the mating posture), when the stockman enters the pen. Hens are inseminated three times in the 10 days prior to the onset of lay and then go to a weekly pattern of insemination.

Egg collection should be done at least five times a day. Only clean eggs should be graded for setting. Turkey eggs are sensitive to storage conditions and should be stored within a temperature range of 13–16°C and relative humidity of 65–70%. Hatchability declines rapidly after 7 days storage.

## General care and handling

Stockmanship and good training of stockmen is paramount to ensure the welfare and productive potential of turkeys as with all other stock. This is especially relevant for young stock during the first week of life. Similarly, obtaining the best egg production, fertility and hatchability is highly dependent on stockmanship, skills and routine in managing both hens and stags.

*The Welfare of Livestock Regulations 1994* require all stock to be inspected *at least* once a day and that stock be cared for by adequately trained staff aware of all the provisions of the relevant welfare code for turkeys (MAFF 1987). The provision of adequate food and water at all times is clearly essential. Maintenance of adequate records ensures flock health is being well monitored and is available for external audit. Monitoring of water consumption through metering may help as the first indicator of ill health in the birds.

### Handling birds

Younger birds must not be exposed to sudden frights which may lead to smothers or injuries. All birds should be handled correctly to avoid injury. Young birds can be handled by grasping both legs with one hand and supporting the breast and wings with the other. Older birds should be grasped with one hand on the shoulder of the wing and holding the legs with the other. Any sick or injured birds should be removed from a pen as soon as possible. If suitable facilities are available, such birds may be removed to a hospital pen where they may be treated, observed and given access to food and water without competition. Where the condition is severe or the bird cannot be readily treated, it should be culled humanely without delay. Most birds are killed by neck dislocation, either by hand or by the use of a number of simple killing devices. FAWC in their *Report on the Welfare of Turkeys (1995)* recommended further research was needed to develop a portable and easy to use device to kill turkeys humanely.

### Catching/transport

Collection of stock for transport to the slaughterhouse gives the potential for injury and damage. Most birds are transported in modular crates, modules having hinged doors or carrying sliding drawers. Birds are usually driven into a small catching area within the house and are then removed manually to modules at the door of the house. This procedure needs to be closely monitored to ensure that all birds are handled correctly and only fit birds are loaded for transport.

Due account of weather conditions during transport should be taken to avoid extremes of temperature in transit. Programming should be efficiently monitored to minimise the time transporters need to wait prior to unloading at the slaughterhouse. The efficiency of all the procedures at catching and handling should be audited at the slaughterhouse by monitoring dead on arrival figures and the incidence of recent injuries.

## Health and welfare

As indicated earlier, a high level of stockmanship is essential to the success of any livestock enterprise. The provision of clean water and fresh food and maintaining stable environment control are paramount in maintaining the health of the flock. Recognising the signs of health and ill health are an integral part of maintaining health and welfare of the birds under the stockman's care. Sick birds should be removed and either culled or treated. Prompt veterinary diagnosis and advice should be sought to pre-empt major disease problems. Post mortem examination in situations of sudden or increased mortality, significant morbidity, or specific clinical signs is essential to aid early diagnosis and, hence, treatment. Preventive medication helps to control parasitic diseases such as coccidiosis and blackhead (histomoniasis). The use of concrete floors greatly reduces the risk of roundworm infestation.

Vaccination against fowl cholera (pasteurellosis) may be necessary, especially in pole barn premises where vermin is often a problem. The most significant viral disease is that of turkey rhinotracheitis (TRT) which affects the respiratory tract and may precipitate secondary bacterial infection. Live vaccines for use in the first week of life can help to reduce the severity of this infection. The other significant respiratory condition is mycoplasmosis, due, in the main, to *Mycoplasma gallisepticum*, is generally introduced onto a site by new stock or possibly personnel. Multi-age sites, ie those which are rarely depopulated and have constant movement of young stock, may be particularly prone to such diseases which, once established, can themselves infect subsequent 'clean' batches of young birds brought onto the premises.

As a result of these factors, effective biosecurity is essential in reducing the risk of the introduction of disease onto premises. Minimising the risk and persistence of disease may be achieved by some, or all, of the following:

1. Restrict access of vehicles and personnel to sites. Only allow essential visitors

2. Spray wheels of feed lorries etc with disinfectant on arrival

3. Provide showering facilities and change of clothes and boots for those entering the site

4. Boots worn by staff and visitors should be easily cleaned, and a brush and foot bath provided outside the entrance to each house or pen

5. Maintain adequate clean water supply to all birds

6. Ensure programmed feed deliveries to maintain supply to the birds. Bulk bins should be easily inspected and cleaned to prevent bridging of food which may become stale or contaminated

7. Maintain optimum litter and environmental conditions in each house

8. Avoid clutter between houses and keep grass and weeds short to remove cover for vermin

9. Provide adequate concrete apron to allow effective removal of litter at depopulation and aid efficient cleaning and disinfection of equipment removed from the house

10. Regularly monitor the health of the flock by inspection, post mortem examinations and other laboratory tests

11. Rectify all problems as soon as they are detected.

12. Maintain accurate records of mortality, feed and water consumption, egg production where appropriate, and all medications administered.

## THE WAY FORWARD

Current methods of turkey production require a high level of stockmanship to succeed. The dedication of effective, enthusiastic and well trained stockmen remains the key to the success of any enterprise.

Consumption of turkey meat, rightly perceived as a wholesome meat with a low fat content, is set to continue to increase. Methods of production which take account of the health and welfare of turkeys are available and with suitable management skills will continue to supply high quality products.

## FURTHER READING

**Farm Animal Welfare Council** 1995 *Report on the Welfare of Turkeys*. PB2033. MAFF: London, UK

**Ministry of Agriculture Fisheries and Food** 1987 *Codes of Recommendation for the Welfare of Livestock: Turkeys (reprinted 1991)*. PB 0077. MAFF: London, UK

**National Research Council** 1994 *Nutrient Requirements of Poultry, 9th edition*. National Academy Press: Washington DC, USA

**Nixey C and Grey T C** (eds) 1989 *Recent Advances in Turkey Science*. Butterworths: London, UK

**Stuart J C** 1993 Disease prevention and control in turkeys. In: Pattison M (ed) *The Health of Poultry,* pp 205-238. Longman: London, UK

# 12 Ducks

## R R Henry

The UK commercial duck producing enterprises are a very small fraction of the total poultry industry, producing only about 12 million meat birds per annum of which three to four million are exported. There is some further processing but most of the 'duckling' is sold either to wholesalers or through supermarket outlets. Most of the large scale duck growing is located in the Eastern area of England, between Yorkshire and Suffolk, with a scattering of smaller, 'backyard' producers across the country but predominantly towards the South. Some producers may import day old duckling from the continent but are most likely to obtain them from the larger UK suppliers in Eastern England. The commercial producer relies on various strains of 'Pekin' duck for most of the meat production, these strains having been bred for rapid weight gain and low feed conversion ratio. The true Aylesbury duckling is no longer a commercially available meat-producing breed but may still be found with specialist rearers.

## BIOLOGY/NATURAL HISTORY

The ducks mainly used for meat (and less commonly, egg) production in the UK originate from the wild Mallard (*Anas platyrhynchos*), a water bird of migratory habit. Because of this latter factor, the bird is genetically adapted to life in environments as disparate as the Arctic Circle and the tropics with all the variation of temperature, food types, light intensity and day-length that this produces. Selection for meat production and egg production over the last twenty to thirty centuries has produced domesticated variants which are physically, physiologically and psychologically better adapted to modern farming environments. These birds have been selected for either high and persistent egg production or rapid growth and economic meat production. As with the fowl there has been selection for pure egg laying strains (mainly the Khaki-Campbell) and pure meat strains (Pekin), but modern farming requirements call for, and have developed, hybrids with various combinations of these characteristics, designed to meet the needs of various countries.

## GENERAL MANAGEMENT CONSIDERATIONS

The day old duckling should weigh approximately 50g. It should appear lively and bright eyed. The navel should be well healed and show no prominent scab.

If duckling are to be sexed this can be done by day-old conventional 'vent sexing' methods which are very much easier in the duck than the fowl or turkey. Even at day old the male duckling has a distinct phallus which can be detected visually in the everted vent. In some Eastern countries sexing is carried out by sense of touch, the male phallus being palpated through the wall of the cloaca. This method can be rapid and accurate but is difficult to learn.

If it is necessary to sex ducks at a later age it is worth remembering that from about five weeks of age onwards the vocalisations of male and female are distinctly different: the female utters the distinct 'quack' associated with the species. The male produces a much more subdued and higher pitched 'squawk'. If sexing on visual characteristics is necessary note that the mature feathered male may show some curled tail feathers and also appears to have a less domed cranium

The birds should grow rapidly and, depending on genotype, achieve approximately 700g by two weeks. There is little sexual dimorphism although it will be noted that males feather grow slightly more slowly than females. This is mainly of significance when it comes to blood sampling.

Until about three weeks of age birds are usually growing in brooder houses. At that time it is possible to continue the grower phase of the meat producing bird in field pens provided that adequate shelter from the weather is available.

Because potential breeder stock may be kept on a strict diet regime it is common, if they are to be field reared, not to move them out until five or six weeks of age. At seven weeks of age meat birds will be approximately 3.0 to 3.5 kg live bodyweight although much depends on genotype and temperature of rearing. Keeping meat birds to a greater age is likely to cause plucking problems at slaughter due to the moult which takes place at around eight weeks.

At sexual maturity the bodyweight will be approximately 4.5 kg (once again depending on genotype). This bodyweight on *ab lib* feeding is achieved at about 12 to 14 weeks. It is advisable to maintain the bird at about 80 per cent of its expected mature bodyweight up to the point of lay for optimal performance, liveability and health. A peak and well sustained egg production of approximately 90 per cent may be anticipated when birds are reared as suggested. Ninety per cent fertility may also be expected with a male to female ratio of 1:5. The first egg may be expected at about 22 weeks of age.

## NUTRITION

Like the fowl and turkey, the ducks have a simple alimentary system but they lack the 'crop' found in the other two species. The duck, however, does have a dilatable region of the lower oesophagus which can be used for temporary food storage. In the wild, ducks are omnivorous and they scavenge whatever they can from the environment. The timing of the breeding cycle is such that young ducklings have access to a plentiful supply of insect life as well as vegetable matter. In the domesticated state a complete formulated food must be provided, for optimum effect, with a range of protein levels from about 20 per cent for the very young duckling to 11 per cent for a ration which will maintain a given bodyweight.

Ducks will adjust their feed intake in relation to the energy level of the feed, taking into consideration the environmental temperatures prevalent at the time. (This presupposes that there is no gross nutritional deficiency driving feed intake to remedy the situation.) The composition of the food for optimum benefit must take into account the energy balance of the duck versus the environment and the necessary intake of protein and micronutrients. There is little published research work on the nutrient requirements of the duck. Most of the recommendations are based on formulations which are found to be successful in commercial practice. (Table 12.1).

Most feed compounders do not produce a specific 'duck diet' but it has been found that in practice a normal chicken or turkey diet is adequate. If these are to be used it should be checked that they do not contain medication which might be disadvantageous to the duck. Because the feed intake of the duck is greater than that of the chicken or turkey in relation to its bodyweight, particularly during the growing phase, the presence of medicaments, which might be beneficial to the other species at their normal intake levels, can prove harmful. There is also a specific species difference in reaction to some medicaments.

**Table 12.1**

| Ration | Starter | Grower | Finisher | Breeder | Developer |
|---|---|---|---|---|---|
| *Metabolisable energy (MJ/Kg)* | 12.1 | 12.1 | 12.4 | 11.3 | 11.8 |
| *Crude Protein %* | 20.0 | 16 | 14 | 18.0 | 13 |
| *Methionine %* | 0.44 | 0.33 | 0.33 | | |
| *Lysine %* | 1.10 | 0.85 | 0.75 | 1.0 | |
| *Calcium %* | 1.0 | 0.8 | 0.6 | 3.0 | 0.6 |
| *Available Phosphorus %* | 0.4 | 0.35 | 0.3 | 0.35 | 0.3 |
| *Salt %* | 0.5 | 0.5 | 0.5 | 0.5 | 0.5 |

Specifically, amprolium, dinitolmide (DOT), aprinocid, nicarbazine and halofuginone have been shown to either cause health problems or slow growth rate. There are no reports known to the author of toxicity from lasalocid, monensin, narasin or salinomycin, given at normal feed inclusion levels.

It is unlikely that in-feed medication will be required routinely for ducks. Coccidiosis is only likely to occur in birds reared in the field and even there it is a fairly rare occurrence, coccidiostats are not normally planned to be included in the food. The starter feed should be given either as a crumb or 3mm pellet.

Good quality pellets should be ensured. Accumulation of dust (or fines) in the feed hoppers can cause problems with meal build-up around the ducks' bills; it also leads to expensive feed wastage. Pellet size for birds over two weeks of age can be up to 4mm in diameter. It is usual, commercially, to use a two or three stage feeding programme (ie starter, grower, finisher) to obtain optimum economic benefit and carcase quality. For birds intended for breeding a starter, developer and breeder ration regime is common.

## Feeding equipment

For growers, feed can be provided either via hoppers or tube feeders; these may be filled automatically from a feed bin where the number of birds makes filling the feeders manually burdensome. A minimum feeder space of 0.5m per 100 birds should be allowed. It is not usual to keep ducks destined for meat to more than about 7½ weeks of age but if this is to be done then the feeder space should be increased by 20 per cent. For breeders on a restricted diet, floor feeding of measured quantities of food gives good control of weight gain and helps to ensure an even uptake of feed. In-lay feed may be offered either on a timed basis using flaps to limit access to the food trough or in measured quantities into hoppers or tube feeders. By whichever method food is offered it is essential to ensure that there is opportunity for all birds to obtain their due ration.

## Water

Water is an essential nutrient. Free access to drinking water should be available at all times. A drinker space of 0.5m/100 birds should be allowed as a

minimum with a 20 per cent increase if birds older than about 7-8 weeks are to be kept. To start ducklings off, flat pans of water in addition to the normal drinker arrangements prove beneficial.

For young ducklings, drinking founts or narrow rimmed chicken bell drinkers are satisfactory. It is necessary to ensure that the young birds find the watering facilities either by dipping the beaks of a proportion of them into the water so that, having been so 'taught', they can act as leaders for the others, or by supplying water in so many containers that it is impossible for the birds not to find it.

For birds older than about three weeks, once again depending on strain and size, a change to the wider turkey bell drinker is advantageous. Nipple drinkers can be used with satisfactory results for birds of any age but it should be ensured that a) the nipples are within reach, and b) that they are in reach in such a way that the birds' necks are almost fully extended in as vertical a position as possible when taking water. Management of nipple line systems is not as easy as the management of individual drinker systems. Water flow rates may need adjusting in very hot weather.

Water troughs are frequently recommended and can work well for larger birds but it should be remembered that young ducks can (and do, occasionally) drown in them. The young duckling has an insatiable curiosity and will get into things from which it cannot get out.

It must be ensured, in field rearing situations, that either the water lines do not freeze up (for instance, by the use of 'trace heating') or that alternative watering facilities are readily available. Under field rearing conditions it will probably be found that the drinkers will have to be moved from place to place on a daily basis because the area surrounding them will become heavily fouled, and very muddy. The duck has a tendency to 'dibble' with its beak in such areas, which can cause drinkers to tip, and cause local flooding.

In some countries in tropical and subtropical areas, pond water is provided in which the birds can swim. This can be useful in helping to reduce the adverse effects of high ambient temperatures but creates a situation with a potential for helminth parasitism and bacterial and viral infections. In temperate climates swimming water is not generally provided for commercially reared birds; this is an apparent welfare disadvantage but has obvious health gains.

Sufficient watering facilities must be provided to permit birds to perform their ablutions but there is apparently no necessity for this to require total immersion of the head as is sometimes reported. Well managed watering facilities will result in clean ducks.

If it is found that watering facilities have to be altered for any reason, even if the change is envisaged as an 'improvement', great care should be taken to introduce the ducks to the new system gradually and carefully. They will not 'instinctively' find water if it is presented in a fashion different from that to which they have become accustomed.

## HOUSING

Up to two weeks of age $0.07m^2$ (0.75sq ft) per bird is desirable. At this time the birds will require some form of heating and it is usual to minimise the cost of this by the provision of well insulated brooder accommodation. Gas brooders are commonly used. Duckling numbers per brooder should be about 66 per cent of their suggested chick capacity. If the birds are to be brooded on wire floors the mesh size should be approximately 2cm (¾ inch) with a wire gauge sufficient to provide good underfoot support to both bird and stockman. A smaller mesh size of appropriate gauge wire tends to clog with faeces. If the birds are to be brooded in tier brooders a smaller mesh may accompany the narrower gauge wire usually used.

Over three weeks of age $0.13m^2$ (1.5 sq ft) per bird should be regarded as minimum for either wire/slatted floor houses or well drained deep litter systems. For wire floors a 2.5cm (1 inch) mesh is appropriate. Where drainage is poor, $0.18m^2$ (2 sq ft) per bird may be necessary. During the grower phase, under large scale conditions, the requirement for good insulation is less important than in the brooding phase. Broiler type housing is in common use but polebarn style accommodation is equally suitable provided that protection from wind can be ensured. Range rearing is common is some areas. The code of practice (see below) suggests 2500 grower birds/hectare (about 1000/acre). The area/bird required will depend on the nature of the soil, particularly its drainage capabilities, and the nature of

the ground cover. It will probably be necessary to move any frequently used equipment (eg feeders) regularly to prevent the land becoming irretrievably fouled. Drinkers will probably have to be moved daily. Weather protection will be required, particularly from high winds and rain but also from summer sun. Breeder birds during the lay period will benefit from an allowance of 0.4 to $0.65m^2$ (5 to 7 sq ft) per bird, depending on the expected ability of the litter to remain dry. The *Codes of Recommendations for the Welfare of Livestock: Ducks* (MAFF 1987) should be consulted for acceptable minima.

One of the objectives of field rearing of growers and replacement breeders is to reduce housing costs. This tends to limit the use of predator-proof fencing, which is expensive to erect and expensive and difficult to move when the ground becomes fouled and the birds have to be moved to fresh land. The normal fencing required to retain the birds *in situ* is about 75cm (30ins) in height, which is no real obstacle to foxes or other predators. The use of electric fencing about 25cm (10ins) outside the duck perimeter fencing, and about the same distance from the ground may be found to be a useful deterrent.

### Litter

For birds up to about 10 days of age the use of sawdust or a pelleted straw material appears optimal but wood shavings or chopped straw can be substituted.

If chopped straw is used it must be of good quality and as free as possible from *Aspergillus* spp and other moulds, to which young duckling are highly susceptible. Long straw should be avoided at this age. For older birds the pelleted straw is too expensive. Softwood shavings or normal straw should be used. Because of the birds' high water use and intake, under normal bird density conditions it will be necessary to spread fresh litter every day or every second day. It is prudent to keep the feeders and drinkers well separated to try to spread the faecal loading on the litter. Feeders, drinkers and other pieces of equipment should be so arranged that duckling cannot become caught in them, under them or around them. If they can, they will.

The disease spectrum to which housed ducks are exposed does not always make discarding litter at the end of a crop as essential for health as is the case for the fowl, but it is usual to remove the litter for cleaning down.

## Temperature

The young duckling appreciates a radiant heat source for about the first 10 to 12 days. If this is supplied and adequate, the environmental temperatures may be reduced to as low as 18°C (65°F). If radiant heat is not supplied then an initial environmental temperature of 32°C (90°F) is required. This should decrease evenly, daily, down to expected ambient temperature by the time the birds are to move out of the brooder accommodation or by about 14 days of age. In either case good ventilation should be maintained but absence of draughts is essential. For a fully feathered duck, the zone of thermal neutrality is around 17°C.

## Ventilation

Ventilation rates of a nominal 0.095m³/kg body wt/min (approx 1.5 cu ft/lb bodyweight/min) would be a satisfactory rule when calculating minimum fan capacity for a new building but ammonia production must also be taken into account. For any given faecal output, this will be found to be greater on straw than on shavings. The atmospheric ammonia level should never be allowed to rise to greater than 10 ppm although any olfactory sense of ammonia should instigate action. On no account should the upper legal limit (25 ppm) for the stockman be considered permissible. Ducks will almost certainly suffer ammonia blindness at this level and a reduction in respiratory tract resistance to infection can be anticipated.

The ventilation systems in use vary greatly, mainly perhaps because they have been taken over from, or modelled on, broiler growing farms. Totally fan ventilated units as set up for broilers are adequate but naturally ventilated units are cheaper to build, cheaper to run and do not require the provision of stand-by generators or other fail-safe devices. When properly run they perform well.

The effect of air movement over the birds should be taken into account when considering their comfort and welfare. Air movement, which in high temperature conditions is frequently referred to, approvingly, as a breeze and in low temperature conditions, disapprovingly, as a draught, causes a significant increase in body heat loss. Whether this is desirable or not will depend on circumstances. In either case it will cause/permit an increase in food consumption under the temperature range likely to be found in the UK and most other areas where duck farming is likely to take place.

## Lighting

As with any species, light plays an important part in physiological development. The duck is essentially a migratory species and appears to tolerate quite happily anything from the 24 hour day of the Arctic circle to the 12 hour day length of tropical areas.

For both grower bird and potential breeder, it appears to be advantageous to permit normal light intensity for the first 24 to 48 hour period; however, Codes of Practice drawn up by MAFF require that a period of reduced illumination be given to accustom birds to accidental 'blackouts' due to power failure. For housed growers, a short period of darkness (about 30 minutes) must be provided to accustom the birds to the dark situation in case of such power failure. For breeders the normal step-down, step-up pattern of daylength (as for the fowl) appears to be satisfactory both for the above purpose and for stimulating the onset of lay.

Since the duck is a twilight feeder under normal conditions its eyesight is geared to low light intensities. Growing birds do not discriminate, behaviourally, between 0.5 and 5 lux. Light intensity may be as low as is convenient for the operatives who are managing the birds. Sufficient light intensity must be available, continuously or on demand, to enable proper inspection of the stock. The light level for layers should probably be about 10 lux overall. The classic definition for adequate light for the fowl (ie that the poultryman should just be able to see to read newsprint) would still seem to be a useful yardstick for background light intensity.

Since ducks can be grown for meat and eggs under tropical conditions there is clearly no upper light intensity limit under normal conditions. Ducks do not appear to be attracted to high light intensities and, unlike the fowl, will not react adversely to 'spot' lighting with 'cannibalism'.

For field reared breeders it can prove beneficial to give the day's ration in the late evening when interference from wild birds is likely to be minimal. This does not inconvenience the duck and ensures that as much as possible of the food supplied reaches the intended customer.

## HANDLING

The primary rule when handling ducks is that they should **never** be caught by the legs. Birds should **always** be caught by the neck, by which structure they may conveniently (for all concerned) be lifted until the weight can be supported by the other hand or by suitable equipment. Small birds may be restrained by placing a hand either over or under the body, care being taken not to press on the thorax or abdomen: excessive pressure in those regions tends to impair respiration. With larger birds, ie once the flight feathers have started to develop, it is necessary to confine the wings to prevent wing tip damage (to the bird) and bruise injury (to the holder). For short periods of time the wings may be held in the fully extended flight position (see blood sampling below). Before handling ducks for any purpose it must be remembered that although food and water are provided *ad lib*, ducks, like humans, are social animals and tend to eat communally. The immediately ingested food is retained in a fusiform extension of the lower oesophagus from which regurgitation may occur if the bird is inverted or if the area is compressed shortly after food (and more particularly food and water) has been taken. This is certainly stressful to the bird, can cause death from inhalation of regurgitated ingesta, and at best, it makes a very unpleasant mess over the careless holder.

It is prudent to remove food one hour prior to handling birds.

## Vaccination

### Subcutaneous injection
Subcutaneous injection is commonly used for vaccines but can also be used for the administration of antibiotics.

The route of choice is the subcutaneous tissue at the base of the neck on the dorsal surface. Smaller birds should be held, with wings restricted, in the hand by one operator while the injector controls the head and neck with one hand, grasps the skin between forefinger and thumb and administers the inoculum into the thus 'tented' subcutaneous tissue at the base of the neck. For larger birds it may be found easier if the birds are held by the wings in full extension to give maximum loose skin at the base of the neck. Irrespective of the manner in which the bird is held, the vaccinator must ensure that the inoculum is not injected into the neck muscle where there is the potential for causing injury, particularly if oil adjuvant vaccines are in use, or laterally where the inoculum may well end up in the oesophagus particularly if this is slightly distended (see Handling).

### Intramuscular Injection
Intramuscular injection may be used for either vaccination or antibiotic administration. The most satisfactory target area is the musculature behind the femur. The bird should be held as for subcutaneous injection (see above). The muscle mass can be felt between finger and thumb of the vaccinator's free hand.

The inoculum should be delivered into the centre of the mass.

### Foot stab vaccination
This method of vaccination is almost totally confined to the application of duck virus hepatitis live vaccines to day old duckling. The vaccine is diluted according to the manufacturer's instructions. An eyed needle (such as a sewing machine needle), is dipped in the vaccine and then stabbed through the foot web of the subject, thus delivering a small amount of virus. Care should be taken to avoid the major blood vessels. The foot should be supported on a suitable pad.

## Blood sampling
Blood sampling is usually carried out either to monitor vaccination effectiveness or to monitor possible disease contact.

The normal sites for withdrawal of blood from the duck are the brachial (upper arm) or saphenous (shin) vein. The former location is to be preferred in birds once flight feathers or, more particularly, the 'under forearm tract' (UFT) feathers have developed. In birds below this age the vein can be difficult/practically impossible to locate (depending on the age of the bird). For these younger birds sampling from the saphenous vein is satisfactory.

Blood sampling from the brachial vein can probably best be carried out by a solo operator. For bleeding from the saphenous vein two operators are preferable, one to restrain the bird in the supine position and the other to take the sample.

For bleeding from the brachial vein it is usually least stressful to the bird to have it assume its normal sitting position on a firm non-slippery surface. It may be found necessary to position smaller birds in lateral recumbency. The bird is restrained by the wings which are raised to full flight extension and the UFT is raised with the fingers (or more usually, thumb) of the restraining hand. Down in the area should be removed by firm wiping with a moistened finger. (Plucking the down between finger and thumb risks tearing the skin.) The vein will be found to run along the line of the base of the feather shafts of the UFT. The needle should be introduced from the cranial end about midway down the upper 'arm'. Blood can then be freely withdrawn. In the smaller bird bled by this method it may be necessary to withdraw the blood in a 'pulse-like' manner to allow blood flow to refill the vein.

To bleed from the saphenous vein it is easiest to have the bird held in the supine position. Blood may be taken from either leg. The length of vein running laterally over the joint should be avoided. This method is applicable, of course, for birds of all ages but in the older bird the skin in that region can become remarkably tough. On the whole, bleeding from the brachial vein is more satisfactory for both the bleeder and the bird where bird size and development permit.

In both cases the vein lies very close to the surface. The needle should be introduced with the bevel (the slope containing the hole) facing downwards and at a very shallow angle to the surface of the skin. This will prevent the overlying vein wall and skin being sucked down to obstruct the blood flow into the needle and also reduces the chance of penetration through the vein into the underlying tissue.

### Livestock monitoring

It is laid down in *Welfare of Livestock (Intensive Units) Regulations 1978* that stock keepers of intensive units must inspect the stock and the equipment on which the stock depend at least once daily. The requirement for those who keep stock on a rather more extensive system should be no less rigorous and an evaluation should be made of the additional risks that the birds may be exposed to. These may include the possible action of predators, and exposure to infectious conditions from feral birds of similar or dissimilar species.

A necessary prerequisite to proper monitoring of stock is a sound knowledge of the normal.

### GENERAL BEHAVIOUR

It should be noted that the duck is a strongly social creature. If birds are exposed to human presence regularly within the first two to three days of life they will become largely imprinted upon the stockman and will permit a closer approach and show less alarm than birds reared in almost total isolation.

Although space requirements for ducks are quoted (see above) it should be noted that it must not be expected that the birds will avail themselves of all the space supplied. They will, at rest, almost always settle in close physical contact with one another in 'rafts' which may, numbers permitting, extend to several hundred birds. This should not be taken to imply that the birds are feeling cold or, as might be the case with the fowl, unwell. This normal tendency to gather together must, however, not be confused with the 'pile-up' of duckling under the age of about ten days which may, before thermoregulation is well established, indicate chilling.

Ducks over about three weeks of age are unlikely to suffer from low temperatures unless these are accompanied by wet and/or windy conditions.

Ducks which are subjected to high temperatures will reduce body temperature by panting and radiating excess heat from feet, beak and under wing regions. Panting will be noted when the environment reaches 27°C at a relative humidity of about 60 per cent and at rather lower temperatures if the birds are very large or the relative humidity is high.

It may well be noted that some birds lie with their legs stretched out behind them in a 'nose-dive' position. In otherwise healthy ducks this is, as might be expected, a sign of relaxed comfort. This should not be confused with the postural abnormalities which may follow bacterial (eg *Riemerella anatipestifer/* streptococcus infection or Clostridial enterotoxaemia (limberneck)).

When disturbed or when on the move for any other reason, ducks tend to move as a group and any one that is, for some reason, separated from the group will show obvious distress, random movement and possibly a plaintive noise until contact with other ducks is re-established.

Young ducks, like the young of many species, tend to run rather than walk. In the first few weeks of life there is a tendency for groups of birds to have a 'run around'. This results in a swirling mass of duckling moving together, frequently in a circular path, for no apparent reason. The activity may change direction or cease as suddenly as it started. The general behaviour pattern resembles that of a shoal of fish or a flock of birds in flight. This pattern of behaviour is not to be confused with panic reaction; however, small or weak birds may be bowled over by their more exuberant mates which can lead to trampling and death. Birds so affected should be culled. As birds mature this behaviour pattern disappears. Unless considerably stimulated, the adult duck tends to have a placid and relatively sedentary lifestyle. The tonal differences between the male and female 'quack' should be noted.

The 'critical distance' (ie how close they will allow a potential threat before taking evasive action) for domesticated ducks varies very considerably with their age and degree of habituation to human contact but it is usually in the region of 4 to 5 metres for birds of greater than about one week of age. Below that age duckling may well approach the stockkeeper, particularly if he/she stands still for a moment or so. Beyond the critical distance healthy birds will show interest in and reaction to any movements the stockman may make. Inside this range there will be a tendency for the birds to crowd and try to escape.

It has been observed that young ducks (3-6 weeks old) spent about 43 per cent of their time sitting, 10 per cent drinking, 17 per cent preening and 2 per cent feeding. Sieving the litter occupied about 15 per cent of the time and standing and walking another 15 per cent. In Pekin ducks it could be shown that preening activity did not differ significantly between birds kept on range or intensive housing, nor when bathing water was or was not supplied.

## BREEDING
### Strategy
Breeding stock should be purchased from specialist breeders, because selection of the lines will already have been carried out with a view to optimising the performance of birds for the purchaser. As with any other 'high tech' purchase it is essential to discover what should be expected and under what conditions these expectations are likely to be met.

The alternative less effective practice is to breed from the existing growing (or laying) stock.

If breeding selections are to be made, it will probably be necessary to identify the offspring of various parents or groups of parents from day old. This can be done using wing-tags, to identify individuals or by foot web slitting (often referred to, incorrectly, as toe-cutting) which is more usually used to identify groups of birds (see marking).

Selection of the breeding stock should be made when they are of an age (for meat birds usually seven weeks) at which their potential to achieve the required conformation has been demonstrated. Several factors will usually be taken into account, including their weight, feed conversion ability, mobility and general health. Their reproductive capacity must be deduced from that of their parents.

### Management
Birds selected for breeding from the grower sheds will probably be over-weight for optimum lay performance. Some form of food rationing will be necessary to bring their bodyweight to that which experience of the strain has shown to be consistent with good egg production.

Birds purchased from a primary breeder should come with a firm set of recommendations with regard to their growth curve (which may or may not be the same for both sexes).

It is usual to rear ducks and drakes together. If this is not to be done for some reason, the drakes should have some female companionship during rearing. If this is not provided, homosexual tendencies are likely to be imprinted with a disastrous effect on subsequent flock fertility.

A male to female ratio of one to five is normal at point of lay for the parents of meat-producing birds and a wider ratio, one to seven or eight for the lighter-bodied egg laying strains.

It is usual to bring birds into lay at about 24 weeks with increments in daylength.

If the birds have been intensively housed the system in use for the fowl will be found to be adequate (ie a reduction in day length for the first three or four weeks to about 8–10 hours and then a step up pattern of about ½ hour a week from 15 weeks of age. In the case of birds reared on natural daylight, the reduction in natural day length will have to be offset by the use of artificial light if year-round egg production is to be achieved.

It is usual to bring birds into lay on daylengths of between 15 and 17 hours. It may be found that initial egg fertility is poor and egg size small, with a relatively large number of 'double-yolkers'. By about peak production, (which should be 90% or over), infertility should have dropped to about 10 per cent. Any marked deviation from the norm for the particular strain of layer usually implies an error in feeding rates, light pattern or both. Egg size for any particular strain of layer will have an optimum for the economic production of first quality day-olds. Egg size can be regulated to some extent by alterations in the amount of feed provided but attempts to reduce egg size beyond a certain point will simply result in a reduction in egg numbers.

For optimum hatch results, eggs should be laid in specially prepared nests. Nests are most commonly kept to floor level with one cubicle to two or three females. A nest size of 40cmx40cm (16x16ins) area with 40cm (16in) high back and sides, one 10cm, front lip and with no roof and no floor (but resting on the litter) appears to work well. The nest construction is traditionally of wood but polypropylene has been found satisfactory and easier to clean. It has the disadvantage of needing a welded construction rather than the less sophisticated nail and batten. Nest colour does not appear to matter.

Nest material should be cheap, (to permit frequent renewal), as contamination-free as possible and comfortable for the bird. Soft wood shavings would appear to be most satisfactory although good quality straw can also be used. The nest material and area should be kept as dry and clean as possible to reduce egg contamination.

Although the bird to nest ratio may seem low, and although it will be noted that many nests are not used for laying whereas others may contain several eggs, a wider ratio is likely to lead to floor laying with a resultant drop in egg hygiene. It will be found that ducks tend to bury their eggs. The nest material should be changed regularly to prevent the unavoidable faecal material from fermenting with the production of moisture and warmth – both antagonistic to the production of clean uncontaminated hatching (or eating) eggs.

Ducks tend to lay their eggs on a 24hr cycle. In the case of very immature birds, or birds approaching the end of their lay cycle, a few birds may have a slower cycle but it is not usual to have to collect eggs more than once daily. More than 95 per cent of the eggs can be expected within 4 hours of the commencement of the 'day'.

Details of hatchery practice are outside the scope of this chapter but, in summary, eggs should be washed, if necessary, and set in incubators kept at 37°C and at a moderate humidity (eg c 33°C wet bulb reading) which will permit an egg weight loss of about half a gram daily. Eggs should be turned regularly during the first 24 to 25 days of incubation. At 24 to 25 days the eggs should be transferred to a hatchery where egg turning should cease. As hatch approaches humidity should be increasing the near saturation point. The hatch will possibly spread over two or three days but most of the ducklings will emerge on the 27th and 28th days.

The problems associated with breeder health are few. The commonest cause of death is probably 'egg peritonitis'. This condition is due to infection of yolk material released into the abdominal cavity at normal ovulation time without the normal physiological process which results in production of an egg being completed. This is a conditions which seems to be most common in females which have been overfed during rearing.

Lameness due to foot injury is commoner on wood shavings litter than on straw. Chronic foot injury may lead to over-mating with resultant injury, mainly to the upper neck region and/or, if the bird survives, to amyloidosis. This is a biochemical dysfunction associated with the overstimulation of the immune system which gives rise to malfunction of various organs, usually one or more of liver, kidney, spleen and heart.

Vent gleet, with or without impaction of an egg in the oviduct, may be a cause of a high cull rate.

Although acceptable stocking rates for breeders have been quoted as being of the order of three to five birds per square metre (see MAFF duck welfare code) strict adherence to these figures for small groups of ducks may result in the females being unable to escape male attention (which in the farmed strains appears to be randomly displayed) with resultant over-mating and resultant injury. Males will give up the chase, frequently without changing to another target, if the females can show a good turn of speed for long enough.

Broodiness is not a significant problem with ducks. Those that occur are easily identified by their aggressive behaviour at egg collection time. It does not appear to be profitable to remove these birds and pen them apart: recovery within the community appears to be equally rapid.

Male birds may suffer damage to the phallus with resultant inability to retract this organ (which in the drake is an intromittent organ, several centimetres in length). Clearly this will affect the ability of the drake to fertilise. The damage usually progresses to abscess formation in the peri-cloacal tissues and may result in the death of the bird. Affected males should be culled as soon as they are identified. Surgical intervention may well save life but does not return function.

Segregation of groups of breeders can usually be achieved by the use of pen partitions of a height of about 75cms. For the lighter, egg laying strains which may still have a weight/wing area ratio that permits flight, clipping of the feathers of one wing is permitted and is effective. Surgical pinioning is illegal (*The Welfare of Livestock (Prohibited Operations) Regulations 1982*).

A lay cycle of between 40 and 45 weeks (depending on strain of bird and commercial considerations) is usual. It is possible to moult birds and bring them back to lay for a second season, if circumstances so demand. In general, ducks can be moulted by reducing their nutrient intake. It is helpful if daylength can, simultaneously, be reduced (to about eight hours). Water should never be withheld. There appears to be no necessity to produce a massive weight reduction, and, indeed this serves only to delay the return to lay unnecessarily.

Although details of the procedure will depend on circumstances it is usually sufficient to reduce the daily ration to about 30g/bird for five days, at which time the ration can be increased to 50g. If day length can be reduced this will help to cause a cessation of lay. Egg production can be expected to cease after about the tenth day. The ration can then be increased

to about 150g/bird which should serve as a maintenance diet for the following three weeks. At that time a gradual increase in food and daylength can be given which should bring the birds back into lay some eight weeks after the commencement of the procedure. During the moult procedure a close eye should be kept for signs of lameness or general distress. Affected birds should be culled. It is more important than ever, during this period, to ensure that food distribution is even. The exact amount to be fed at various times will depend on the nature of the food (standard breeder, holder rations or even whole wheat), the ambient temperature and the initial condition of the birds.

The advantages of moulting are that it permits a rather longer total lay length than would otherwise be the case. It may also help in arranging egg production to meet some foreseen, or unforeseen, fluctuation in demand. The egg size in the second lay will be almost instantly of 'settable' weight.

The disadvantage of the procedure is that accommodation is occupied unproductively for about two months.

### Marking

The 'pin type' wing tag can be inserted at day old into the web of skin stretching across the front edge of the 'elbow'. Care must be taken to pierce only the skin web. The main problems which may be met are:
1. The tag becomes caught or the wing passes through the loop of the tag
2. The tag comes out (a tag in each wing might be worth consideration)
3. The puncture becomes infected. This is relatively uncommon in a properly placed tag
4. The tags become outgrown and start to become embedded in the wing tissue.

Replacement of the first pin tags with larger tags will be necessary, at about three weeks of age.

The birds should be checked a few days after tagging to ensure that mishaps have not occurred.

Some small-scale amateur duck breeders toe punch or slit the webs of the feet for identification purposes. The discomfort to the birds seems minimal if the job is properly done.

Web slitting is carried out at day old using a sharp knife or scissors. Two webs are available on each foot. The cuts should be made well into the web to ensure that they do not become indistinguishable with age. Excessive cutting should be avoided and, on the whole, it is preferable to avoid the outer web if possible as this helps to reduce the strain on the outer toes as the birds gain weight. It is possible to make punch holes in the webs instead of slits but these do have a tendency to close over which can make reliable identification difficult or impossible.

## Emergency slaughter

However good the environment and health of the stock, there will be occasions when welfare or other considerations necessitate the slaughter of a small number of birds. This is best achieved by cervical dislocation. This may be done either by holding the bird by the legs at chest height in one hand and, with the other hand placed behind the skull at the back of the neck, pressing rapidly downwards while tilting the birds head back, or by resting the bird's breast on the bent knee and then dislocating the neck as above. It may prove helpful to keep the arm straight and thrust down from the shoulder.

## DISEASES

The duck is normally regarded as the least prone of farmed poultry species to infectious diseases, but inevitably some outbreaks do occur. These are most common on continuous production sites, or where managemental faults have arisen.

## Viral Infections

*Duck Virus Hepatitis (DVH)*
There are three unrelated conditions of this name identified, where necessary, by the numbers 1, 2 and 3.

*Duck Virus Hepatitis 1*. An enterovirus infection of duckling under about four weeks of age are most susceptible. Up to 95 per cent mortality has been reported. Birds over this age can be infected without being obviously affected. They will excrete virus. Control is by vaccination either of parents (to provide maternal passive immunity to the duckling) or the day old duckling itself. This condition is endemic in some areas of the country on certain farms.

*Duck Virus Hepatitis 2*. There have been two or three localised outbreaks in the South East of England,

spread over several years. Control is by vaccination using an emergency vaccine produced by the Central Veterinary Laboratory at Weybridge. The disease pattern resembles that of DVH 1 but the susceptible age range is up to about six weeks.

*Duck Virus Hepatitis 3*. This is so far confined to the USA. It is a picorna-virus unrelated to DVH 1, with a rather lower mortality rate in affected flocks.

*Duck Virus Enteritis*. This condition can affect any age of susceptible duckling but since the source is most commonly migratory wild waterfowl the condition tends to be commoner in birds old enough to have access to common ground/water. With poor hygiene the condition can, of course, be carried into housed stock. Control is by vaccination.

## Bacteria

*Riemerella anatipestifer* infection, until recently more commonly known as *Moraxella anatipestifer* infection and, prior to that *Pasteurella anatipestifer* infection, is the cause of a septicaemia condition (New Duck Disease, Duckling Serositis) which may cause mortality and carcase condemnation in grower flocks. The condition is one which usually arises following stress or poor hygiene, in grower birds between three and five weeks of age. It may merge in to a colisepticaemia.

Control of outbreaks using either potentiated sulphonamides or amoxycillin is possible. Emergency vaccines can be prepared. A commercial vaccine is being evaluated for possible licensing.

*Escherichia coli* infection can cause a septicaemia condition, colisepticaemia, which is considered as always being secondary to some other infection (commonly *R. anatipestifer* in ducks) or to some managemental stress. Medication may prove helpful although there are frequently doubts as to the economics of such action, unless the precipitating factor(s) can first be removed.

Streptococcal septicaemia, commonly due to a Group D streptococcus can occur, usually at about 10 days of age, the time when artificial heat has just been removed and the birds crowd together for warmth. There is lung congestion and oedema, with cyanosis of the beak and toenails; the spleen may be enlarged. Mortality is usually low. If treatment is considered necessary, amoxycillin proves effective.

*Erysipelothrix rhusiopathiae* infections can occur but these are relatively uncommon. They tend to leave a legacy of chronic infection in spite of medication and vaccination.

### Fungi

Aspergillosis is a fungal infection of bronchi, lung and airsacs due to *Aspergillus fumigatus*. The term 'aspergillosis' is loosely used to describe infections of the above type which may, on investigation, turn out to be due to other fungi (eg *Absidia, Rhizopus*). The source of these infections is commonly mouldy straw which may, in spite of looking to be of good quality, carry a heavy mould spore burden. Mouldy feed residues, particularly in the warm brooder areas, are another potential source of infection.

Aflatoxicosis is one of the better known toxic conditions associated with preformed fungal toxins from *Aspergillus flavus* in the food. The duck is particularly susceptible to this toxin.

## LEGISLATION

Various items of legislation are applicable to those who farm poultry (including ducks) whether this is on a large scale or small.

*The Animal Health and Welfare Act of 1984,* which is an update of *The Slaughter of Poultry Act 1967* governs the welfare of poultry at slaughter wherever and on whatever scale it takes place.

*The Slaughter of Poultry (Humane Conditions) Regulations Act 1984* governs the slaughter of poultry for a commercial purpose.

*The Welfare of Livestock (Intensive Units) Regulations 1978* requires that stock keepers examine their stock and the equipment associated with them at least once daily.

*The Welfare of Livestock (Prohibited Operations) Regulations 1982* makes illegal the surgical castration of poultry and the impeding of flight by operations on wing tissues (other than the clipping of feathers).

*The Welfare of Animals (Transport) Order 1997* now largely controls the conditions under which animals can be transported. The rules and regulations are complex - see MAFF 1998 for details.

The *Codes of Recommendations for the Welfare of Livestock: Ducks* (MAFF 1987) outlines basic welfare requirements. A copy of this booklet must be available to all relevant stock keepers.

## THE WAY FORWARD

The development of the duck industry has meant that fewer flocks are now maintained on free range and more in naturally ventilated poultry buildings. Mechanical litter spreaders are being developed. It is important to ensure that excessive dust is not produced during the process. It is unlikely that ducks will be caught using the newly developed 'harvesting' systems for broilers. This is because the duck is able to see in near darkness and takes evasive action.

## LITERATURE

**Farrel D J and Stapleton P** (eds) 1985 *Duck Production Science and World Practice.* University of New England: Armidale, NSW

**Gooderham K R** 1996 Diseases of ducks. In: Jordan F T W and Pattison M (eds) *Poultry Diseases, 4th edition*, pp 415-421. Saunders: London, UK

**Ministry of Agriculture Fisheries and Food** 1987 *Codes of Recommendations for the Welfare of Livestock: Ducks (reprinted 1991).* PB 0079. MAFF: London, UK

**Ministry of Agriculture Fisheries and Food** 1998 *Guidance on the Welfare of Animals (Transport) Order 1997.* PB 3766. MAFF: London, UK

**Scott M L and Dean W F** 1991 *Nutrition and Management of Ducks.* M L Scott of Ithaca, NY

# 13 Quail production

## B Hodgetts

## THE INDUSTRY

Quail originated in the Far East where they are still highly popular. In the UK, both quail meat and eggs are seen as luxury foods and quite difficult to obtain. On the Continent quail are more sought-after and large scale production is commonplace.

The UK industry is rather precarious, being very susceptible to cheap imported birds. Detailed production figures are difficult to obtain, but an estimate would be 12,000-15,000 meat birds produced per week. In the UK the market for quail eggs for consumption is very limited.

## BREEDING AND GENETICS

The most used breed is the Japanese quail (*Coturnix coturnix Japonica*). In its wild state it weighs about 150g maturing at six weeks and laying 240 eggs in 12 months. In recent years geneticists have improved the meat production characteristics of the bird. Stocks now have a much heavier mature bodyweight of around 230g. This gain has, however, been at the expense of its egg production capabilities, which have declined to about 200 eggs.

It is possible to distinguish male and female quail from about three weeks of age. The male is cinnamon to dull chestnut on the upper throat and lower breast region, whereas the female is much lighter coloured on the throat and upper breast and the feathers are long and pointed. The female also has characteristically black stippled feathers on the breast.

### Selection of breeders

It is important to take future breeding stocks from strains with good egg production, growth rate and fertility. Breeders select stock based on health, uniformity and body size. Too much emphasis in a breeding programme on one factor, such as growth rate, can lead to a fall in other reproductive traits, such as fertility and egg production. A structured breeding programme is advisable to maximize genetic improvement and reduce the undesirable effects of inbreeding. A simple breeding programme is one where the total breeding stock is spread into six or eight families. The female progeny of each family remains with that family to form the next generation while the male progeny move to another family. In the first generation, the males move one step in the cycle, in the second generation two steps and in the third generation three steps and so on. For example, with eight families A-H, in the first generation males from family 'A' go with females family 'B'. In the second generation males from family 'A' go to females from family 'C' and in the third generation males from family 'A' go to females of family 'D' and so on. This system should produce a degree of hybrid vigour while avoiding inbreeding.

In practice, mating ratios of 2-3 females to each male is the most satisfactory. Quail will mature between 6 and 7 weeks of age and breed well for 10-12 months. However, it is common to replace the males part way through the cycle to arrest a decline in fertility. Females lay about 200-220 hatching eggs, with 85-90 per cent fertility.

## ENVIRONMENT AND HOUSING

Breeding stock appear to perform best when in a warm environment; a temperature of 20-21°C is desirable. For optimum fertility and egg production quail require a light intensity of 15-35 lux and a photoperiod of 16-17 hours. Breeders prefer to house their stock in cages rather than on litter because it produces cleaner eggs. Commercial cages for nine females and three males measure 41.5cm deep, 60cm wide and 20cm high. The cages normally have wooden floors and 15mm square section wire mesh with a 7˚ slope so that eggs roll to the front. Producers should house breeders on litter at a density of 55 birds/m$^2$.

It is vital to insulate the building that houses either breeders or growers so that it conserves the body heat of the birds. It is also important to have a ventilation system designed to maintain a balance between a fresh atmosphere and the temperature most appropriate for economic growth and egg production. The maximum ventilation rates are about 0.94m$^3$ sec$^{-1}$/hour per 1000 adult quail and 0.47m$^3$ sec$^{-1}$ for growing stock; these values should ensure that during the summer months, the ventilation system removes excess heat and the noxious waste gases, such as ammonia. Minimum ventilation rate should not be less than 0.06m$^3$ sec$^{-1}$ per 1000 birds.

### Incubation

Quail eggs weigh about 14g. Shells are generally very thin and easy to crack. Probably because the shells are thin, eggs lose moisture very easily during storage and hatchability suffers as a result. It is usual to experience a decline of 3 per cent hatchability for each day of storage. The humidity in the egg store must be high, 75–80 per cent relative humidity (RH), to limit this moisture loss. Storage temperatures should be between 16 and 18˚C. It is therefore important to avoid storing eggs for longer than seven days.

It is possible to adapt most chicken incubators for quail eggs using special plastic quail egg trays. Incubation takes about 17 days, but egg age and the type of breeding stock influences this. Some authorities advise that incubation humidity should be slightly higher than that for chickens (30.6˚C wet bulb instead of 27.5˚C). This is probably to control the water loss from the thin shelled quail egg.

Age of the parent stock greatly affects hatchability and fertility. Maximum hatchability occurs from 8–24 week-old birds, after which there is a rapid decline.

### Brooding and Growing

It is possible to brood and rear quail in either cages or on litter. Cage housing provides better control of the birds but initial capital costs are greater. Growing quail on the floor should have 150 birds/m$^2$ for the first 14 days reducing to 100 birds/m for the rest of the growing period. Light intensity should be quite bright (ie 60–70 lux) at day old to help the chicks find food and water. Reducing intensity to 10 lux or less by three weeks does not appear to adversely effect production. To accustom the birds to possible power failures, it is important to give them regular periods of dark. Producers normally provide 23 or 23½ hours of light for the first 2–3 days, reducing it to 14 hours by 14 days. Should cannibalism occur, producers often use a reduction in the light intensity to help control the problem.

Brooding temperatures at day one should be 35–36˚C, gradually reducing to 20˚C by three weeks of age. It is important to avoid draughts during the brooding stage.

## PRODUCTION OF EATING EGGS

Producers of quail eggs usually house 12 adult females in a cage measuring 45cm by 60cm. They grade the eggs to discard large, small and cracked ones. From maturity to 30 weeks of age female quail will produce about 100 eggs.

## FEEDING AND NUTRITION

Breeding quail have similar nutritional requirements to breeding pheasants and a typical game breeder ration or light hybrid breeder ration is perfectly satisfactory. They will consume approximately 25g/bird/day if on a diet containing 20 per cent protein and 11.5MJ of metabolisable energy per kg, provided on an *ad libitum* basis. Clean water should always be available. Occasionally the stockperson should give the birds small quantities of chick-size flint grit. When egg production is high, limestone grit or oyster shell should be provided about once a week, sprinkled on top of the food.

Farmers can place feed in box lids or egg trays for the first few days and introduce normal feeders by 14 days. Initially the birds require baby chick drinkers, but later the stockperson weans the chicks onto nipple or cup drinkers. Quail chicks are prone to drowning; placing marbles in the drinkers for the first weeks reduces the chance of this happening.

During the brooding and growing period the stockperson usually feeds the quail *ad libitum* turkey rations in crumb form. The birds are supplied from day old to 14 days with a turkey superstarter; they follow this with a turkey starter, which they remain on until slaughter at six weeks or 40 days. Total food consumption to slaughter at 40 days will be about 500g/bird, producing a bodyweight of around 230g. As food conversion deteriorates rapidly after 40 days it is not worth growing them beyond this age.

A light hybrid chicken layer ration is sufficient for birds producing eggs for human consumption.

## HEALTH AND DISEASE

Coturnix quail are susceptible to many common diseases of chickens, turkeys and game birds. Some of the most common are:

1. Bacterial diseases such as salmonellosis, *Clostridia, Escherichia coli* and *Staphylococcus* infections often due to poor farm hygiene
2. Respiratory diseases due to *Aspergillus fumigatus, Mycoplasma* are often complicated by secondary bacterial infections
3. Parasitic diseases, such as coccidiosis, blackhead and worms.

Successful treatment and prevention of disease depend upon correct diagnosis. Consult, when necessary, veterinary surgeons who have experience with avian diseases.

## SLAUGHTER

The *Welfare of Animals (Slaughter or Killing) Regulations 1995* require that the slaughter or killing of quail must not cause any avoidable excitement, pain or suffering. Decapitation, neck dislocation, or stunning by electronarcosis followed by bleeding are the recognised methods of killing. Those who slaughter quail on-farm can use neck dislocation or decapitation provided they have the necessary skills and knowledge. Those using stunning by electronarcosis, whether on-farm or in a slaughterhouse, must hold a slaughtermans licence.

## LEGISLATION

There are many rules and regulations surrounding the production and marketing of quail. These concern the welfare of the stock, transport of live birds and procedures for humane slaughter. There are also new regulations for the handling and storage of quail meat. For a current list of requirements, it is advisable that producers contact their local Ministry of Agriculture office who will provide full information.

## THE WAY AHEAD

The future for quail in the UK can only develop if the product receives more exposure and correct marketing. There is a good consistent demand for quail in France and Belgium, and some more enterprising UK producers see exporting as the way forward. There is scope for applying selective breeding techniques to improve egg numbers and growth rate. Many production techniques that the chicken and turkey industry use to handle eggs and improve hatchability apply equally well to quail. There is a need to apply hygienic control measures to control disease, and to bring standards of quail production into line with other sections of the poultry industry.

# 14 Guineafowl

## E Ruth I Biswas

## THE PLACE OF GUINEAFOWL IN THE LIVESTOCK INDUSTRY

Though present as small, mostly non-commercial flocks on farms and country estates the world over, guineafowl play a very minor role in poultry production. During the early 1990s around 70 million keets (chicks) for fattening were hatched annually within the EEC. France, Italy and Belgium contributed about 70, 30 and 1 per cent respectively with a small seasonal contribution from Spain, Greece and Germany. In 1994 production of keets as compared with broiler chicks amounted only to 5.4, 5.3 and 0.7 per cent in these countries. However, the number of keets hatched exceeded that of goslings throughout these countries, although in France and Italy producers now hatch goslings throughout the year. In Belgium keets outnumbered turkey poults for fattening.

Guineafowl have a lean, faintly gamey meat that has been royal food from ancient times, and the eggs have always been regarded as a delicacy. Nowadays an eight-week-old bird will produce a dressed carcase of 1kg and many guineafowl hens can produce at least 150 eggs in 7–9 months of lay. Slower growing strains (there are no separate breeds) seem ideally suited for free-range management, a practice that consumers now increasingly demand. High production costs – the guineafowl's main drawback – should, however, be offset by the value of the gourmet end product which also has less fat and a higher meat to bone ratio than chicken, but unfortunately is not readily available in the UK.

## DESCRIPTION AND DOMESTICATION

Guineafowl are recognised by their white pearled grey plumage, interspersed or sometimes substituted by whole black, white, lilac, blue or buff shades. They are gallinaceous birds native to Africa. A variety of feathered crowns, crests, bony casques as well as brightly coloured patches of naked skin or fleshy wattles decorate their head and neck and separate the four genera and numerous species. The striking appearance of some has made them valuable game birds – the friendly but sensitive Vulturine guineafowl being the best example.

The domesticated species is the helmet guineafowl *Numida meleagris*. The EEC marketing regulations require its scientific name to bear the suffix *domesticus* in common with the other domestic poultry and, as the name suggests, it has a bony casque or helmet. It is the most numerous and widely distributed in nature. Although most are grey with distinctive white pearls throughout, albino, lavender and black varieties and mixtures occur. These varieties have been increased by selection so that some shades now predominate in some countries (for example, white is predominant in Russia). A study of the colour inheritance has made sex-linked day-olds possible.

### Sex differentiation

Guineafowl are monogamous and the sexes are so similar that breeding failures in small flocks, even in Africa, are often due to wrong identification. Comb and spurs are absent. The best way of separating the sexes in the field is by the call. Females utter a repeated two syllable call: 'come-back, come-back'; while males use a single 'back, back' call. If excited, females may also adopt a one-syllable shriek but a male never uses a two-syllable call. Vent sexing at day old is not recommended but is possible when drafting potential breeders by about 6 months. At this age the distance between sternum and cloaca in the female is greater than in the male. This can be shown by holding the bird in one hand and measuring with the other hand placed across the abdomen. This can be followed up when the point of lay is approaching by comparing the distance between the pin bones on either side of the cloaca. As in the domestic hen, the pin bones are wider apart, easier to feel and more supple, than in the male.

There are other more subtle ways to differentiate the sex of guineafowls. Although the young male may be marginally heavier than the female, the weight of the well-fed improved female tends to surpass that of the male from 9–10 weeks onwards. When comparing breeders, in the male the helmet grows stronger and more upright and the wattles develop thicker edges. In addition, slight differences in the caruncle and the length of the keel and shanks have been reported. By checking several of these features at different times it is possible to achieve an accuracy of above 95 per cent.

## NATURAL HISTORY AND BEHAVIOUR

The guineafowl has proven to be highly adaptable wherever it has been taken, ranging from the tropics to Siberia. Its natural habitat is woodland savanna with ground level cover for its nest and trees for roosting at night. In its native Africa, insects, leaves, grass seeds and tubers, especially of sedges, appear to form the guineafowls' diet. The temporary abundance of this nutritious diet may partly trigger the onset of the short breeding activity during and after the seasonal rains. In common with other wild animals naturally exposed to prolonged dry season food shortages, reproductive organ functioning depends more critically on adequate nutrition than does the survival of the young produced.

The guineafowl is a most easily tamed wild bird but even after generations of selection and productive improvement, guineafowl retain some element of their wild nature. They have a nervous disposition, an urge to sleep on high perching places, a gregarious way in moving about and a liking for communal nests. When pursued by predators, guineafowl prefer to escape by running first and taking to flight only as a last resort, aiming for trees or high structures nearby. Management practices should make allowances for their behaviour by providing high perches and communal nestboxes.

## PRODUCTION

The figures in this section should only be taken as a rough guide as there may be considerable variation in practice due to strain of bird, environment (especially ambient temperature), feed and equipment. Producers should request advice on feed and management from the breeder when purchasing stock. Where proprietary guineafowl feed mixtures are not available it is possible to modify broiler starter or turkey feeds, according to the age of the guineafowl.

Guineafowl eggs are smaller than chicken eggs and more pointed at the narrow end, with the colour ranging from white to dark brown, sometimes with speckling. The shell is thicker than in other domestic birds and the eggs are famous for their long shelf life, although hatchability reduces markedly beyond a week. In the course of genetic improvement for other parameters the eggs have become heavier and the shell thinner. The weight ranges from 30–45g but it is not advisable to use eggs below 38–40g for incubation. Clutch size ranges from 8–18 but if eggs are removed continually a female may lay some 50 in a season and this can be increased further by enclosure and feeding. A guinea hen can cover 12–15 eggs and incubation takes 26–28 days. In small scale production a bantam or the common broody, which can cover well over 20 eggs, is preferable to the guinea hen for hatching and care of the keets. Under such conditions hatchability often exceeds that in the incubators that breeders use for large scale production. Incubation is at 37°C and at relative humidity (RH) of 55–60 per cent. Candling can be done at 10 and 23 days and the fertile eggs transferred to the hatcher at 24 days while reducing the temperature gradually to 36.4°C and increasing RH to 98–100 per cent. During the final 2 days a change from continuous to intermittent ventilation facilitated hatching.

Newly hatched keets are very sensitive to cold and damp so it is essential to keep them at 37°C for the first few days. As they feather quickly, heat should be provided beyond one month only if the ambient temperature is below 50°C. The rate of stepping down the temperature should depend on the keets' behaviour. Standard liveweight for slaughter is 1.3kg which will yield a carcase of around 1kg. In the unimproved guineafowl this will require 15–18 weeks of feeding. In the early 1970s, before serious improvement had modernized guineafowl, it took 10–12 weeks to obtain these weights: now it is possible in 8 weeks. Mature wild guineafowl may weigh 1.5kg and improved ones 2.5kg.

Meat birds are fed *ad libitum* while future breeders are raised more slowly to achieve adequate fertility and stamina. The sexes are reared separately by using sex-linked crosses or determining the sex early and marking the birds with coloured wing bands or leg rings. Light is stepped down to 8 hours and feed restricted in quantity and crude protein content by completion of breeding. Dimmers are used to make all light changes gradual. During the 5th month stimulation of the male is started by slowly stepping up light to 14 hrs, followed by the females during the 6th month. In the 7th month breeders are selected, 4 females to each male, and housed in their breeding pens to start breeding by 30 weeks.

It is very important to select the best birds as any sub-standard male would be subdued and fertility level lowered because of the permanent bonding habit and the fact that guineafowl cocks are not renowned for fertility. In deep litter systems, females readily accept nestboxes on ground level. Large scale breeding units always use artificial insemination.

Guineafowl hens will lay in the absence of cocks - in case egg production alone is planned. The point of lay is sometimes reported before 25 weeks but under adverse condition in the wild, may be delayed to a year.

Guineafowl are very secretative about mating which is therefore rarely observed and needs confirmation by candling the eggs in the second week of incubation. Most eggs appear to be laid from late morning to early afternoon. Laying continues for about 9 months in housed birds. Improved strains produce 150-180 eggs and sometimes even 200 eggs in the year. Hens can be kept for a second year of production after a moult. In Africa, as under free range conditions in temperate countries, breeding is seasonal – late spring to early autumn in Europe. All - the-year breeding is easily achieved by housing and management. Mortality during each stage of life is usually below 5% as guineafowl are very hardy.

## HOUSING, FEEDING AND SPECIAL MANAGEMENT ASPECTS

Guineafowls are inquisitive, nervy and tend to be noisy with repeated use of 'come back' calls. It is therefore best to site them at some distance from the farm house and major roads. The birds soon get used to their handlers and do best under conditions of continuity. For this reason it is important to keep keets intended for fattening under an 'all in all out system'. Guineafowl can be aggressive in the establishment of their pecking orders and can be brutal to weaklings: perches should be provided from the second month onwards so that birds can remove themselves from the social competition on the pen floor. The sound of sudden rain on corrugated roofing may cause panic. It is therefore advisable to round off the corner of the pens with high narrow mesh netting to avoid trampling and suffocation in case of a stampede. Guineafowl like to pick fibrous material; the litter for young keets should therefore be free from long fibres that could cause compaction when swallowed. In deep litter systems it is recommended to provide a sandbath area to allow dustbathing. During the first few days it is important to introduce an hour of darkness in the 24-hour lighting to accustom the keets in case of a power failure. As keets start flying very early, the brooder surrounds confining them should not be less than 1m high.

Guineafowl are wasteful feeders and should be given crumbs or pellets rather than mash. Long feed troughs with an inwardly facing lip or a partial top cover also help to cut down wastage; similarly it is advisable to raise the trough as the birds grow to force them to reach up for feeding. Suspended tubular feeders and drinking founts are best as they can be readily raised and the birds are not so likely to misuse them as perches. Keets drown easily in chick-type water founts, so these should be partially filled with pebbles.

Feather pecking does not normally occur under good management unless a bird has been injured by accident.

Weekly sample weighings are recommended as a check on the feeding and general success; for this the use of a catching crate causes least disturbance of the flock. Handlers should catch the birds with both hands cupped securely over the wings to prevent flapping. The birds are quite amenable to handling as a standard routine. For the first few weeks they can be put straight on a weigh pan. Later it may be necessary to immobilize them by twisting their wings gently to a locked position or suspending them with a soft cloth strap by their shanks from a spring balance.

## NUTRITION

Although guineafowl will find enough food to survive under most free range conditions, adequate protein containing sufficient methionine, cystine and lysine is crucial to fast growth. Crude protein levels should be at 24 per cent for the first four weeks, falling then to about 20 for intensively reared meat birds and reducing further to 15–16 per cent for extensively kept meat birds and future breeders. The initial energy levels need to start about 3000KCal/kg with a gradual reduction in the second month. As conversion ability rapidly deteriorates from nine weeks, with ratios rising to above 3:1, producers sell the standard broiler type guineafowl before this happens.

When enclosed, regular supplies of fresh leafy food and also grit should be given.

## HEALTH AND WELFARE

Although very hardy, guineafowl need the customary prophylactic and curative treatment usual for domestic fowls with particular attention to control coccidiosis and parasitic worms. It is vital never to give the drugs monensin, elancoban or stenorol to guineafowls, and with other drugs it is necessary to seek expert advice. As guineafowls have a fascination for long fibres and strings they sometimes suffer from intestinal impaction or become crop bound. In the latter case oil and water should be given and the crop gently massaged to help move the food.

Flocks of guineafowl should not contain mixed ages or be mixed with other types of poultry - partly for hygiene and also because of the guineafowl intolerance of newcomers. The regular rotation of outside pens is essential for the control of gut parasites and houses must be thoroughly disinfected after each period of 10-14 days.

The flightiness of guineafowl can be a real problem when outside runs are used for extensive rearing as high fences or fully netted aviaries are very expensive. Feather clipping is allowed but does not stay effective for long. There is no code of recommendations for the welfare of guineafowl but

where they are farmed on agricultural land they are covered by *The Welfare of Livestock (Prohibited Operations) Regulations 1982* which prohibit cutting the tip of one wing (pinioning) to impede flight.

## MARKETING

In Britain guineafowl meat is mainly sold as oven-ready fresh or frozen entire birds. The majority are fastgrown so-called 'industrial' broilers. The demand for heavier, older extensively reared birds has been met by the EEC Marketing Regulations which cover all domestic poultry and which allows quality labelling for 'extensive indoor or barn reared' and three levels of 'free range' birds of slow-growing strains. These have usually been bred for meat conformation and high food conversion ability. Precise conditions have been laid down for food, accommodation, stocking rate and minimum slaughter age to separate the product from the standard broilers though both are classed as 'young guinea fowl' as the tip of their sternum is still flexible, ie not yet ossified.

Minimum slaughter age for 'barn reared' is 82 days at a stocking rate of less than 25kg total liveweight (ie 15-16 birds) per m². The same applies for 'free range' for which there must have been access during at least half the lifetime, to open air runs allowing 1m² per bird, while slaughter age for 'traditional free range' must be at least 94 days and the feed formula for fattening must contain at least 70% cereal. However, in the case of guineafowl, open-air runs may be replaced by a perchery with a floor surface at least double that of the house and a height of at least 2m. Perches of at least 10cm length per bird must be available. This concession avoids the problem of impeding flight to keep the birds in.

Old breeders with an inflexible ossified top of the sternum, will be sold as 'guineafowl'.

## THE WAY AHEAD

In Britain the guineafowl has received little official attention although breeders have achieved excellent improvement of stock. It is regrettable that teaching establishments rarely keep guineafowls or include them on the syllabus. The modern guineafowl will still breed with its wild counterpart; this presents a vast natural gene pool for research. New strains bred for eggs or meat under a variety of managements are on the way. The industry needs to cut production costs generally and to improve the performance of free range flocks. Britain regularly imports fresh guineafowl carcases; so there is a case for encouraging increased home production.

The breeding of the more ornamental crested and vulturian guineafowl, for stocking nature parks and zoos, is also worth investigation. Some of these have been previously reared in captivity.

## FURTHER READING

**Ancel A and Girard H** 1992 Egg shell of the domestic guineafowl. *British Poultry Science 33:* 993-1001

**Ayeni J S O, Olomu J M and Aire T A (eds)** 1983 *The Helmet Guineafowl (numida meleagris galeata Pallas) in Nigeria - Selected papers from State of Knowledge Workshop on the grey breasted helmet guinea fowl.* Kainji Lake Research Institute: New Bussa, Nigeria

**Beaumont C, Sauveur B, Plouzeau M, Jamenot P and Duval E** 1993 Breeding and genetics of guineafowl: current state and future prospects. In: Gavora J S (ed) *Proceedings of the 10th International Symposium on Current Problems of Avian Genetics, Nitra, Slovakia* pp 79-84. Slovac Technical University: Bratislava, Slovakia

**Best P** 1974 A sunday treat for the French. *Poultry International 13:* 16-24

**European Commission** 1987-1996 Guineafowls eggs placed in incubation and chicks hatched for fattening by member countries. *Animal Production Quarterly Statistics.* European Commission Publications Office: Luxembourg

**Feltwell R** 1980 Developments in guineafowl production. *Poultry International 19:* 24-32

**Hastings Belshaw R H** 1985 *Guineafowl of the World.* Nimrod Book Services: Liss, UK

**Hoesen C Van and Stromberg L** (eds) 1975 *Guineafowl - a book on Guinea Culture.* Stromberg Publishing Company: Fort Dodge, USA

**Jamenot P, Galor S A and Seigneurim F** 1994 La Pintade an unknown product. *World Poultry - Misset 10(9):* 10-13

**Mongin P and Plouzeau M** 1984 Guineafowl. In: Mason I L (ed) *Evolution of Domesticated Animals,* pp 322-325. Longman: London, UK

**Oguntona T** 1986 The performance of the guineafowl (*Numida meleagris*) keet on various dietary energy levels. *Journal of Animal Production Research 6:* 15-20

**Okaemne A N** 1983 Disease conditions in guineafowl production in Nigeria. *World's Poultry Science Journal 39(3):* 179-183

**Smetana P** 1974 Some aspects of guineafowl production. *Proceedings 1974 Australasian Poultry Science Convention, Hobart 302-306*

# 15 Fish farming
## A E Wall

Controlled fish culture probably originated many centuries ago in China and the Far East. The culture of salmonoid fish is relatively more recent, beginning during the 19th Century to enable the stocking of fish into rivers for anglers.

The farming of Rainbow trout (*Onychorynchus mykiss*) for human consumption was pioneered in Denmark. This practice has developed throughout the UK during this century, resulting in the relatively stable production we see today. Atlantic salmon (*Salmo salar*) farming using sea cages to ongrow fish for the market, is a relatively recent development which has grown rapidly in Scotland, Norway and elsewhere, over the last 30–35 years, and continues to expand.

The development of these industries in the UK has corresponded with a demand for the relatively high value fresh and smoked products the industries produce, both within the UK and in Northern Europe. Increasing production and more widespread availability and public awareness has led to a gradual but substantial reduction in the market price, and although the European market continued to increase steadily, the rapid increase in production, particularly in Atlantic salmon, has led to a search for new markets, mainly in North America and Japan.

Currently, the UK exports in the region of 60 per cent of its total Atlantic salmon production. Production methods and efficiencies have improved dramatically in tandem with reducing market prices, in order that the industry maintains its profitability.

## UK FARMED FISH PRODUCTION

UK production of Atlantic salmon and Rainbow trout from 1991–1994 are summarised in Table 15.1.

While rainbow trout production remains relatively stable, Atlantic salmon production rose by a further 20–30 per cent in 1995. This was achieved mainly by improved survival and performance, rather than a straightforward increase in numbers of fish. The majority of salmon and trout production in the UK is sold fresh or smoked.

The total UK production figure for 1998 is projected to be over 100,000 tonnes.

Pilot schemes and small scale projects for the farming of other species, such as turbot, halibut, carp and tilapia, are also part of the UK aquacultural effort, although these species are in the main hatched and reared for export to warmer water for ongrowing, or grown to satisfy limited niche consumer markets. As such, they do not contribute significantly to national aquacultural output in terms

**Table 15.1.    UK production of Atlantic Salmon and Rainbow trout from 1991-1994.**

| | Production (metric tonnes) | | | |
|---|---|---|---|---|
| | 1991 | 1992 | 1993 | 1994 |
| *Atlantic salmon* | 40,500 | 46,100 | 48,500 | 56,000 |
| *Rainbow trout* | 14,500 | 14,500 | 14,500 | 15,500 |

of tonnage. Rainbow, brown and sea trout are also grown for the restocking of angling rivers.

With the constant pressure on market prices, most effort to improve quality of the final product has been focused on fish husbandry and especially feeding. Production engineering to maintain high standards of water quality and quantity has led to improvements in the environment of the fish. Some selection and stock manipulation programmes are in use, to improve health status and production efficiency. These include:

- Programmes to select genetic stock lines for particular purposes (eg low or late maturation, increased growth and efficiency of food conversion)
- Production of triploid (having three sets of chromosomes instead of the usual diploid two) and all female stocks, to reduce the production and quality implications of maturation. Female fish usually mature later than their male counterparts; the later maturing female is desirable before sexual maturation causes a decrease in flesh quality.
  Triploid fish, usually females, do not reach sexual maturity before harvest. These fish are sterile and so escapees cannot endanger the native gene pool by breeding with wild stocks.
- Manipulation of photoperiod and temperature regimes, in order to allow extension of the spawning season, and in Atlantic salmon to induce early or late smoltification.

These programmes allow the industry to exert greater control over production parameters so that fish of appropriate size and quality can be supplied to the market all year round to meet market demands. Because of the relative youth of the salmonoid industry, and the relatively long production cycle, breeding and selection programmes are at an early stage and take time to develop.

## BIOLOGY AND REPRODUCTIVE PHYSIOLOGY

### In nature
Atlantic salmon can be found naturally in suitable rivers bordering the Atlantic in the northern hemisphere. Most populations are migratory from fresh to sea water and back again in order to complete their life cycle.

Because the natural range of Atlantic salmon is so great, conditions under which they develop in fresh water also varies greatly, particularly light and temperature. This results in very different growth rates and time spent in fresh water before migration to sea.

In the wild, at the onset of sexual maturity salmon return and collect in the coastal waters and estuaries before moving upstream, towards the spawning beds where, in the majority of cases, they have spent their first few months of life. Fish can begin returning to the rivers up to 12 months before spawning commences, usually from October to January. Spawning takes place where suitable conditions usually in tributaries of the main river, can be found. In order that the eggs have the best chance of survival they need to be laid in clean gravel of the correct size and depth with no silt, so that the river water is able to flow through the substrate as well as over it. The hen will then excavate a trench or 'redd' using her fins to displace gravel downstream. Once the redd is the correct size the male moves alongside the female and fertilises the eggs as they are laid. At this point it is sometimes possible for a precocious male parr to sneak in and fertilise the eggs before the adult male. All the eggs are not laid at once but in a succession of redds moving upstream, the material from the excavation of one redd covering the previous one downstream; the final redd containing very few eggs. Spawning is usually completed in two or three days but can take longer at lower temperatures. On average the female will produce about 1200 eggs per kilo body weight.

Development time for the egg varies according to water temperature but in degree days it takes approximately 250 from green to eyed egg and a further 250 from eyed egg to hatch. The newly hatched fish, or alevin, at this stage continues to feed on the reserves on the yolk sac and for another 250 degree days until it is fully absorbed, and until the mouth and digestive system complete their development. During this time the alevin remains amongst the gravel which provides it with support and protection. As the last of the yolk sac is absorbed the fish moves out of the gravel and begins to feed. Zooplankton now form the basis for the diet of the young fry but as the fish grows the diet changes to become almost exclusively aquatic insect larvae. The parr spend between one and eight years in fresh water before undergoing physiological and behavioural changes prior to migration into the sea in May or June. This variation in the time taken to migrate to sea is due to the length of the summer feeding period available to the fish. The average for fish in the British Isles would probably be two years with a survival rate of about 5–10 per cent from egg to smolt.

Once the fish leave fresh water much less is known about where they go or on what they feed. It is probable that movements of the fish are to a large extent dependent on where the food species are, salmon having been caught along the edge of the Arctic pack ice and into the Labrador Sea. Some populations remain relatively close to home; for example the Swedish, Finnish and Russian stocks in rivers which drain into the Baltic tend to remain there. It is thought that initially they feed on shrimp and krill but as they grow, fish are increasingly included in their diet, particularly the fattier species.

Salmon may return to spawn in their native river after only 1 winter at sea; most of these fish are male and known as grilse. The majority return after two or three winters at sea but some can remain at sea for up to five years. During the spawning migration to their home rivers the fish utilise the fat reserves in their flesh for gonad development. During this development the pigment in the flesh is also remobilised and is used particularly in egg production, resulting in a reduced colour in the flesh of the fish. The fish, particularly the male, undergo external changes when a large hooked lower jaw or kype develops. Once spawned only about 5 per cent of the fish will survive as kelts to spawn again on another occasion.

**On the farm**

Atlantic salmon are now extensively cultivated in a number of countries around the world. In order that this can be achieved profitably a number of changes have been made to their natural lifecycle; the fundamental change for farmed fish being that they are now confined within various tanks and/or cages for their entire life. It is therefore possible, with the correct feeding and manipulation of their environment, to produce a market fish of about 3kg in approximately two and a half years where it would

have probably taken at least four years to grow to this size in the wild.

Manipulation of the salmon begins as soon as the egg is stripped from the hen, with the selection of milt used to fertilise it. The farmer can either use milt (sperm) from his own stock or from sources outside his farm. Studies have indicated that each salmon river has it own distinct genetic stock. Thus, stocks taken from different rivers have slightly different characteristics, which can be selected for if desirable, using, for example, milt from early or late maturing stocks.

Many of the following decisions have the size of fish and time of harvest in mind, so that a continuous production of optimally sized fish for market is possible. The farmer has to decide under what conditions to cultivate the eggs. Should they be chilled, heated or left to develop under ambient water conditions? This will determine when they hatch, so that it is possible to obtain batches of fry at different development points from the same eggs. As the fry grow and become parr it is also possible, once they have reached a critical size, to manipulate their photoperiod in order to induce early smoltification. Using this method smolts can be produced that are ready for transfer to sea in less than a year, traditionally known as S½ smolts. Once smolted the fish are transferred to sea water. This may be pumped ashore into tanks, or more commonly put into cages at sea.

Broodstock are usually selected either by line breeding or more commonly by mass selection. In mass selection the broodstock are usually chosen by selecting the fish which grow largest in the shortest time. They receive no special attention other than that after selection they may be kept at a lower stocking density than market fish and at the onset of maturity receive a supplemented diet.

Survival rates in farmed stocks are much higher than those in the wild, with probably a 10 per cent loss from egg to fry and a further 10 per cent from fry to smolt. Survival at sea in farmed salmon is now in the region of 90-95% of the smolt input. There has been a significant improvement over the last few years mainly due to sound husbandry practices.

At first the basic information necessary for the culture of salmonoids came from studies of wild fish

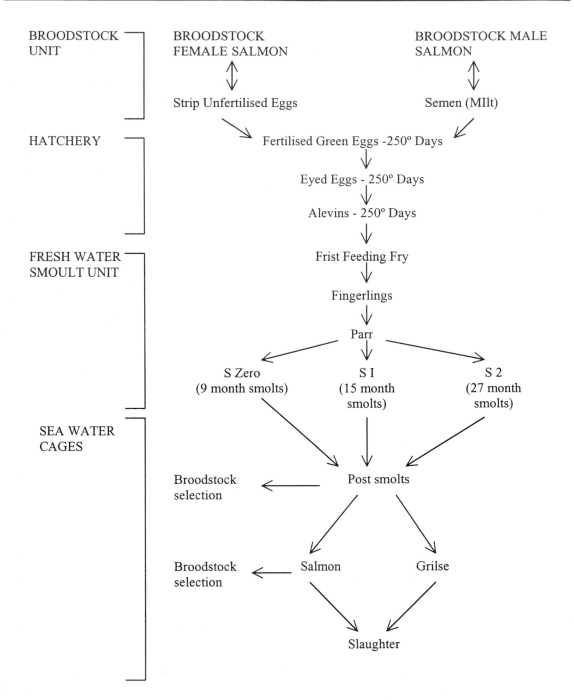

Figure 15.1    **Broodstock salmon are usually stripped in November/December. However the fish are now being photo manipulated to produce 'Out of Season' fish. This facilitates the production of out of season or S - Zero smoults which are also photomanipulated. These fish are useful in ironing out the peaks and troughs of the production cycle.**

but now more information is being gained from fish studied in captivity, which has led to a greater understanding of the fishes life cycle.

## FEED AND NUTRITION

Most UK salmonoid farmers now use commercial 'dry' pelleted diets, formulated by large feed companies, throughout the production cycle. In the past, the use of 'moist diets', largely consisting of trash fish, was widespread although the inefficient utilisation of such diets by fish, and the risk of spreading disease onto fish farms, has almost completely eliminated this practice. Much research and development has taken place in dry feed production, resulting in diets that now enable farmers to achieve rapid growth rates throughout the production cycle, at feed conversion rates that can be close to 1:1. Feed remains the largest variable cost to the farmer, and as such salmonoid farmers need to maintain highly efficient conversion rates to remain profitable. The feed is mainly derived from fish meal containing protein and oil. Most farmed fish are carnivorous and are not capable of metabolising large amounts of carbohydrates. Fish meal is an expensive source of protein, and alternate sources are being investigated especially single cell proteins. This is necessary work, as removing fish from the sea to feed farm fish may not in the long term be sustainable.

Oil levels from which the fish derive most of their energy source (high energy diets) have risen over the last few years to up to 30 per cent of the total diet. These oils have a protein sparing effect, leaving the available proteins to be used for body growth rather than as an energy source. Rapid growth using high energy diets does have consequences on flesh quality at harvest, an area which is currently under some scrutiny from within the industry following reaction from processors and consumers.

The fish are fed either by hand or automatically from hoppers, the food dropping into the cages or being blown in, or by a combination of the two. It is important that some part of the daily ration is hand fed as this is often the only chance that the stockman has to observe his fish.

It is usual to feed larger fish in 'meals' rather than continuously adding feed in small amounts. This method of introducing large amounts of pellets in a short time gives all the fish a supply of feed – rather than just the larger, more aggressive ones and so will reduce biomodality of growth. Smaller fish, especially fry and first feeders, are usually fed *ad lib.* via automatic feeders.

### Withholding food

Fish may starve for a period of time for various reasons. Prior to a bath treatment, handling or vaccination, fish may be given no food for 2–3 days. They may sometimes be starved for a period of time in the face of a bacterial infection. This strategy can be partially effective by decreasing the metabolic rate of the fish and reducing the bacterial loading in the water. More controversially, fish are often starved for up to two weeks prior to slaughter to improve flesh quality. There is much debate as to whether this prolonged period is necessary – indeed some recent work has shown that a lower oil diet in the months prior to slaughter has a most beneficial effect on carcase composition.

Notwithstanding the debate on flesh quality, there remains the thorny issue of whether a two week starvation period affects the welfare of fish which have been used to daily feeding.

### ENVIRONMENT

Most fish spend all their life in water. This has the advantage of providing the fish with support, making it relatively weightless. However, as the fish is so intimately surrounded by its environment any adverse changes in water quality will be much more serious than similar changes in terrestrial animals. These adverse changes may be caused by external factors such as pollution in the water body, or related to the method of fish husbandry. For example, water quality will deteriorate rapidly in the presence of excess food or faeces from the fish farming operation.

Some species of fish will tolerate poorer water quality than others. Typically carp can survive and grow well in water conditions that would kill salmonoid species. Related to this is the ability of eels to grow well at much higher stocking densities than most other species of farmed fish, partly due to their tougher/thicker skin and also because of their ability to tolerate relatively poor water quality and lower oxygen levels. Individual parameters to measure water quality can rarely be made in isolation.

**Figure 15.2**      **Salmon in sea cages being fed by hand.** This method provides a good spread of feed and ensures all fish get a ready supply of pellets. As cage sizes enlarge and the amount of feed that needs to be given each day increases, more and more farms are using automated feeding systems. This presents new challenges especially ensuring that all fish are given an adequate ration and the observation of the stocks is not neglected.

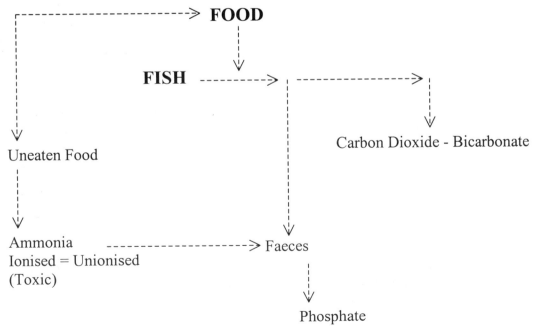

**Figure 15.3**      **Food, faeces and phosphate relationships.**

Temperature, oxygen, ammonia, carbon dioxide, suspended solids and pH all modify each other in their effects on the fish.

## Temperature

The body temperature of the fish shadows by 1–2˚C that of the water temperature. Any increase in water temperature will elevate the metabolic rate of the fish, which will in turn increase appetite and excretory products, and so the demand for oxygen will be correspondingly greater. Conversely, at higher temperatures the oxygen-carrying capacity of the water decreases. This is exacerbated by the elevated levels of food and faeces, which will further increase the demand for oxygen (the biological oxygen demand). Generally a dissolved concentration which falls below 6ppm and/or 70% saturation for cold water species can be considered dangerous.

This conflict of maximising growth at high water temperatures is a tightrope which the fish farmer must often walk. The problem can be highlighted in rainbow trout, which are much greedier than Atlantic salmon. Whereas Atlantic salmon will often naturally reduce their feeding levels at very high temperatures (usually over 19˚C), rainbow trout will eat to excess. At these high temperatures and low oxygen levels the increased demand for oxygen by the trout to metabolise this food cannot be met and death may ensue.

## Carbon dioxide

High levels of carbon dioxide (>10ppm), especially at low pH, can cause problems with oxygen uptake. Nephrocalcinosis can occur when calcium carbonate formed by the dissociation of the gas in the water is deposited in the kidney.

## Ammonia

The non-ionised form is toxic, causing gill, liver and kidney damage. This toxicity is increased at higher temperatures and higher pH. The build-up of ammonia is more likely in recirculation systems, where the biological filtering is inadequate and fails to convert the ammonia into nitrite. Ammonia can also reach toxic levels from pollution of the incoming water, usually caused by agricultural run-off.

## pH

The optimum pH for most species is between 6.5 and 8.5; outside this range direct toxic effects can occur and stress levels will be high. Low, acidic pH will cause primary gill damage and is more common than excess alkalinity.

The condition of the water will largely determine the health or otherwise of the fish. Poor water quality associated with high levels of suspended solids can cause gill diseases predisposing to other conditions such as bacterial and parasitic diseases.

Increasing the stocking density will lead to more competition for food, resulting in a bimodal size population. It will also reduce water quality due to the increased demand for available oxygen and the increased levels of food and faeces.

## Recirculation systems

In some fresh water systems the limiting factor in biomass of fish produced is the small volume of water available. However, it is possible to re-use water in a recirculation system using only 10 per cent or less of new water from the outside source.

These systems involve mechanical filtration, biological filters and usually sterilisation of the water. This system must be accompanied by a high level of stockmanship to avoid the build up of pathogens and waste products. Such systems especially with the use of biological filters' precludes the use of any chemical or antibacterial therapy as this would destroy the bacteria present in the filter.

## Oxygenation and aerators

These systems are used mainly for emergencies such as when fish are crowded or there are water shortages, high temperatures, chemical treatments and during transport. Both methods will provide additional oxygen to the water, but aerators will have the additional advantage of removing ammonia from the water with the air bubbles.

## Food and faeces

The levels of food and faeces in the water discharged from the farm are measured indirectly by the local or national river purification boards. These regulatory bodies control the discharge of effluent from a fish farm into the water course.

The quality of most fish foods has increased over the years. Some of the main problems associated with the build up of phosphates, suspended solids and ammonia (see Figure 15.3 on page 284) have been corrected by changing the formulation of the diets.

## FISH KEEPING SYSTEMS

Aquaculture systems, compared with terrestrial ones, are usually three dimensional. Fish do not just live side by side but above and below one another. This can make observation of fish stocks particularly difficult. Indeed it is quite possible that some fish in a large marine cage are never seen until slaughter.

### Fresh water systems

*Hatcheries*

Salmon and trout eggs are usually reared in trays. The eggs are poured into these aluminium trays in a layer one to two deep. The trays have a perforated aluminium base through which the hatched fish, or alevins, can swim, ending up in the fibreglass tank below the egg tray. These trays are usually stacked in layers, the water flowing over the eggs and down into the egg tray below. The eggs are kept in the dark or in very subdued lighting. As the trays can easily be removed for inspection, any dead eggs, which can be readily identified by their whitish opacity, can be picked out.

In larger hatcheries eggs may be incubated in large jars up to a metre high. The water is introduced at the bottom, rising through the column of eggs. While this method can save space and time, dead eggs cannot be removed, leading to fungal infections in the adjacent healthy eggs. The use of prophylactic antifungal agents is usually necessary to maintain this system.

Halibut and turbot eggs are planktonic and are incubated while suspended in the water column. Here, good water quality is very important; often this water is filtered and sterilised before being used again in a recirculation system.

### First feeding and fry rearing systems

The salmonoid species produce large eggs which in turn produce large alevins. The development of these larvae is relatively uncomplicated. After swimming through the egg tray the alevins settle on the bottom of the tank, where they remain relatively immobile until most of the yolk sac is absorbed. Some substrate, either gravel or more usually a plastic imitation grass turf (eg Astroturf) is used to give these fish protection at this stage. Once the fish have absorbed most of the yolk sac they will start to 'swim up' through the water column in search of food. The timing of the introduction of food to these fish is critical and requires a high level of stockmanship. Feed them too soon and the waste food builds-up with a reduction of water quality and consequent gill disease. Leave it too late and the fish do not have the energy to apprehend or metabolise the food – ending up as 'pin heads' – typically a long thin body with a big head.

Turbot and halibut larvae require live food for their first feeding. Hatchery units usually include facilities for the culture of live food such as rotifers, atremia and copepods, which are themselves fed on algae specially grown in the hatchery. The rearing of the larvae is difficult ad the techniques used are still being developed. The usual large losses of the eggs and larvae of these species is compensated for by the very large number of eggs produced by the female.

### Fingerling and smolt systems

These systems can vary depending on the species involved as well as the type of water available.

Most salmon are reared from fingerling to smolt either in circular tanks or in fresh water cages. The tanks used vary in diameter from 2m to 10m with a peripheral inlet of water. The angle of the inlet pipe can be varied to control the speed of water flow in the tank. The outlet is a screened central pipe where any mortalities can be collected. These tanks are usually shaded to remove the danger of sunburning and reduce the levels of stress. Such tanks can have a high capital cost compared with fresh water cages but offer the advantage of a more controlled environment, where monitored water flows, observations of stocks, chemical treatments and mortality removal are easily carried out.

Fresh water cages are used in smolt production, often for the 6–9 months before smolt transfer. Notwithstanding some of the difficulties of managing this system, these fish often grow extremely well and adapt more quickly to the marine cages than their tank-reared counterparts.

Rainbow trout are usually on-grown in raceways or rectangular earth ponds. In both these systems water is introduced at one end with the outlet being at the other, operating a continuous flow-through system. (The use of static water systems does not produce enough high water quality for salmonids - although this system is used for channel catfish and carp.)

It can be difficult with raceways and earth ponds to get a good flushing effect. Failure to achieve this can result in an accumulation of faeces and uneaten rotting food. Raceways are often constructed of concrete which can abrade the fins and skin. Earth ponds can provide a suitable habitat for the intermediate hosts of whirling disease and eye fluke parasites. Fallowing of these ponds or treating the empty pond may be necessary to break the cycle of infections.

Generally earth ponds and raceways work well in growing rainbow trout. Rearing Atlantic salmon in similar conditions prior to sea-water entry has given equivocal results.

**Marine systems**

Marine systems of farming sea bass and gilt-head bream in the Mediterranean and cod in Northern Europe have been developed which are similar to the on-growing marine systems used for Atlantic salmon.

These systems rely on a floating walkway below which is suspended a bag-net enclosing the fish. The walkway can be 12m to 20m square, octagonal or round. The depth of the suspended net is very variable but usually will vary from 5 to 20m deep. The mesh of the net is usually sufficiently open to allow a good water exchange without allowing fish to escape. The cages are often enclosed in another outer net – the predator net – which is used to prevent

**Figure 15.4      Salmon cages.**

attacks from marine predators and diving birds. There is usually another predator net over the top of the cage to prevent herons and aerial birds removing or damaging the fish. The nets are individually weighted at the corners to prevent distortion which would reduce the volume of the cage. These cages may be grouped together in blocks. They are moored to the seabed with anchors and large concrete blocks. Water exchange and hence water quality relies on tides and currents. This can be compromised if the net becomes heavily fouled with weed. Generally the water quality in sea cages is satisfactory.

Observations of all the fish in a sea cage is not possible without the use of divers. The farmer must rely on the behaviour of those fish visible near the surface to give an indication of the status of fish at deeper levels. Indeed, it is often the presence of dead fish which will alert the farmer to a possible problem.

Mortality collection in sea cages is difficult – involving the need for divers, the use of removable internal nets at the bottom of the cage or, more recently, an airlift system which lifts fish up to the surface from a sack in the base of the cage (Figure

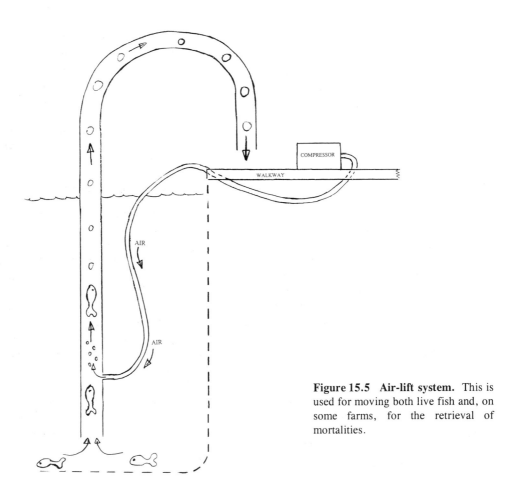

**Figure 15.5 Air-lift system.** This is used for moving both live fish and, on some farms, for the retrieval of mortalities.

15.5). Mortality removal on a daily or very regular basis is essential to monitor the health of the stocks as well as removing a source of disease from the healthy fish (see Figure 15.7 on page 293). The last few years has shown an increased awareness in the need for prompt removal of mortalities by whatever means.

## MANAGEMENT OF PARTICULAR GROUPS

### Smolt transfer to sea water

Atlantic salmon smolts are traditionally transferred from fresh water to salt water in April or May, although adjustment to the photoperiod has now resulted in smolts going to sea at most times of the year. This dramatic osmoregulatory change is achieved more gradually in wild fish, where the fish will adjust to the hyperosmotic sea water in the estuary.

Farmed fish, however, are transferred to sea water suddenly and so it is important that they are ready to adapt to this change. Smolts ready for the sea will become more silvery, with black tips to their fins and they will lose their parr marks. Analysis of the blood chloride levels of a few of the fish after they have been in full strength sea water for 72 hours can be used to assess whether the majority of the population is ready for transfer. Blood samples can be taken from anaesthetised fish or immediately after stunning. Blood can be withdrawn into a syringe from the caudal vein between the anus and the tail. The use of vacuum blood collection methods will often collapse the blood vessels in a small fish.

Fish should be graded before sea water transfer to remove any parr, otherwise the parr will die in sea water, usually during the six weeks after transfer. These losses should be minimised for economic and welfare considerations.

### Broodstock management

As the adult fish becomes sexually mature the appetite decreases and the pigment naturally occurring in the flesh is transferred to the egg. The fish becomes darker in colour, often with a reddening of the abdomen. Salmonoid species produce relatively few eggs, whereas some flat fish such as turbot will yield over a million per g bodyweight. Thus, relatively large numbers of salmon and trout will need to be kept as broodstock to ensure adequate numbers of eggs.

Selection of these broodstock fish cannot be made on fast growth alone as this method will tend to select only the larger males. At the stage of broodstock selection 2–3 times the number finally needed will be held back from the fish ready for slaughter. Later in the year, as the secondary sex characteristics become more apparent, the number can be reduced. Usually, to achieve adequate fertilisation as well as genetic diversity one male is used to fertilise two females.

Fish are 'stripped' of the eggs or sperm (milt) under anaesthesia. This is usually done by manually squeezing the abdomen gently in an anterior-posterior direction. Other methods involve the use of gas introduced into the abdominal cavity and the increase in pressure forcing out the eggs and sperm. This latter method is only performed in dead or deeply anaesthetized fish. The milt is mixed with the eggs and any excess milt is washed off.

Most broodstock are very large fish and can be difficult to handle out of water. Generally speaking manual removal of the gametes cannot be achieved in the unanaesthetized fish without causing damage and stress. For this reason fish are stripped either under anaesthesia or immediately after death.

## GRADING AND HANDLING

Grading fish is important to maintain uniformity of size in a population, otherwise the biggest, most aggressive feeding fish will eat most of the food. There is evidence to show that the smallest fish in a population are inhibited from feeding even in the presence of excess feed. Also, in salmon hatcheries, parr must be separated from potential smolts ready for transfer to the sea.

Grading and handling are especially necessary when fish are small and fast growing. At this time it can easily be achieved with the minimum of stress and damage. As fish become larger and heavier the physical effects of handling and grading can become more serious. These larger fish are more likely to suffer mucus loss, skin abrasions, bruising and eye damage, which may be routes of entry for systemic bacterial infections.

Most systems growing larger fish such as the sea water phase on salmon farms will try to keep grading and handling of these large fish to a minimum. It may be necessary to separate the sexually maturing grilse from the non-maturing salmon after 15–18 months in the sea. This stress-inducing procedure often initiates an outbreak of disease and so an all-in, all-out system is becoming increasingly popular.

Size grading is carried out by pumping or netting fish out of the water and over a series of adjustable bars through which the smaller fish will fall. Grilse grading entails manual selection of fish over a table with two or more chutes. The maturing grilse are separated into one pen for early slaughter - while the non-maturing salmon will be on-grown through the summer and autumn.

Any method of handling fish is potentially stressful and carries the risk of external damage to the fish. Small fish can be manually netted out of water without causing any damage. Care must be taken to limit the number of fish in a net. The fish at the bottom of a full net will be much more likely to be damaged than one at the top. Manually removing larger fish from the water by netting is not carried out, due to operator fatigue and the serious external damage that can occur to the fish. These bigger fish are usually handled by pumping fish or more recently using an air-lift system. These methods can be very efficient causing minimal damage to the fish (Figure 15.5).

## TRANSPORT

Live fish can be transported in tanks, usually on lorries or trucks. Helicopters are being increasingly used. This expensive method can be cost effective by reducing transport time involved. Fish are usually starved for two days prior to transport to reduce the biological loading of the water and metabolic rate of the fish. This is important as not only have oxygen levels to be adequate but ammonia levels should be kept low. Oxygenation and/or aeration is necessary when transporting fish.

Changes in temperature are common during fish transport and if possible they should be kept to a minimum. Insulated tanks and the addition of ice if the temperature rises are helpful.

Fish should be checked regularly during transport. Most losses are due to failure of the oxygenation system and so oxygen monitoring is essential.

## SLAUGHTER

Prior to slaughter most fish will be starved for a period of time. During this time the gut contents will be eliminated, thus reducing the contamination to the carcase during gutting. Sometimes food is withheld for longer periods of up to two weeks in the hope of improving flesh quality.

It is important that adequate training of personnel in slaughter techniques is carried out. Whatever method is used it is important that the fish are rendered insensible instantaneously and as efficiently as possible. This method should be irreversible.

### Salmon

Methods used include:

1.  Stunning with or without bleeding. Using a wooden or plastic club a single hard blow or series of short sharp blows is delivered just in front of the eyes. This method is effective in producing sufficient immediate brain damage when carried out properly. It may be followed by cutting all four gill arches to ensure rapid bleeding. Although the operators can easily become tired using this method, it can be the most humane killing system available for salmon.

2.  Carbon dioxide narcosis. The fish are placed in a bath of sea water previously saturated with gaseous carbon dioxide. This saturation can only be determined in practice by measuring the fall in pH. Narcosis is attained after 1–2 minutes. There is a period of excitement and head shaking when the fish are first placed in the bath. In practice, the large number of fish being slaughtered at any one time on a commercial farm makes it difficult to maintain high levels of carbon dioxide.

    If these fish are returned to sea water they will recover. For this reason bleeding after carbon dioxide narcosis is essential.

    Even though this method may have some limitations, such as the time taken to reach insensibility and the initial excitement phase, at the moment it is probably the most efficient method available.

3.  Bleeding. This method is usually used after prior stunning or carbon dioxide narcosis. It is especially important if the fish are to be smoked. The gill arches are cut with a sharp knife

**Figure 15.6**     **Salmon being moved to the killing area. The fish are netted from the cage where they have been crowded (1) and moved to a water bath saturated with carbon dioxide where they become narcotised (2). The fish are then exsanguinated by severing the gill arches. Fish should not remain crowded for too long prior to killing otherwise skin abrasions and poor fish quality may occur.**

- ideally on both sides - to allow rapid and effective exsanguination.

On a small number of farms this method is used alone without prior stunning or narcosis. There is an increase in activity and muscular spasms which may last up to four minutes. In the opinion of the author this method is unacceptable.

**Trout**

Methods used include:

1. Electric shock. A pulse of direct electric current causes instantaneous immobilisation of the fish and appears to be satisfactory from a humane point of view. However, this method can cause spinal fractures and haemorrhages into the muscle. There is the added disadvantage that great care needs to be taken to ensure operator safety.

2. Ice slurry. After the fish are removed from the water they are placed directly into an ice and water slurry at about 2°C. These fish become torpid due to the cold and eventually death is by anoxia. This may take some time as the oxygen requirements at these temperatures is low.

3. Severing the spinal cord at the cervical area. Even if the blood supply to the brain is interrupted fish are tolerant to prolonged hypoxia and so immediately sensibility may not be achieved.

There is room for improvement in methods of trout slaughter.

**Flat fish**

Methods used include:

1. Carbon dioxide narcosis
2. Stunning. This is very difficult without damaging or destroying the eyes
3. Severing the spinal cord. This may not cause immediate insensibility as fish are tolerant to prolonged hypoxia.

**HEALTH, DISEASE AND WELFARE**

**Health and welfare**

The last few years has seen a considerable improvement in fish health in the UK – especially in Atlantic salmon. This has been brought about mainly by adopting sound farming principles well recognized in terrestrial farming. Most salmon farmers will now fallow sea sites for up to 3–6 months before the introduction of the new stock. Single year batches are also the norm so further reducing the spread of disease from one class to another. These two measures have been largely responsible for the dramatic decline in bacterial, parasitic and viral diseases.

New technologies have been developed which facilitate the removal of dead fish, especially from sea cages. This has helped to reduce the spread of disease.

Stocking densities have also been reduced on most salmon farms, so any disease spread in the populations will be less rapid. Low densities also decrease the need for thinning out and grading the fish later on – a stressful procedure which is likely to precipitate an outbreak of bacterial infection.

Rainbow trout farmed in fresh water are usually kept at much higher stocking densities. This species is better adapted to higher densities than salmon, with its tougher integument and greater resistance to surface abrasion. Even so, these higher stocking levels are often reflected in poorer fish quality and scale loss, commonly seen in rainbow trout.

Most fish farmers have accepted the need for careful and patient observation of fish stocks. Often they have had to be convinced that, like the shepherd leaning over the gate looking at his flock, time spent noting the behaviour and appearance of his fish is not a waste of time.

Detailed records will be necessary, not only to document any previous disease in the stocks, but also to record any prophylactic chemical treatments, changes in behaviour and weather conditions that may cause changes in water quality.

Consideration should be given to:
– the behaviour of the fish (are they shoaling, using the whole water column, crowding around the inlet or outlet pipes)
– the external appearance of the fish (ragged fins, boils or furuncles, blindness, excessive opercular (respiratory) movements, colour changes, excess mucus or fungus present, predator damage)
- feeding response (vigorously, not at all, spitting out food)
– the fish or mixed sizes in one population may indicate faulty feeding, metabolic disease or overcrowding sizes.

**Diseases**
*Bacterial infections*
Furunculosis, enteric redmouth and vibriosis have been of the main cause for large mortalities in farmed fish. Although effective vaccines are now available to control these diseases, the importance of good husbandry and hygiene is paramount. Overcrowded fish living in poor environmental conditions will often succumb to these infections even after vaccinations. These bacterial diseases can sometimes be controlled using oral antibiotics incorporated into the food. The build up of bacterial resistance to these drugs has been very rapid in some cases, which is a cause for concern, especially as the number of antibiotics licensed for use in fish is limited.

Bacterial kidney disease (BKD) causes variable mortalities in both salmon and trout. This disease must be controlled mainly by the production of disease-free stock. There is no vaccine available and the use of antibiotics is not generally effective.

*Viral infections*
Infectious pancreatic necrosis virus can cause mortalities in first feeding and yolk sac fry in both salmon and trout.

Salmon smolts entering the sea can succumb in large numbers but other causes such as inadequate smoltification, transport and handling damage must also be considered. The isolation of the virus, however, is not necessarily confirmation of the cause of mortalities – some strains of the virus can often be non-pathogenic and appears to be ubiquitous in the sea.

Pancreas disease, recently found to be caused by a Toga virus, causes destruction of the exocrine pancreatic tissues, leading to anorexic moribund salmon in the seawater phase. Losses are variable – usually below 10 per cent. There is often a small proportion of fish which do not recover sufficiently to resume feeding. These fish should be removed and humanely disposed of.

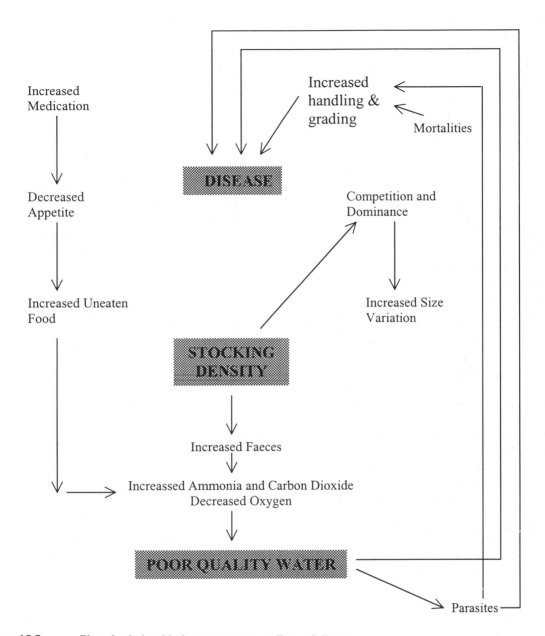

**Figure 15.7**       **Pivotal relationship between water quality and disease.**

*Parasites*

Newly hatched salmon and trout in fresh water are often parasitized with protozoal ectoparasites. These are more common at higher temperatures and when fish are overcrowded. Good hatchery management will incorporate regular microscopic investigations for these organisms before large numbers cause problems.

Lice are a major cause of loss of production, and mortality in salmon in the sea. Infestation with large numbers can lead to skin and scale loss resulting in deep ulcerated areas around the head. Treatments usually consist of an organosphosphate, pyrethroid or hydrogen peroxide bath. This disease is probably the most important from a welfare and production point of view. As in other diseases, fallowing sites and separation of year groups has reduced the impact of this parasite.

## Therapy

Fish may be treated orally using medication incorporated into the feed pellet. Bath treatments are common for fungus and ectoparasite infections. If adequate oxygenation is maintained, these treatments in fresh water are usually quite safe. Bath treatments in the sea, especially for sea-lice, can be potentially dangerous as the fish may panic and burrow into the corners, causing large losses. Novel in-feed sea-lice treatments are being developed which should remove the need for some bath treatments.

Both salmon and trout are regularly vaccinated against the major bacterial diseases. These vaccines are usually administered by intraperitoneal injection or by immersion.

The use of any medication for fish is restricted to those compounds for which the Regional Water Purification Board will grant a licence – the so-called Water Discharge Consents. So the fact that the veterinary surgeon or farmer may wish to use a certain drug will matter little if no discharge consent has been granted. This restriction is made worse in that the number of medications available for the treatment of fish – already small – is gradually being eroded by new legislation. These constraints on effective treatments are a serious cause for concern. There are very few licensed antibiotics and antiparasitic treatments available.

A number of the chemicals currently used as bath treatments are not licensed and are unlikely to become so. The health and welfare of farmed fish could be seriously jeopardised in the future by this lack of effective medications.

## THE WAY AHEAD

The rapid growth in salmon and trout farming in the UK in the last few decades will continue, and there will be new developments such as the farming of turbot, halibut and cod. The general decline in wild fisheries will ensure a ready market for these fish: the main challenge will be developing the technology for rearing and growing fish with such diverse life histories.

The use of high levels of fish meal in most fish feeds may not be sustainable in the long term. Development of single cell proteins from yeasts and bacteria as well as the use of modified plant proteins will be necessary to ensure sustainable and ready supplies of fish food.

Finally, over the last 20 years there has been an acceptance by farmers, scientists and veterinary surgeons that the fish kept on fish farms need to be treated in a humane and compassionate manner. As new technologies emerge, so also do new welfare considerations; challenges which need to be constantly addressed.

## FURTHER READING

**Brown L** (ed) 1993 *Aquaculture for Veterinarians. Fish Husbandry and Medicine*. Pergammon Press: Oxford, UK

**Bruno D W, Alderman D J, Schlefeldt H J** *What Should I Do? The Practical Guide to the Marine Fish Farmer*. The European Association of Fish Pathologists. (ISBN 09526292 62)

**Farm Animal Welfare Council** 1996 *Report on the Welfare of Farmed Fish*, PB 2765. MAFF: London, UK

**Ferguson H W** 1989 *Systemic Pathology of Fish*. Iowa State University Press: Ames, USA

# Acronyms and Abbreviations

| | |
|---|---|
| ACCESS | a computer-controlled eating stall system (for calves) |
| AFRC | Agricultural and Food Research Council (now BBSRC) |
| AHV-1 | Alcelaphine herpesvirus -1 |
| AI | artificial insemination |
| ARC | Agricultural Research Council (changed firstly to AFRC and then to BBSRC) |
| ART | avian rhinotracheitis |
| ATB | Agricultural Training Board (now Lantra National Training Organisation Ltd) |
| | |
| BBSRC | Biotechnology and Biological Sciences Research Council |
| BCMS | British Cattle Movement Service |
| BKD | bacterial kidney disease (of fish) |
| BSE | bovine spongiform encephalopathy |
| BSI | British Standards Institute |
| BUT | British United Turkey (company) |
| | |
| CAB | Commonwealth Agriculture Bureaux |
| CE | Council of Europe |
| CJD | Creutzfeldt-Jakob Disease |
| CP | crude protein |
| CTS | Cattle Tracing Service |
| | |
| DCP | digestible crude protein |
| DE | digestible energy |
| DFD | dark firm and dry (meat) |
| DIY | do-it-yourself |
| DM | dry matter |
| DUP | digestible undegradable protein |
| $DVH_1$ | duck virus hepatitis - 1 |
| $DVH_2$ | duck virus hepatitis - 2 |
| | |
| EC | European Community |
| *E. coli* | *Escherichia coli* |
| EDS | egg drop syndrome |
| EEC | European Economic Community (now EC/EU) |
| ELISA | enzyme-linked immunosorbent assay |
| ERDP | effective rumen-degradable protein |
| ESF | electronic sow feeders |
| EU | European Union |

| | |
|---|---|
| FAWC | Farm Animal Welfare Council (formerly the Farm Animal Welfare Advisory Committee) |
| FCR | food conversion ratio (kilogram of food per kilogram of liveweight gain) |
| FME | fermentable metabolizable energy |
| | |
| GATT | General Agreement on Tariffs and Trade |
| GE | gross energy |
| | |
| ha | hectare |
| HSA | Humane Slaughter Association |
| HSE | Health and Safety Executive |
| | |
| IAM | individual animal model |
| IB | infectious bronchitis (of birds) |
| IgA | immunoglobulin A |
| IgG | immunoglobulin G |
| ILT | infectious laryngotracheitis (of birds) |
| IU | international units |
| | |
| KKCF | kidney, knob and channel fat (beef cattle) |
| | |
| MAFF | Ministry of Agriculture Fisheries and Food |
| MCF | malignant catarrhal fever |
| M/D | see MJ ME/kg DM |
| MDC | Milk Development Council |
| ME | metabolizable energy |
| MEm | metabolizable energy requirement for maintenance |
| MJ DE/kg | megajoules of digestible energy per kilogram (of liveweight) |
| MJ ME/kg DM | megajoules of metabolisable energy per kg of dry matter (abbreviated to M/D) |
| MLC | Meat and Livestock Commission |
| MMB | Milk Marketing Board (functions now largely transferred to MDC, NDC and a number of commercial organisations) |
| MOET | multiple ovulation and embryo transfer |
| MP | metabolizable protein |
| | |
| ND | Newcastle disease |
| NDC | National Dairy Council |
| NFU | National Farmers Union |
| NOAH | National Office of Animal Health |
| NPN | non-protein nitrogen |
| NSA | National Sheep Association |
| | |
| ODR | oestrus detection rate |
| OHV - 2 | ovine herpesvirus - 2 |
| | |
| PIDA | Pig Industry Development Authority (now incorporated into MLC) |
| ppm | parts per million |
| PRID | progesterone releasing intravaginal devices |
| PSE | pale soft exudative (meat) |

| | |
|---|---|
| PTA | predicted transmitting ability |
| | |
| RDP | rumen digestible protein |
| RH | relative humidity |
| RSPCA | Royal Society for the Prevention of Cruelty to Animals |
| | |
| SOAFD | Scottish Office Agriculture and Fisheries Department (now SOAEFD) |
| SOAEFD | Scottish Office Agriculture, Environment and Fisheries Department |
| SSC | somatic cell content (count) |
| | |
| TB | tuberculosis |
| TBC | total bacterial count |
| TMR | total mixed ration |
| TR | turkey rhinotracheitis |
| TTT (3T rule) | training, taming, tempo (handling of deer) |
| | |
| UDP | undegradable protein |
| UFAW | Universities Federation for Animal Welfare |
| UK | United Kingdom of Great Britain and Northern Ireland (ie England, Scotland, Wales and Northern Ireland) |
| UKROFS | United Kingdom Register of Organic Food Standards |
| | |
| VFA | volatile fatty acids |
| VHD | viral haemorrhagic disease (of rabbits) |
| | |
| WTO | World Trade Organisation |

# Index